# THE
# INTERNATIONAL SERIES
# OF
# MONOGRAPHS ON PHYSICS

## SERIES EDITORS

# INTERNATIONAL SERIES OF MONOGRAPHS ON PHYSICS

128. R. E. Raab, O. L. de Lange: *Multipole theory in electromagnetism*
127. A. Larkin, A. Varlamov: *Theory of fluctuations in superconductors*
126. P. Goldbart, N. Goldenfeld, D. Sherrington: *Stealing the gold*
125. S. Atzeni, J. Meyer-ter-Vehn: *The physics of inertial fusion*
124. C. Kiefer: *Quantum Gravity*
123. T. Fujimoto: *Plasma Spectroscopy*
122. K. Fujikawa, H. Suzuki: *Path integrals and quantum anomalies*
121. T. Giamarchi: *Quantum physics in one dimension*
120. M. Warner, E. Terentjev: *Liquid crystal elastomers*
119. L. Jacak, P. Sitko, K. Wieczorek, A. Wójs: *Quantum Hall systems*
118. J. Wesson: *Tokamaks Third edition*
117. G. Volovik: *The universe in a helium droplet*
116. L. Pitaevskii, S. Stringari: *Bose–Einstein condensation*
115. G. Dissertori, I. G. Knowles, M. Schmelling: *Quantum chromodynamics*
114. B. DeWitt: *The global approach to quantum field theory*
113. J. Zinn-Justin: *Quantum field theory and critical phenomena, Fourth edition*
112. R. M. Mazo: *Brownian motion—fluctuations, dynamics, and applications*
111. H. Nishimori: *Statistical physics of spin glasses and information processing—an introduction*
110. N. B. Kopnin: *Theory of nonequilibrium superconductivity*
109. A. Aharoni: *Introduction to the theory of ferromagnetism, Second edition*
108. R. Dobbs: *Helium three*
107. R. Wigmans: *Calorimetry*
106. J. Kübler: *Theory of itinerant electron magnetism*
105. Y. Kuramoto, Y. Kitaoka: *Dynamics of heavy electrons*
104. D. Bardin, G. Passarino: *The Standard Model in the making*
103. G. C. Branco, L. Lavoura, J. P. Silva: *CP violation*
102. T. C. Choy: *Effective medium theory*
101. H. Araki: *Mathematical theory of quantum fields*
100. L. M. Pismen: *Vortices in nonlinear fields*
99. L. Mestel: *Stellar magnetism*
98. K. H. Bennemann: *Nonlinear optics in metals*
97. D. Salzmann: *Atomic physics in hot plamas*
96. M. Brambilla: *Kinetic theory of plasma waves*
95. M. Wakatani: *Stellarator and heliotron devices*
94. S. Chikazumi: *Physics of ferromagnetism*
91. R. A. Bertlmann: *Anomalies in quantum field theory*
90. P. K. Gosh: *Ion traps*
89. E. Simánek: *Inhomogeneous superconductors*
88. S. L. Adler: *Quaternionic quantum mechanics and quantum fields*
87. P. S. Joshi: *Global aspects in gravitation and cosmology*
86. E. R. Pike, S. Sarkar: *The quantum theory of radiation*
84. V. Z. Kresin, H. Morawitz, S. A. Wolf: *Mechanisms of conventional and high $T_c$ superconductivity*
83. P. G. de Gennes, J. Prost: *The physics of liquid crystals*
82. B. H. Bransden, M. R. C. McDowell: *Charge exchange and the theory of ion–atom–collision*
81. J. Jensen, A. R. Mackintosh: *Rare earth magnetism*
80. R. Gastmans, T. T. Wu: *The ubiquitous photon*
79. P. Luchini, H. Motz: *Undulators and free-electron lasers*
78. P. Weinberger: *Electron scattering theory*
76. H. Aoki, H. Kamimura: *The physics of interacting electrons in disordered systems*
75. J. D. Lawson: *The physics of charged particle beams*
73. M. Doi, S. F. Edwards: *The theory of polymer dynamics*
71. E. L. Wolf: *Principles of electron tunneling spectroscopy*
70. H. K. Henisch: *Semiconductor contacts*
69. S. Chandrasekhar: *The mathematical theory of black holes*
68. G. R. Satchler: *Direct nuclear reactions*
51. C. Møller: *The theory of relativity*
46. H. E. Stanley: *Introduction to phase transitions and critical phenomena*
32. A. Abragam: *Principles of nuclear magnetism*
27. P. A. M. Dirac: *Principles of quantum mechanics*
23. R. E. Peierls: *Quantum theory of solids*

# Multipole Theory in Electromagnetism

*Classical, quantum, and symmetry aspects, with applications*

R. E. RAAB
O. L. DE LANGE

*School of Chemical and Physical Sciences, University of Natal, Pietermaritzburg, South Africa*

CLARENDON PRESS · OXFORD
2005

phys

0185709633

# OXFORD
UNIVERSITY PRESS

Great Clarendon Street, Oxford OX2 6DP

Oxford University Press is a department of the University of Oxford.
It furthers the University's objective of excellence in research, scholarship,
and education by publishing worldwide in

Oxford New York

Auckland Bangkok Buenos Aires Cape Town Chennai
Dar es Salaam Delhi Hong Kong Istanbul
Karachi Kolkata Kuala Lumpur Madrid Melbourne Mexico City Mumbai
Nairobi São Paulo Shanghai Taipei Tokyo Toronto

Oxford is a registered trade mark of Oxford University Press
in the UK and in certain other countries

Published in the United States
by Oxford University Press Inc., New York

© Oxford University Press 2005

The moral rights of the author have been asserted
Database right Oxford University Press (maker)

First published 2005

A catalogue record for this title is available from the British Library

Library of Congress Cataloging in Publication Data
(Data available)

ISBN 0 19 856727 8

10 9 8 7 6 5 4 3 2 1

Printed in Great Britain
on acid-free paper by
Biddles Ltd, Kings Lynn, Norfolk

CL4
3/21/05

# PREFACE

*It is a capital mistake to theorize before one has data.*
Arthur Conan Doyle
(*Scandal in Bohemia*)

Multipole expansions in electrostatics, magnetostatics, and electrodynamics provide a useful and powerful method of characterizing charge and current distributions, and the associated electromagnetic potentials and fields. When applied to macroscopic electrodynamics, and used in conjunction with quantum mechanics and relevant space–time properties and symmetries of physical quantities, multipole theory enables one to describe certain macroscopic electromagnetic phenomena in terms of the underlying microscopic structure — molecules for gases and unit cells for crystals. For example, one can relate birefringences (natural and induced), dichroisms, and reflectivities to polarizabilities and multipole moments of the microscopic unit. Where phenomena require that both electric and magnetic contributions be considered, as in certain transmission effects in crystals, multipole theory provides a way of circumventing the otherwise intractable nature of the problem (Section 5.13).

Although there are several excellent books on electromagnetic theory, none concentrates exclusively on multipole theory; this despite the considerable research that has been reported on applications, and even the formulation, of multipole theory during the last few decades. This circumstance provides the motivation for the present book.

It might be thought that the formulation and applications of multipole theory in macroscopic electromagnetism are, in principle, straightforward and lacking in surprises or undue subtlety. The first indication to the contrary occurs in linear constitutive relations for the electromagnetic response fields $\mathbf{D}$ and $\mathbf{H}$ obtained directly from multipole theory. The dynamic material constants (permittivity, permeability, and magnetoelectic coefficients) in these relations have the surprising and unphysical feature that, for contributions beyond electric dipole order, they are not translationally invariant (independent of the choice of origin of coordinates). A further surprise follows: despite this defect, one can use multipole theory to construct a satisfactory theory of transmission effects. This happy circumstance does not extend to reflection effects, for which existing multipole theory yields fundamentally unphysical results when taken beyond electric dipole order, notably reflected intensities that are not translationally invariant.

These difficulties arise in the following way. Multipole theory expresses observables in terms of quantities which are, in general, origin dependent: namely, polarizabilities (and sometimes also multipole moments). Therefore, in the description of origin-independent observables, the theory should combine various polarizabilities (and multipole moments) in such a manner that the overall expression is origin independent. It does this successfully for transmission and scattering phenomena (Chapter 5), but not for the dynamic material constants (Chapter 4) or reflection phenomena (Chapter 6). Thus multipole theory in macroscopic electromagnetism presents a rather puzzling picture.

Our purpose in writing this monograph has been three-fold. First, we intend it as a detailed introduction to classical, quantum-mechanical, and symmetry aspects of multipole theory in electromagnetism. We consider both a charge distribution in vacuum and, by extension, bulk matter (Chapters 1 to 3). This extension involves an averaging technique which yields the macroscopic multipole moment densities; we do not discuss the details of this technique because there are comprehensive treatments in the literature.

Second, we provide an account of some of the successes and failures of the direct application of multipole theory to macroscopic media. This is presented in Chapters 4 to 6 which deal with constitutive relations and transmission, scattering, and reflection phenomena. These chapters should be of interest and value to the novice and experienced researcher alike, since the existing extensive literature is rather fragmented, sometimes misleading, and occasionally incomplete, making it difficult to assess the state of this subject. For instance, one will search the literature in vain for any demonstration, such as that of Section 6.8, of the unphysical nature of the multipole theory of reflection from crystal surfaces. In our selection of topics for inclusion in Chapters 4 to 6 we have omitted certain standard applications of multipole theory, such as those in electrostatics and magnetostatics, because these are available in several texts on electromagnetism.

Our third purpose in writing this monograph is to present an alternative to the unphysical standard formulation of multipole theory for macroscopic media. We exploit the non-uniqueness of the response fields $\mathbf{D}$ and $\mathbf{H}$ in Maxwell's macroscopic equations to construct a transformation theory for these fields (Chapter 7). This transformation theory is designed to restore the translational invariance of the theory, thereby changing unphysical multipole constitutive relations into unique, physically acceptable relations. It also modifies expressions for macroscopic multipole moment densities in a desirable manner and restores the Post constraint for the magnetoelectric tensors (Chapter 8). In Chapter 9 we show that this transformation theory retains the good features of the standard formulation of multipole theory, as exemplified by its application to transmission phenomena, while it removes the undesirable features of this theory, such as its unphysical consequences for reflection phenomena.

The required background for reading this monograph is a knowledge of introductory electromagnetic theory (including Maxwell's equations in vacuum), and introductory non-relativistic quantum mechanics (in the Dirac formalism and including perturbation theory). A reader who is already acquainted with multipole theory, and who is interested in the unphysical aspects of this theory and their resolution, may commence reading at Chapter 4. To assist the reader, a glossary of symbols is provided at the end of the book.

The content of this book, in part or in whole, should appeal to a broad spectrum of scientists: to physics students wishing to advance their knowledge of multipole theory beyond the treatment offered in texts on electromagnetism; to optical and molecular physicists, researchers in ellipsometry, physical and theoretical chemists who conduct research related to multipole effects; to solid state physicists studying the effects of static and dynamic fields in condensed matter; to engineers with interests ranging from microwave effects in synthetic chiral and other materials to free-space radiation; and to applied mathematicians developing theories of anisotropic materials.

The advent of increasingly powerful computers and computational techniques has brought a most interesting development. Recent *ab initio* numerical calculations by several researchers have shown that the accurate evaluation of rather complicated polarizabilities and hyperpolarizabilities is now feasible. These studies provide strong support for the multipole approach (Sections 5.11 and 5.12). It is likely that the combination of experiment, theory, and computational research will lead to further progress in the understanding of complex electromagnetic systems. We therefore hope that our book will stimulate interest in this topic among students of physics, chemistry, and mathematics, and that it will assist computational physicists and chemists wishing to work in this area to acquire the necessary background in multipole theory.

Each field of endeavour has a beginning, even if serendipitous. For one of us (RER) there were the immediate inspiration and solid foundation provided by Professor David Buckingham, CBE, FRS, to his first research student in the field of multipoles and their applications. David's ongoing interest in the molecular physics group at the University of Natal[1] has been a source of much encouragement. Some of the material in this book draws on the collaborative research undertaken over many years with Dr. Elizabeth Graham and Professor Clive Graham and a succession of excellent research students. Their contributions are gratefully acknowledged.

*Pietermaritzburg, South Africa*                                              R. E. R.
May 2004                                                                      O. L. de L.

---

[1] The University of Natal merged with the University of Durban-Westville on 1 January 2004 to become the University of KwaZulu-Natal.

# CONTENTS

**1  Classical multipole theory**                                    1

  1.1   Multipole expansion for the potential of a finite static
        charge distribution                                         1
  1.2   Dependence of electric multipole moments on origin          5
  1.3   Permanent and induced multipole moments                     6
  1.4   Force and torque in an external electrostatic field         7
  1.5   Potential energy of a charge distribution in an
        electrostatic field                                         8
  1.6   Multipole expansion for the vector potential of a finite
        distribution of steady current                             10
  1.7   Dependence of magnetic multipole moments on origin         12
  1.8   Force and torque in an external magnetostatic field        13
  1.9   Potential energy of a current distribution in a
        magnetostatic field                                        14
  1.10  Multipole expansions for the dynamic scalar and vector
        potentials                                                 15
  1.11  The far- and near-zone limits                              17
  1.12  Macroscopic media                                          18
  1.13  Maxwell's macroscopic equations: multipole forms for
        $\mathbf{D}$ and $\mathbf{H}$                              23
  1.14  Discussion                                                 25
  1.15  Primitive moments versus traceless moments                 27
      1.15.1 A charge distribution                               27
      1.15.2 Macroscopic media                                   28
  References                                                       29

**2  Quantum theory of multipole moments
and polarizabilities**                                              32

  2.1   Semi-classical quantum mechanics                           32
  2.2   Electrostatic perturbation                                 33
  2.3   Buckingham's derivation of electrostatic multipole
        moments                                                    37
  2.4   Magnetostatic perturbation                                 38
  2.5   Time-dependent fields: standard gauge                      40
  2.6   Time-dependent fields: the Barron–Gray gauge               45
  2.7   Polarizabilities for harmonic plane wave fields            47
  2.8   Absorption of radiation                                    50
  2.9   Additional static magnetic polarizabilities                51
  2.10  Symmetries                                                 52

2.11  Macroscopic multipole moment and polarizability
          densities                                                          53
2.12  Phenomenology of the wave–matter interaction                          54
References                                                                   56

3   **Space and time properties**                                           59
3.1   Coordinate transformations                                            59
3.2   Vectors                                                               60
3.3   Cartesian tensors                                                     62
3.4   Time reversal                                                         65
3.5   The space and time nature of various tensors                         67
3.6   Symmetry and property tensors                                         70
3.7   Origin dependence of polarizability tensors                          75
3.8   A pictorial determination of symmetry conditions                     79
3.9   Discussion                                                            82
References                                                                   82

4   **Linear constitutive relations from multipole theory**                 84
4.1   Constitutive relations                                                84
4.2   Origin independence                                                   86
4.3   Symmetries                                                            86
4.4   The "Post constraint"                                                 90
4.5   Comparison with direct multipole results                             92
        4.5.1  Electric dipole order                                        92
        4.5.2  Electric quadrupole–magnetic dipole order                   93
        4.5.3  Electric octopole–magnetic quadrupole order                 94
4.6   Discussion                                                            95
References                                                                   98

5   **Transmission and scattering effects: direct multipole
    results**                                                              100
5.1   The wave equation                                                    100
5.2   Intrinsic Faraday rotation in a ferromagnetic crystal               103
5.3   Natural optical activity                                             106
5.4   Time-odd linear birefringence in magnetic cubics                    110
5.5   Optical properties in the Jones calculus                            111
5.6   Gyrotropic birefringence                                            112
5.7   Linear birefringence in non-magnetic cubic crystals (Lorentz
          birefringence)                                                   115
5.8   Intrinsic Faraday rotation in magnetic cubics                       118
5.9   The Kerr effect in an ideal gas                                      120
5.10  Forward scattering theory of the Kerr effect                        124
5.11  Birefringence induced in a gas by an electric field
          gradient: forward scattering theory                             127
        5.11.1 Forward scattering by a molecule                           128

|  | 5.11.2 Induced moments | 129 |
|  | 5.11.3 Forward scattering by a lamina | 130 |
|  | 5.11.4 The electrostatic field | 131 |
|  | 5.11.5 Radiated field for linearly polarized light | 132 |
|  | 5.11.6 Field-gradient-induced birefringence | 133 |
|  | 5.11.7 Comparison between theory and experiment | 135 |
| 5.12 | Birefringence induced in a gas by an electric field gradient: wave theory | 136 |
| 5.13 | Discussion | 140 |
|  | References | 141 |

**6 Reflection effects: direct multipole results** — 145
| 6.1 | Reflection and the reflection matrix | 145 |
| 6.2 | The principle of reciprocity | 147 |
| 6.3 | Equations of continuity | 150 |
| 6.4 | Matching conditions in multipole theory | 153 |
| 6.5 | The reflection matrix for non-magnetic uniaxial and cubic crystals | 156 |
| 6.6 | Solutions of the wave equation | 162 |
| 6.7 | Reflection coefficients | 165 |
| 6.8 | Tests of translational and time-reversal invariance | 168 |
| 6.9 | Discussion | 169 |
|  | References | 170 |

**7 Transformations of the response fields and the constitutive tensor** — 172
| 7.1 | Gauge transformations of the 4-vector potential | 172 |
| 7.2 | "Gauge transformations" of response fields | 173 |
| 7.3 | Faraday transformations | 174 |
| 7.4 | Transformations of linear constitutive relations in multipole theory | 174 |
|  | References | 177 |

**8 Applications of the gauge and Faraday transformations** — 178
| 8.1 | Electric dipole order | 178 |
| 8.2 | Electric quadrupole–magnetic dipole order, non-magnetic medium | 180 |
| 8.3 | Electric quadrupole–magnetic dipole order, magnetic medium | 183 |
| 8.4 | Discussion | 186 |
|  | References | 189 |

**9 Transmission and reflection effects: transformed multipole results** — 191
| 9.1 | The wave equation and transmission | 191 |

9.2   Reflection from non-magnetic uniaxial and cubic
      crystals                                                      192
9.3   Explicit results for non-magnetic uniaxial crystals          195
9.4   Explicit results for non-magnetic cubic crystals             197
9.5   Tests of translational and time-reversal invariance          198
9.6   Reflection from antiferromagnetic $Cr_2O_3$: first
      configuration                                                 199
9.7   Reflection from antiferromagnetic $Cr_2O_3$: second
      configuration                                                 202
9.8   Comparison with experiment for $Cr_2O_3$                      205
9.9   Uniqueness of fields                                          206
9.10  Summary                                                       206
      References                                                    210

A   Transformations involving J                                    211

B   Magnetostatic field                                            213

C   Magnetostatic force                                            214

D   Magnetostatic torque                                           215

E   Integral transformations                                       216

F   Origin dependence of a polarizability tensor                   218

G   Invariance of transformed tensors                              220

Glossary of symbols                                                221

Index                                                              229

# 1

## CLASSICAL MULTIPOLE THEORY

*And mighty folios first, a lordly band,*
*Then quartos, their well-order'd ranks maintain,*
*And light octavos fill a spacious plain;*
*See yonder, ranged in more frequent rows,*
*A humbler band of duodecimos.*
George Crabbe
(*The library*)

In this chapter we present an introduction to classical multipole theory. This is first discussed for electrostatics (Sections 1.1–1.5) and magnetostatics (Sections 1.6–1.9). Multipole expansions of the scalar and vector potentials for time-dependent charge and current distributions are derived (Section 1.10), together with their far- and near-zone limits (Section 1.11). Macroscopic media are considered in Section 1.12, where the macroscopic multipole moment densities are introduced, and it is shown how bound charge and current densities can be expressed in terms of these multipole moment densities. This leads to a discussion of Maxwell's macroscopic equations and expressions for the response fields **D** and **H** in terms of multipole moment densities (Section 1.13). The use of primitive versus traceless moments is discussed in Section 1.15.

Multipole expansions are represented by infinite series such as (1.3), (1.47), (1.76), (1.78), (1.118), and (1.119). In this book, the highest multipole order to which we present explicit results is electric octopole–magnetic quadrupole. There are two reasons for this. First, this is the highest order to which physical effects have been studied (Section 2.12 and Chapter 5). Second, working to this order is useful in elucidating aspects of the theory (Chapter 4).

### 1.1 Multipole expansion for the potential of a finite static charge distribution

We consider a finite, continuous distribution of charge in vacuum, and choose an origin of coordinates $O$ inside the distribution. The charge inside an infinitesimal volume element $dv$ is $\rho(\mathbf{r})\, dv$ where $\rho$ is the charge density and $\mathbf{r}$ is the position vector of $dv$ (see Fig. 1.1). Let $P$ be a field point with position vector $\mathbf{R}$. The electrostatic potential at $P$ is

$$\Phi(\mathbf{R}) = \frac{1}{4\pi\varepsilon_0} \int_V \frac{\rho(\mathbf{r})\, dv}{|\mathbf{R} - \mathbf{r}|}, \qquad (1.1)$$

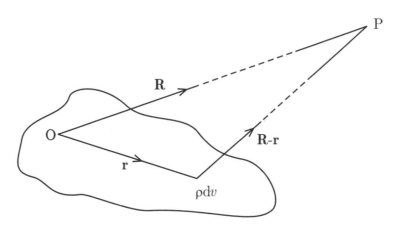

FIG. 1.1. Coordinates and notation for a finite, continuous charge distribution.

where, for a distribution of volume $V$, we have taken the zero of potential at infinity. Next, we consider a distant field point $(R \gg r)$ and expand

$$|\mathbf{R} - \mathbf{r}|^{-1} = (R^2 - 2\mathbf{R} \cdot \mathbf{r} + r^2)^{-1/2}$$

$$= \frac{1}{R}\left(1 + \frac{1}{R^2}\left[r^2 - 2\mathbf{R} \cdot \mathbf{r}\right]\right)^{-1/2}$$

$$= \frac{1}{R} + \frac{\mathbf{R} \cdot \mathbf{r}}{R^3} + \frac{3(\mathbf{R} \cdot \mathbf{r})^2 - R^2 r^2}{2R^5} + \frac{5(\mathbf{R} \cdot \mathbf{r})^3 - 3R^2(\mathbf{R} \cdot \mathbf{r})r^2}{2R^7} + \cdots,$$

$$\tag{1.2}$$

where in the last step we have used the binomial expansion and grouped terms in powers of $\mathbf{r}$.

From (1.1) and (1.2) we obtain the multipole expansion for the electrostatic potential at $P$ relative to the origin $O$

$$\Phi(\mathbf{R}) = \frac{1}{4\pi\varepsilon_0}\left[\frac{q}{R} + \frac{R_i}{R^3}p_i + \frac{3R_i R_j - R^2\delta_{ij}}{2R^5}q_{ij}\right.$$

$$\left. + \frac{5R_i R_j R_k - R^2(R_i\delta_{jk} + R_j\delta_{ki} + R_k\delta_{ij})}{2R^7}q_{ijk} + \cdots\right]. \tag{1.3}$$

Here $\delta_{ij}$ is the Kronecker delta function ($\delta_{ij} = 1$ if $i = j$, $\delta_{ij} = 0$ if $i \neq j$) and $q$, $p_i$, $q_{ij}$, $q_{ijk}, \ldots$ are electric multipole moments, relative to the origin $O$, defined as follows:

$$q = \int_V \rho(\mathbf{r})\, dv \tag{1.4}$$

is the zeroth moment, or electric monopole moment,

$$p_i = \int_V r_i\, \rho(\mathbf{r})\, dv \tag{1.5}$$

is the first moment, or electric dipole moment,

$$q_{ij} = \int_V r_i r_j\, \rho(\mathbf{r})\, dv \tag{1.6}$$

is the second moment, or electric quadrupole moment,

$$q_{ijk} = \int_V r_i r_j r_k \rho(\mathbf{r})\, dv \tag{1.7}$$

is the third moment, or electric octopole moment, and so on. In (1.3) we have introduced a notation which will be used throughout this book: a subscript $i, j, \ldots$ ($= 1, 2,$ or $3$) denotes a component of a Cartesian tensor, and a repeated subscript implies summation from 1 to 3. In (1.3) we have written the expansion explicitly to electric octopole order. (A reader who is unfamiliar with Cartesian tensors may find it helpful to refer to Section 3.3.)

For a finite, discrete charge distribution with $N$ charges $q^{(\alpha)}$ ($\alpha = 1, 2, \ldots, N$) located at positions $\mathbf{r}^{(\alpha)}$ one starts with

$$\Phi(\mathbf{R}) = \frac{1}{4\pi\varepsilon_0} \sum_{\alpha=1}^{N} \frac{q^{(\alpha)}}{|\mathbf{R} - \mathbf{r}^{(\alpha)}|}, \tag{1.8}$$

instead of (1.1). (Here we have, for clarity of notation, used a superscript $\alpha$ to label the charges: an alternative notation which is frequently used in the literature assigns two subscripts to a vector—one labelling the particle, the other its Cartesian component [1, 2].) The result is the expansion (1.3) with

$$q = \sum_{\alpha=1}^{N} q^{(\alpha)} \tag{1.9}$$

$$p_i = \sum_{\alpha=1}^{N} q^{(\alpha)} r_i^{(\alpha)} \tag{1.10}$$

$$q_{ij} = \sum_{\alpha=1}^{N} q^{(\alpha)} r_i^{(\alpha)} r_j^{(\alpha)} \tag{1.11}$$

$$q_{ijk} = \sum_{\alpha=1}^{N} q^{(\alpha)} r_i^{(\alpha)} r_j^{(\alpha)} r_k^{(\alpha)} \tag{1.12}$$

for the first four electric multipole moments. An alternative method of deducing (1.9)–(1.12) uses the Dirac delta function: this function is defined by $\delta(\mathbf{r}) = 0$ if $\mathbf{r} \neq 0$ and

$$\int \delta(\mathbf{r})\, dv = 1$$

if the region of integration includes the origin. If we replace $\rho(\mathbf{r})$ in (1.4)–(1.7) with $\sum_{\alpha=1}^{N} q^{(\alpha)} \delta(\mathbf{r} - \mathbf{r}^{(\alpha)})$ for a discrete distribution, we obtain (1.9)–(1.12).

The electric field $\mathbf{E} = -\boldsymbol{\nabla}\Phi$ can be found from (1.3). Writing $E_i = -\partial\Phi/\partial R_i$ we obtain

$$E_i(\mathbf{R}) = \frac{1}{4\pi\varepsilon_0}\left[\frac{R_i}{R^3}q + \frac{3R_iR_j - R^2\delta_{ij}}{R^5}p_j\right.$$
$$\left. + \frac{3}{2R^7}\left\{5R_iR_jR_k - R^2(R_i\delta_{jk} + R_j\delta_{ki} + R_k\delta_{ij})\right\}q_{jk} + \cdots\right]. \quad (1.13)$$

The expansions (1.3) and (1.13) show that:

(i) Each multipole contributes separately to the potential and field at an external point.

(ii) Each multipole behaves as if it were located at the origin. This is because its contribution to the potential and field contains, apart from the moment itself, only $\mathbf{R}$, the displacement of the field point $P$ from the origin $O$.

(iii) The potential and field due to each multipole depend not only on its moment and on its distance $R$ from the field point, but in general also on its orientation relative to $\mathbf{R}$. For example, the dipole contribution in (1.3) and (1.13) contains $R_ip_i = Rp\cos\theta$, where $\theta$ is the angle between $\mathbf{R}$ and $\mathbf{p}$.

(iv) The $R$-dependences of the potential and field, respectively, are charge: $R^{-1}$ and $R^{-2}$; dipole: $R^{-2}$ and $R^{-3}$; quadrupole: $R^{-3}$ and $R^{-4}$, etc.

(v) At sufficiently large distances (distance $\gg$ dimensions of charge distribution), the potential and field are dominated by the leading non-vanishing term in $1/R$, all higher-order terms being negligible. Associated with the leading non-vanishing term is a unique multipole moment independent of choice of origin (see Section 1.2). Thus, at a distant point the $SO_4^=$ ion behaves like a charge, HCl like a dipole, $CO_2$ like a quadrupole, $CH_4$ like an octopole, etc., even though each has higher moments. If the distance to the field point is not sufficiently large, then higher-order moments contribute as well.

(vi) The potential and field external to a neutral spherically symmetric charge distribution are zero.

We will refer to (1.4)–(1.7) and (1.9)–(1.12) as primitive moments to distinguish them from traceless moments which can be constructed for all terms beyond the electric dipole. For example, a commonly used definition of a traceless electric quadrupole moment is [1, 2]

$$\Theta_{ij} = \frac{1}{2}(3q_{ij} - q_{kk}\delta_{ij}). \quad (1.14)$$

Since $\delta_{ii} = 3$ we see that $\Theta_{ii} = 0$ as desired. According to (1.14), the quadrupole contribution to the potential (1.3) can be written as

$$\frac{1}{4\pi\varepsilon_0}\frac{3R_iR_j - R^2\delta_{ij}}{2R^5}\frac{1}{3}(2\Theta_{ij} + q_{kk}\delta_{ij}) = \frac{1}{4\pi\varepsilon_0}\frac{3R_iR_j - R^2\delta_{ij}}{3R^5}\Theta_{ij}.$$

Thus the term involving the trace $q_{kk}$ in (1.14) does not contribute to the potential, and one can use either the primitive moment $q_{ij}$ or the traceless moment $\Theta_{ij}$

to calculate the quadrupole contribution to the potential, and hence the electric field. Now $q_{ij}$ is symmetric and in general possesses six independent components, whereas the traceless property of $\Theta_{ij}$ means that it possesses five independent components. This reduction in the number of components is a consequence of Laplace's equation [3].

Higher-order traceless moments can be constructed in a similar manner, for example the traceless electric octopole moment [1,4]

$$\Omega_{ijk} = \frac{1}{2}(5q_{ijk} - q_{ill}\delta_{jk} - q_{jll}\delta_{ki} - q_{kll}\delta_{ij}). \tag{1.15}$$

Again, the traces of the primitive moment $q_{ijk}$ in (1.15) do not contribute to the potential (1.3), and one can use the traceless moment (1.15) to evaluate the octopole contribution to the potential.

The above definitions of traceless multipole moments are those used in molecular and crystal physics. Alternative definitions occur in the literature, such as [5] $\Theta_{ij} = q_{ij} - \frac{1}{3}q_{kk}\delta_{ij}$: with this definition the electrostatic potential (1.3) and field (1.13) are invariant under the replacement $q_{ij} \to \Theta_{ij}$. This property applies to electric multipoles of arbitrary order, and is indicative of the fact that for a pole of order $2^\ell$ ($\ell = 0, 1, 2, \ldots$) in electrostatics, the electric multipole moment can be represented by a symmetric, traceless tensor of rank $\ell$ having $2\ell + 1$ independent components rather than the $\frac{1}{2}(\ell + 1)(\ell + 2)$ independent components of the corresponding primitive moment [5]. Thus in electrostatics, primitive moments beyond the dipole provide more components than are necessary. A similar property applies in magnetostatics (Section 1.6).

A convenient formulation of the theory of traceless multipole moments can be given in terms of spherical harmonics. We do not discuss this formulation here for two reasons. First, detailed accounts abound in the literature [3], [5–7]. Second, for the material presented in this monograph, it is more convenient to work in a Cartesian basis. For example, the Cartesian basis is more suitable for the derivation of macroscopic electrodynamics in multipole theory (Sections 1.10–1.13); for use in quantum theory of multipole moments and polarizabilities (Chapter 2); and to determine the origin dependence of polarizability tensors (Section 3.7). Use of this basis is also favoured in tables of space–time symmetries of property tensors (Section 3.6).

We will return to the topic of primitive and traceless multipole moments again in Section 1.15, particularly in relation to their role in macroscopic electromagnetism.

## 1.2   Dependence of electric multipole moments on origin

In general, electric multipole moments beyond the monopole depend on the choice of origin. Consider, for example, an electric quadrupole moment and an origin $\bar{O}$ displaced by $\mathbf{d}$ from $O$. The position vector of an element of charge $\rho\, dv$ relative to $\bar{O}$ is $\bar{\mathbf{r}} = \mathbf{r} - \mathbf{d}$. Thus the quadrupole moment (1.6) relative to $\bar{O}$ is

$$\bar{q}_{ij} = \int_V (r_i - d_i)(r_j - d_j)\rho\,dv$$
$$= q_{ij} - p_j d_i - p_i d_j + q d_i d_j, \tag{1.16}$$

where $q$ and $p_i$ are the moments (1.4) and (1.5). (Equation (1.16) also holds for the quadrupole moment (1.11) of a discrete charge distribution.) Thus the quadrupole moment is independent of an arbitrary shift of origin only if the monopole and dipole moments $q$ and $p_i$ are zero. This may be generalized: only the leading non-vanishing electric multipole moment is independent of the choice of origin of coordinates. The matter of origin dependence of certain physical quantities plays an important role in this book (see Chapters 3–9). For discussion of the origin dependence of electric quadrupole moments in relation to an experiment to measure these moments, see Section 5.11.

## 1.3 Permanent and induced multipole moments

The permanent multipole moments possessed by the static charge distribution $\rho(\mathbf{r})$ considered so far are denoted by

$$p_i^{(0)}, \quad q_{ij}^{(0)}, \quad q_{ijk}^{(0)}, \quad \dots, \tag{1.17}$$

to distinguish them from the multipole moments in the presence of an applied electric field. Consider first a uniform applied electric field. The charges in the distribution will be displaced to new equilibrium positions different from their field-free positions. The corresponding multipole moments (the total moments) are denoted by

$$p_i, \quad q_{ij}, \quad q_{ijk}, \quad \dots \tag{1.18}$$

The differences

$$p_i - p_i^{(0)}, \quad q_{ij} - q_{ij}^{(0)}, \quad q_{ijk} - q_{ijk}^{(0)}, \quad \dots \tag{1.19}$$

are termed the induced electric multipole moments—dipole, quadrupole, octopole, . . . .

For a weak applied field one may assume that a given induced moment is proportional to the field. Then, for an anisotropic charge distribution,

$$p_i - p_i^{(0)} = \alpha_{ij} E_j \tag{1.20}$$
$$q_{ij} - q_{ij}^{(0)} = \mathfrak{a}_{ijk} E_k \tag{1.21}$$
$$q_{ijk} - q_{ijk}^{(0)} = \mathfrak{b}_{ijkl} E_l, \text{ etc.} \tag{1.22}$$

In these $\alpha_{ij}$, known as the (dipole) polarizability, is a constant of proportionality between the $i$th component of the induced dipole moment and the $j$th component of the uniform field that induces the moment. A similar role is played by $\mathfrak{a}_{ijk}$, termed the quadrupole polarizability (or quadrupolarizability), and $\mathfrak{b}_{ijkl}$, the octopole polarizability.

For a strong uniform field $p_i$ may be expanded in powers of the field

$$p_i = p_i^{(0)} + \alpha_{ij}E_j + \frac{1}{2}\beta_{ijk}E_jE_k + \frac{1}{6}\gamma_{ijkl}E_jE_kE_l + \cdots . \qquad (1.23)$$

This was used by Buckingham and Pople [8] in a theory of the Kerr effect, in which $\beta_{ijk}$, $\gamma_{ijkl}$, ... were collectively termed hyperpolarizabilities [9]. Like $\alpha_{ij}$ they are independent of the field and are thus properties of the unperturbed charge distribution. Similar expansions may be made for $q_{ij}$, $q_{ijk}$, .... At this stage equations (1.20)–(1.23) have only a phenomenological basis but they may be formally derived by means of quantum-mechanical time-independent pertur-bation theory (see Chapter 2). In much of this book only a linear response to a field, as in (1.20)–(1.22), will be assumed. An exception is the theory of the Kerr effect in Chapter 5.

A non-uniform electrostatic field also distorts a charge distribution, thereby inducing multipole moments. To electric octopole order the total moments may be expressed as (see Section 2.2)

$$p_i = p_i^{(0)} + \alpha_{ij}E_j + \frac{1}{2}a_{ijk}\nabla_kE_j + \frac{1}{6}b_{ijkl}\nabla_l\nabla_kE_j + \cdots \qquad (1.24)$$

$$q_{ij} = q_{ij}^{(0)} + \mathfrak{a}_{ijk}E_k + \frac{1}{2}d_{ijkl}\nabla_lE_k + \cdots \qquad (1.25)$$

$$q_{ijk} = q_{ijk}^{(0)} + \mathfrak{b}_{ijkl}E_l + \cdots . \qquad (1.26)$$

The tensors $\mathfrak{a}_{ijk}$, $\mathfrak{b}_{ijkl}$, $a_{ijk}$, ... are also properties of the undistorted distribu-tion. They are referred to collectively as polarizabilities. If a charge distribution is unaffected by an applied field, it is said to be non-polarizable or rigid.

In (1.24)–(1.26) it is assumed that the field is slowly varying inside the dis-tribution; the field and its gradients are evaluated at the origin at which the point multipoles are located in the distribution. Again these expressions may be derived by means of quantum-mechanical perturbation theory, and this also yields explicit expressions for the polarizability tensors (Chapter 2), from which any intrinsic symmetry of tensor subscripts and relationships between certain tensors may be deduced. Thus (see Sections 2.2 and 2.7)

$$\begin{cases} \alpha_{ij} = \alpha_{ji}, \quad a_{ijk} = a_{ikj} = \mathfrak{a}_{jki} \\ b_{ijkl} = b_{ijlk} = b_{ikjl} = \mathfrak{b}_{jkli} \\ d_{ijkl} = d_{klij} = d_{jikl}. \end{cases} \qquad (1.27)$$

## 1.4 Force and torque in an external electrostatic field

We consider a discrete charge distribution in an electrostatic field which is slowly varying. In the expression for the total force on the distribution

$$F_i = \sum_{\alpha=1}^{N} q^{(\alpha)}E_i(\mathbf{r}^{(\alpha)}) \qquad (1.28)$$

we make a series expansion of the field about the origin $O$

$$E_i(\mathbf{r}^{(\alpha)}) = E_i + (\nabla_j E_i)r_j^{(\alpha)} + \frac{1}{2}(\nabla_k \nabla_j E_i)r_j^{(\alpha)}r_k^{(\alpha)}$$
$$+ \frac{1}{6}(\nabla_l \nabla_k \nabla_j E_i)r_j^{(\alpha)}r_k^{(\alpha)}r_l^{(\alpha)} + \cdots, \tag{1.29}$$

where the field and its gradients are evaluated at the origin. Using (1.29) in (1.28) and the multipole moments (1.9)–(1.12), we obtain

$$F_i = qE_i + p_j\nabla_j E_i + \frac{1}{2}q_{jk}\nabla_k\nabla_j E_i + \frac{1}{6}q_{jkl}\nabla_l\nabla_k\nabla_j E_i + \cdots. \tag{1.30}$$

Note that the multipole moments in (1.30) are the total moments (1.18). The same result (1.30) can be derived for a continuous distribution.

In a similar manner, the total torque on the distribution is

$$\begin{aligned}N_i &= \sum_{\alpha=1}^{N}(\mathbf{r}^{(\alpha)} \times \mathbf{F}^{(\alpha)})_i\\&= \sum_{\alpha=1}^{N}\varepsilon_{ijk}r_j^{(\alpha)}q^{(\alpha)}E_k(\mathbf{r}^{(\alpha)})\\&= \varepsilon_{ijk}\left[p_j E_k + q_{jl}\nabla_l E_k + \frac{1}{2}q_{jlm}\nabla_m\nabla_l E_k + \cdots\right]. \end{aligned}\tag{1.31}$$

Here $\varepsilon_{ijk}$ is the Levi–Civita tensor

$$\varepsilon_{ijk} = \begin{cases} 1 & ijk = \text{any even permutation of }1\,2\,3 \\ -1 & ijk = \text{any odd permutation of }1\,2\,3 \\ 0 & \text{if any two subscripts are equal.} \end{cases} \tag{1.32}$$

The torque on a quadrupole (the term in $q_{jl}$ in (1.31)) is the basis of an experiment due to Buckingham to measure directly the electric quadrupole moment of a molecule [10,11]. (See also Section 5.11.) This experiment has been successfully performed in a number of laboratories [12–15].

## 1.5   Potential energy of a charge distribution in an electrostatic field

We deduce an expression for the potential energy $W$ of a charge distribution in an external electrostatic field in terms of the multipole moments of the distribution. The potential energy is given by minus the work done by the electrostatic force on the charges in bringing the distribution from outside the field into its configuration within the field, that is,

$$W = -\int_i^f \mathbf{F} \cdot d\mathbf{r}, \tag{1.33}$$

where the limits $i$ and $f$ denote the initial and final configurations. We assume that the electric field varies slowly in the distribution and use the expansion (1.30). Also, for an electrostatic field $\nabla \times \mathbf{E} = 0$ and therefore

$$\nabla_j E_i = \nabla_i E_j = \partial E_j / \partial r_i. \tag{1.34}$$

Using (1.30) and (1.34) in (1.33) we have

$$W = -\int_i^f \left[ -q\frac{\partial \Phi}{\partial r_i} + p_j \frac{\partial E_j}{\partial r_i} + \frac{1}{2}q_{jk}\frac{\partial}{\partial r_i}\nabla_k E_j + \frac{1}{6}q_{jkl}\frac{\partial}{\partial r_i}\nabla_l\nabla_k E_j + \cdots \right] dr_i$$

$$= \int_i^f \left[ q\, d\Phi - p_i\, dE_i - \frac{1}{2}q_{ij}\, d(\nabla_j E_i) - \frac{1}{6}q_{ijk}\, d(\nabla_k\nabla_j E_i) + \cdots \right], \tag{1.35}$$

where $\Phi$ is the electrostatic potential. In (1.35) the potential, field, and its derivatives are evaluated at the origin. As in (1.30), the multipole moments are the total moments. They are given by

$$p_i = -\frac{\partial W}{\partial E_i}, \quad q_{ij} = -2\frac{\partial W}{\partial(\nabla_j E_i)}, \quad q_{ijk} = -6\frac{\partial W}{\partial(\nabla_k\nabla_j E_i)}, \quad \text{etc.} \tag{1.36}$$

The quantum-mechanical expressions corresponding to (1.36) are derived in Section 2.3.

For a rigid (non-polarizable) distribution the multipole moments in (1.35) are constant, equal to the permanent moments in (1.17). Then the integrals in (1.35) can be performed to yield

$$W = q\Phi - p_i^{(0)}E_i - \frac{1}{2}q_{ij}^{(0)}\nabla_j E_i - \frac{1}{6}q_{ijk}^{(0)}\nabla_k\nabla_j E_i + \cdots. \tag{1.37}$$

In the special case of a neutral rigid distribution in a uniform field this reduces to

$$W = -\mathbf{p}^{(0)} \cdot \mathbf{E}. \tag{1.38}$$

Equation (1.37) corresponds to various rigid multipoles interacting with the potential, the field, or field gradients. Thus a charge interacts with the potential $\Phi$ with energy $q\Phi$; a dipole interacts with the field with energy $-p_i^{(0)}E_i$; a quadrupole interacts with the field gradient with energy $-\frac{1}{2}q_{ij}^{(0)}(\nabla_j E_i)$, and so on. These various energies allow one to determine the approximate interaction energy between any two distant atoms or molecules, because the leading multipole of the one interacts with the relevant field or field gradient established at it by the leading multipole of the other. For example, HCl and $N_2$ interact at a distance through the dipole of HCl (this being its leading multipole) experiencing the field set up by the leading moment of $N_2$, namely its quadrupole. Alternatively, the quadrupole of $N_2$ experiences the field gradient due to the HCl dipole. The two approaches yield the same result for the interaction energy.

## 1.6  Multipole expansion for the vector potential of a finite distribution of steady current

We consider a finite distribution of steady currents in vacuum. The vector potential of the distribution is given by

$$\mathbf{A}(\mathbf{R}) = \frac{\mu_0}{4\pi} \int_V \frac{\mathbf{J}(\mathbf{r})\, dv}{|\mathbf{R} - \mathbf{r}|}, \tag{1.39}$$

where $\mathbf{R}$ is the position vector of the field point $P$, $\mathbf{r}$ is the position vector of the volume element $dv$, $\mathbf{J}(\mathbf{r})$ is the current density at $dv$, and $V$ is the volume of the distribution (see Fig. 1.2). Using the expansion (1.2), we can expand (1.39) as follows

$$A_i(\mathbf{R}) = \frac{\mu_0}{4\pi} \left[ \frac{1}{R} \int_V J_i\, dv + \frac{R_j}{R^3} \int_V J_i r_j\, dv + \frac{3R_j R_k - R^2 \delta_{jk}}{2R^5} \int_V J_i r_j r_k\, dv + \cdots \right]. \tag{1.40}$$

(Note that for the electrostatic expansion (1.3) we have written terms to electric octopole order, whereas in the corresponding magnetostatic expansion (1.40) we have done so only to magnetic quadrupole order; this is discussed in Sections 1.10 and 1.14.)

The first term in (1.40) is zero because for steady conditions the continuity equation is $\nabla \cdot \mathbf{J} = 0$ and hence

$$\nabla_j(r_i J_j) = J_i. \tag{1.41}$$

Using (1.41) and Gauss' theorem we have

$$\int_V J_i\, dv = \int_S r_i J_j n_j\, ds = 0. \tag{1.42}$$

Here $ds$ is an element of area on the surface $S$ enclosing $V$, and $\mathbf{n}$ is the outward normal to $ds$. In (1.42) we have used the boundary condition $\mathbf{J} \cdot \mathbf{n} = 0$ on $S$.

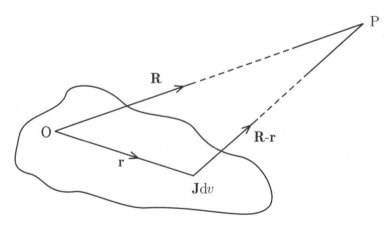

FIG. 1.2. Coordinates and notation used for a finite, continuous current distribution.

With regard to the other terms in (1.40), it is not convenient to define the magnetic multipoles in terms of the integrals in this expansion. These are first transformed in a manner that facilitates comparison with electrostatics and with quantum-mechanical expressions (Chapter 2): it is shown in Appendix A that for magnetostatics (i)

$$R_j \int_V J_i r_j \, dv = (\mathbf{m} \times \mathbf{R})_i, \tag{1.43}$$

where

$$\mathbf{m} = \frac{1}{2} \int_V \mathbf{r} \times \mathbf{J} \, dv \tag{1.44}$$

and (ii)

$$(3R_j R_k - R^2 \delta_{jk}) \int_V J_i r_j r_k \, dv = \varepsilon_{ijl}(3R_k R_l - R^2 \delta_{kl}) m_{jk}, \tag{1.45}$$

where

$$m_{ij} = \frac{2}{3} \int_V (\mathbf{r} \times \mathbf{J})_i r_j \, dv. \tag{1.46}$$

Using (1.42), (1.43), and (1.45) in (1.40) we have the desired expression

$$A_i(\mathbf{R}) = \frac{\mu_0}{4\pi} \left[ \frac{1}{R^3}(\mathbf{m} \times \mathbf{R})_i + \frac{1}{2R^5}\varepsilon_{ijl}(3R_k R_l - R^2 \delta_{kl})m_{jk} + \cdots \right]. \tag{1.47}$$

Here $\mathbf{m}$ given by (1.44) is the magnetic dipole moment of the steady current distribution, and $m_{ij}$ given in (1.46) is the magnetic quadrupole moment. The magnetic field $\mathbf{B} = \nabla \times \mathbf{A}$ of the steady currents can be obtained from (1.47). The calculation is outlined in Appendix B and it yields

$$B_i(\mathbf{R}) = \frac{\mu_0}{4\pi} \left[ \frac{3R_i R_j - R^2 \delta_{ij}}{R^5} m_j \right.$$
$$\left. + \frac{3}{2R^7} \left\{ 5R_i R_j R_k - R^2(R_i \delta_{jk} + R_j \delta_{ki} + R_k \delta_{ij}) \right\} m_{jk} + \cdots \right]. \tag{1.48}$$

It is interesting to compare the dipole and quadrupole orders of the electrostatic scalar potential (1.3) with the corresponding orders for the magnetostatic vector potential (1.47). These result in similar expressions for the dipole contributions, and for the quadrupole contributions, to the electrostatic field (1.13) and the magnetostatic field (1.48) in terms of the respective multipole moments.

It is also useful to express the magnetic multipoles in discrete form. To do so we use the relation $\mathbf{J} \, dv = \mathbf{u} \, dq$, where $\mathbf{u}$ is the velocity of the charge $dq$. Then (1.44) becomes

$$\mathbf{m} = \frac{1}{2} \int \mathbf{r} \times \mathbf{u} \, dq \longrightarrow \frac{1}{2} \sum_{\alpha=1}^{N} q^{(\alpha)} \mathbf{r}^{(\alpha)} \times \mathbf{u}^{(\alpha)} \tag{1.49}$$

for a distribution of $N$ discrete charges. In terms of the angular momentum of a particle $\mathbf{l}^{(\alpha)} = \mathbf{r}^{(\alpha)} \times m^{(\alpha)}\mathbf{u}^{(\alpha)}$, where $m^{(\alpha)}$ is the mass of the particle, this is

$$\mathbf{m} = \sum_{\alpha=1}^{N} \frac{q^{(\alpha)}}{2m^{(\alpha)}} \mathbf{l}^{(\alpha)}. \tag{1.50}$$

Similarly, the magnetic quadrupole moment (1.46) can be expressed as

$$m_{ij} = \sum_{\alpha=1}^{N} \frac{2q^{(\alpha)}}{3m^{(\alpha)}} l_i^{(\alpha)} r_j^{(\alpha)}. \tag{1.51}$$

The quadrupole moment (1.46) is a primitive moment (Section 1.1). One can also construct traceless, symmetric magnetostatic moments, such as the quadrupole moment [16]

$$\Gamma_{ij} = \frac{1}{2}(3m'_{ij} - m'_{kk}\delta_{ij}), \tag{1.52}$$

where $m'_{ij} = \frac{1}{2}(m_{ij} + m_{ji})$ is the symmetric primitive moment. Using (1.52) one can express the quadrupole contribution to the field (1.48) in terms of the traceless moment $\Gamma_{ij}$: the trace $m_{kk}$ of the primitive moment does not contribute. This result can be extended to arbitrary multipole order, and hence for the magnetostatic field, primitive moments beyond dipole provide more components than are necessary [5]. In particular, for a pole of order $2^\ell$ $(\ell = 0, 1, 2, \dots)$ in magnetostatics, the magnetic multipole moment can be represented by a symmetric, traceless tensor of rank $\ell$ having $2\ell + 1$ independent components [5]. Note that unlike the electrostatic case (Section 1.1), the quadrupole term in the potential (1.47) cannot be expressed solely in terms of the symmetric, traceless moment (1.52), although it can be expressed in terms of the unsymmetric, traceless moment $\frac{1}{2}(3m_{ij} - m_{kk}\delta_{ij})$.

## 1.7  Dependence of magnetic multipole moments on origin

Magnetic multipole moments, like their electric counterparts (Section 1.2), are in general origin dependent. As an example consider the magnetic quadrupole moment (1.46) with respect to an origin $\bar{O}$, that is,

$$\bar{m}_{ij} = \frac{2}{3} \int_V [(\mathbf{r} - \mathbf{d}) \times \mathbf{J}]_i (r_j - d_j) \, dv, \tag{1.53}$$

where $\mathbf{d}$ is the position vector of $\bar{O}$ with respect to $O$. Using (1.42) and (1.44) in (1.53) we have

$$\bar{m}_{ij} = m_{ij} - \frac{2}{3}m_i d_j - \frac{2}{3}\varepsilon_{ikl} d_k \int_V J_l \, r_j \, dv. \tag{1.54}$$

Consider first the integral in (1.54): of the nine elements of this integral, the three diagonal elements are zero because of (A.15) in Appendix A, and the six off-diagonal elements are zero when the magnetic dipole moment $m_i$ in (A.15) is zero. Thus, according to (1.54) the condition for the magnetic quadrupole moment $m_{ij}$ to be invariant for an arbitrary displacement $\mathbf{d}$ of the origin is that the magnetic dipole moment $m_i$ should be zero. Presumably, this is an example of a general result, already established for electric moments (Section 1.2), that only the leading non-vanishing magnetic moment is origin independent.

## 1.8 Force and torque in an external magnetostatic field

Consider a finite distribution of steady currents in a static magnetic field $\mathbf{B}(\mathbf{r})$. The force on an element of charge $dq$ moving with velocity $\mathbf{u}$ is

$$d\mathbf{F} = dq\,\mathbf{u} \times \mathbf{B}$$
$$= \mathbf{J} \times \mathbf{B}\,dv. \tag{1.55}$$

The total force on the distribution is therefore

$$F_i = \varepsilon_{ijk} \int_V J_j\,B_k\,dv. \tag{1.56}$$

We assume a slowly varying field and make a series expansion about the origin $O$

$$B_k(\mathbf{r}) = B_k + (\nabla_l B_k)r_l + \frac{1}{2}(\nabla_m \nabla_l B_k)r_l r_m + \cdots. \tag{1.57}$$

Using (1.57) and (1.42) in (1.56) we have

$$F_i = \varepsilon_{ijk}\left[(\nabla_l B_k)\int_V J_j\,r_l\,dv + \frac{1}{2}(\nabla_m \nabla_l B_k)\int_V J_j\,r_l r_m\,dv + \cdots\right]. \tag{1.58}$$

In Appendix C it is shown how to express (1.58) in terms of the magnetic multipole moments (1.44) and (1.46). The result is

$$F_i = m_j \nabla_j B_i + \frac{1}{2}m_{jk}\nabla_k \nabla_j B_i + \cdots. \tag{1.59}$$

This result is similar to the corresponding electrostatic formula (1.30).

For the total torque on the distribution

$$\mathbf{N} = \int \mathbf{r} \times d\mathbf{F} \tag{1.60}$$

we first use (1.55) to express this as

$$\mathbf{N} = \int_V \mathbf{r} \times (\mathbf{J} \times \mathbf{B})\,dv. \tag{1.61}$$

Thus

$$N_i = \varepsilon_{ijk}\,\varepsilon_{klm} \int_V r_j J_l B_m\,dv. \tag{1.62}$$

In (1.62) we use the expansion (1.57). Then

$$N_i = \varepsilon_{ijk}\,\varepsilon_{klm}\int r_j J_l\,[B_m + (\nabla_n B_m)r_n + \cdots]\,dv. \qquad (1.63)$$

In Appendix D it is shown how to express (1.63) in terms of the magnetic dipole and quadrupole moments. The result is

$$N_i = \varepsilon_{ijk}\left[m_j B_k + \frac{1}{2}(m_{jl} + m_{lj})\nabla_l B_k + \cdots\right]. \qquad (1.64)$$

This may be compared with the corresponding result (1.31) for the torque on a charge distribution in an electrostatic field. We see that the dipole terms are analogues of each other, as are the quadrupole terms. (In comparing the latter, recall that $q_{jl}$ in (1.31) is symmetric and therefore equal to $\frac{1}{2}(q_{jl} + q_{lj})$.)

## 1.9   Potential energy of a current distribution in a magnetostatic field

The calculation is similar to that in Section 1.5 for a charge distribution in an electrostatic field. We assume that the external magnetic field varies slowly in the current distribution. The potential energy is given by (1.33) and (1.59). We also have $\nabla \times \mathbf{B} = 0$ because the sources of the external field are outside the distribution and the currents are steady. Thus $\nabla_j B_i = \nabla_i B_j$ in (1.59), and (1.33) yields

$$\begin{aligned}
W &= -\int_i^f \left[m_j \frac{\partial B_j}{\partial r_i} + \frac{1}{2}m_{jk}\frac{\partial}{\partial r_i}\nabla_k B_j + \cdots\right]dr_i\\
&= -\int_i^f \left[m_i\,dB_i + \frac{1}{2}m_{ij}\,d(\nabla_j B_i) + \cdots\right]. \qquad (1.65)
\end{aligned}$$

Here the field and its derivatives are evaluated at the origin, and the multipole moments are the total moments. From (1.65)

$$m_i = -\frac{\partial W}{\partial B_i}, \qquad m_{ij} = -2\,\frac{\partial W}{\partial(\nabla_j B_i)}, \qquad \text{etc.} \qquad (1.66)$$

For a rigid (non-polarizable) distribution the multipole moments in (1.65) are constants, equal to the permanent moments $m_i^{(0)}$, $m_{ij}^{(0)}$, .... Then (1.65) yields

$$W = -m_i^{(0)} B_i - \frac{1}{2}m_{ij}^{(0)}\nabla_j B_i + \cdots. \qquad (1.67)$$

This is the magnetostatic counterpart of (1.37).

## 1.10   Multipole expansions for the dynamic scalar and vector potentials

Up to this point we have restricted ourselves to electrostatic and magnetostatic properties. We now consider a finite distribution of time-dependent charge and current densities $\rho(\mathbf{r}, t)$ and $\mathbf{J}(\mathbf{r}, t)$. These give rise to time-dependent scalar and vector potentials at a field point $\mathbf{R}$ relative to an origin of coordinates $O$ inside the distribution. In the Lorenz gauge [17, 18]

$$\Phi(\mathbf{R}, t) = \frac{1}{4\pi\varepsilon_0} \int_V \frac{\rho(\mathbf{r}, t - |\mathbf{R} - \mathbf{r}|/c)\, dv}{|\mathbf{R} - \mathbf{r}|} \tag{1.68}$$

and

$$\mathbf{A}(\mathbf{R}, t) = \frac{\mu_0}{4\pi} \int_V \frac{\mathbf{J}(\mathbf{r}, t - |\mathbf{R} - \mathbf{r}|/c)\, dv}{|\mathbf{R} - \mathbf{r}|}. \tag{1.69}$$

Here $c$ is the speed of light in vacuum, $V$ is the volume of the distribution, and $\rho$ and $\mathbf{J}$ are the charge and current densities at the retarded time $t - |\mathbf{R} - \mathbf{r}|/c$ at a microscopic volume element $dv$ with position vector $\mathbf{r}$ relative to $O$.

We consider a distant field point $(R \gg r)$ and use the binomial theorem to expand $|\mathbf{R} - \mathbf{r}|$. This enables us to write the retarded time at $dv$ as

$$t - |\mathbf{R} - \mathbf{r}|/c = t' + \Delta t, \tag{1.70}$$

where

$$t' = t - R/c \tag{1.71}$$

is the retarded time at the origin $O$, and

$$\Delta t = \frac{1}{c}\left[\frac{R_i}{R} r_i + \frac{R_i R_j - R^2 \delta_{ij}}{2R^3} r_i r_j \right.$$
$$\left. + \frac{3R_i R_j R_k - R^2(R_i \delta_{jk} + R_j \delta_{ki} + R_k \delta_{ij})}{6R^5} r_i r_j r_k + \cdots \right]. \tag{1.72}$$

Consider first the scalar potential (1.68) and make a Taylor series expansion of $\rho(\mathbf{r}, t' + \Delta t)$ about $t'$, that is,

$$\rho(\mathbf{r}, t' + \Delta t) = \rho(\mathbf{r}, t') + \frac{\partial \rho(\mathbf{r}, t')}{\partial t'}\Delta t$$
$$+ \frac{1}{2!}\frac{\partial^2 \rho(\mathbf{r}, t')}{(\partial t')^2}(\Delta t)^2 + \frac{1}{3!}\frac{\partial^3 \rho(\mathbf{r}, t')}{(\partial t')^3}(\Delta t)^3 + \cdots. \tag{1.73}$$

From (1.72) and (1.73), and using the abbreviated notation $\rho = \rho(\mathbf{r}, t')$, we have

$$\rho(\mathbf{r}, t' + \Delta t) = \rho + \dot\rho\frac{R_i}{cR} r_i + \left[\dot\rho\frac{R_i R_j - R^2 \delta_{ij}}{2cR^3} + \ddot\rho\frac{R_i R_j}{2c^2 R^2}\right] r_i r_j$$
$$+ \left[\frac{1}{6cR^5}\left(\dot\rho + \frac{R}{c}\ddot\rho\right)\{3R_i R_j R_k - R^2(R_i \delta_{jk} + R_j \delta_{ki} + R_k \delta_{ij})\}\right.$$
$$\left. + \dddot\rho\frac{R_i R_j R_k}{6c^3 R^3}\right] r_i r_j r_k + \cdots, \tag{1.74}$$

where a dot indicates a partial derivative with respect to time. From (1.68), (1.70), (1.74), and (1.2) we find

$$
\begin{aligned}
\Phi(\mathbf{R}, t) = \frac{1}{4\pi\varepsilon_0} \Bigg[ & \frac{1}{R} \int_V \rho\, dv + \frac{R_i}{R^3} \int_V r_i \left( \rho + \frac{R}{c}\dot{\rho} \right) dv \\
& + \frac{3R_iR_j - R^2\delta_{ij}}{2R^5} \int_V r_ir_j \left( \rho + \frac{R}{c}\dot{\rho} \right) dv + \frac{R_iR_j}{2c^2R^3} \int_V r_ir_j\ddot{\rho}\, dv \\
& + \frac{5R_iR_jR_k - R^2(R_i\delta_{jk} + R_j\delta_{ki} + R_k\delta_{ij})}{2R^7} \int_V r_ir_jr_k \left( \rho + \frac{R}{c}\dot{\rho} \right) dv \\
& + \frac{6R_iR_jR_k - R^2(R_i\delta_{jk} + R_j\delta_{ki} + R_k\delta_{ij})}{6c^2R^5} \int_V r_ir_jr_k\ddot{\rho}\, dv \\
& + \frac{R_iR_jR_k}{6c^3R^4} \int_V r_ir_jr_k\dddot{\rho}\, dv + \cdots \Bigg].
\end{aligned}
\tag{1.75}
$$

In terms of the electric multipole moments given by (1.4)–(1.7) with $\rho(\mathbf{r})$ replaced by $\rho(\mathbf{r}, t')$, the expansion (1.75) is

$$
\begin{aligned}
\Phi(\mathbf{R}, t) = \frac{1}{4\pi\varepsilon_0} \Bigg[ & \frac{q}{R} + \frac{R_i}{R^3}\left( p_i + \frac{R}{c}\dot{p}_i \right) + \frac{3R_iR_j - R^2\delta_{ij}}{2R^5}\left( q_{ij} + \frac{R}{c}\dot{q}_{ij} \right) \\
& + \frac{R_iR_j}{2c^2R^3}\ddot{q}_{ij} + \frac{5R_iR_jR_k - R^2(R_i\delta_{jk} + R_j\delta_{ki} + R_k\delta_{ij})}{2R^7}\left( q_{ijk} + \frac{R}{c}\dot{q}_{ijk} \right) \\
& + \frac{6R_iR_jR_k - R^2(R_i\delta_{jk} + R_j\delta_{ki} + R_k\delta_{ij})}{6c^2R^5}\ddot{q}_{ijk} + \frac{R_iR_jR_k}{6c^3R^4}\dddot{q}_{ijk} + \cdots \Bigg].
\end{aligned}
\tag{1.76}
$$

Note that for static phenomena (1.76) reduces to the electrostatic potential (1.3).

Consider next the expansion of the vector potential [19]: from (1.69)–(1.72) and (1.2), and using the same procedure as above for the scalar potential, we obtain

$$
\begin{aligned}
A_i(\mathbf{R}, t) = \frac{\mu_0}{4\pi} \Bigg[ & \frac{1}{R} \int_V J_i\, dv + \frac{R_j}{R^3} \left( \int_V J_ir_j\, dv + \frac{R}{c} \int_V \dot{J}_ir_j\, dv \right) \\
& + \frac{3R_jR_k - R^2\delta_{jk}}{2R^5} \left( \int_V J_ir_jr_k\, dv + \frac{R}{c} \int_V \dot{J}_ir_jr_k\, dv \right) \\
& + \frac{R_jR_k}{2c^2R^3} \int_V \ddot{J}_ir_jr_k\, dv + \cdots \Bigg].
\end{aligned}
\tag{1.77}
$$

Here $J_i = J_i(\mathbf{r}, t')$ and $t'$ is the retarded time at the origin given by (1.71). In Appendix A it is shown how to express the integrals in (1.77) in terms of magnetic and electric multipole moments, thus leading to

$$A_i(\mathbf{R}, t) = \frac{\mu_0}{4\pi} \left[ \frac{1}{R}\dot{p}_i + \frac{R_j}{R^3} \left\{ \left( \frac{1}{2}\dot{q}_{ij} - \varepsilon_{ijk}m_k \right) + \frac{R}{c} \left( \frac{1}{2}\ddot{q}_{ij} - \varepsilon_{ijk}\dot{m}_k \right) \right\} \right.$$
$$+ \frac{3R_j R_k - R^2 \delta_{jk}}{2R^5} \left\{ \left( \frac{1}{3}\dot{q}_{ijk} - \varepsilon_{ijl}m_{lk} \right) + \frac{R}{c} \left( \frac{1}{3}\ddot{q}_{ijk} - \varepsilon_{ijl}\dot{m}_{lk} \right) \right\}$$
$$\left. + \frac{R_j R_k}{2c^2 R^3} \left( \frac{1}{3}\dddot{q}_{ijk} - \varepsilon_{ijl}\ddot{m}_{lk} \right) + \cdots \right]. \tag{1.78}$$

Note that in (1.76) and (1.78) the multipole moments are located at an arbitrary origin in the distribution and are at the retarded time at the origin; that is, they are given by (1.5)–(1.7), with $\rho(\mathbf{r})$ replaced by $\rho(\mathbf{r}, t')$, and by (1.44) and (1.46) with $\mathbf{J}(\mathbf{r})$ replaced by $\mathbf{J}(\mathbf{r}, t')$. For a time-independent distribution (1.78) reduces to the magnetostatic potential (1.47). One can check that, as required, the Lorenz condition $\boldsymbol{\nabla} \cdot \mathbf{A} + c^{-2}\dot{\Phi} = 0$ is satisfied by (1.76) and (1.78).

A noteworthy feature of the vector potential (1.78) is the appearance of combinations of both magnetic and electric multipole moments. In particular, if one expands the vector potential to magnetic quadrupole order, as in (1.78), then it is necessary to retain terms of electric octopole order in (1.78) and in the expansion (1.76) for the scalar potential.

## 1.11 The far- and near-zone limits

We consider charge and current densities $\rho(\mathbf{r}, t)$ and $\mathbf{J}(\mathbf{r}, t)$, and therefore multipole moments, which vary harmonically with time as $\exp(-2\pi ict/\lambda)$. (This presupposes an underlying Fourier analysis and therefore involves no loss of generality [20].) In the far zone the distance to the field point is much greater than the wavelength ($R \gg \lambda$) and terms in $R^{-1}$ dominate in (1.76) and (1.78). Thus the far-zone potentials are

$$\Phi(\mathbf{R}, t) = \frac{1}{4\pi\varepsilon_0 R} \left[ q + \frac{R_i}{cR}\dot{p}_i(t') + \frac{R_i R_j}{2c^2 R^2}\ddot{q}_{ij}(t') + \frac{R_i R_j R_k}{6c^3 R^3}\dddot{q}_{ijk}(t') + \cdots \right] \tag{1.79}$$

$$A_i(\mathbf{R}, t) = \frac{\mu_0}{4\pi R} \left[ \dot{p}_i(t') + \frac{R_j}{cR} \left\{ \frac{1}{2}\ddot{q}_{ij}(t') - \varepsilon_{ijk}\dot{m}_k(t') \right\} \right.$$
$$\left. + \frac{R_j R_k}{2c^2 R^2} \left\{ \frac{1}{3}\dddot{q}_{ijk}(t') - \varepsilon_{ijl}\ddot{m}_{lk}(t') \right\} + \cdots \right]. \tag{1.80}$$

Equations (1.79) and (1.80) are the multipole forms of the scalar and vector potentials for the far zone, given explicitly to electric octopole–magnetic quadrupole order. The corresponding electric and magnetic fields are given by $\mathbf{E} = -\boldsymbol{\nabla}\Phi - \dot{\mathbf{A}}$ and $\mathbf{B} = \boldsymbol{\nabla} \times \mathbf{A}$. In evaluating these we use (1.71) in

$$\nabla_i \left[ \frac{1}{R} f_j(t') \right] = -\frac{R_i}{R^3} f_j(t') + \frac{1}{R} \frac{df_j(t')}{dt'} \frac{\partial t'}{\partial R_i}$$
$$\approx -\frac{R_i}{cR^2} \dot{f}_j(t') \tag{1.81}$$

in the far zone. Thus

$$E_i(\mathbf{R}, t) = \left(\frac{R_i R_j}{R^2} - \delta_{ij}\right) \dot{A}_j(\mathbf{R}, t) \tag{1.82}$$

and

$$B_i(\mathbf{R}, t) = -\varepsilon_{ijk} \frac{R_j}{cR} \dot{A}_k(\mathbf{R}, t), \tag{1.83}$$

so that

$$\mathbf{B} = \frac{\mathbf{R} \times \mathbf{E}}{cR}. \tag{1.84}$$

From (1.82) and (1.80) the electric field in the far zone is given explicitly as

$$E_i(\mathbf{R}, t) = \frac{1}{4\pi\varepsilon_0 c^2 R} \left[ \left\{ \frac{R_i R_j}{R^2} - \delta_{ij} \right\} \left\{ \ddot{p}_j(t') + \frac{R_k}{2cR} \dddot{q}_{jk}(t') + \frac{R_k R_l}{6c^2 R^2} \dddot{q}_{jkl}(t') + \cdots \right\} \right.$$
$$\left. - \varepsilon_{ijk} \left\{ \frac{R_k}{cR} \ddot{m}_j(t') + \frac{R_k R_l}{2c^2 R^2} \dddot{m}_{jl}(t') + \cdots \right\} \right]. \tag{1.85}$$

Next, we consider the opposite limit, the near zone ($R \ll \lambda$). In this limit, time-derivative terms such as $(R/c)\dot{p}_i$ in (1.76) are of order $(R/\lambda)p_i$ and therefore negligible compared to the corresponding term $p_i$ in this expression. Similarly, $(R/c)\dot{q}_{ij} \ll q_{ij}$ and so on. Thus we can write down the near-zone expression for $\Phi$ simply by neglecting time-derivative terms in (1.76). This yields

$$\Phi(\mathbf{R}, t) = \frac{1}{4\pi\varepsilon_0} \left[ \frac{1}{R} q(t') + \frac{R_i}{R^3} p_i(t') + \frac{3R_i R_j - R^2 \delta_{ij}}{2R^5} q_{ij}(t') \right.$$
$$\left. + \frac{5R_i R_j R_k - R^2 (R_i \delta_{jk} + R_j \delta_{ki} + R_k \delta_{ij})}{2R^7} q_{ijk}(t') + \cdots \right], \tag{1.86}$$

which has the same form as the electrostatic potential in (1.3). Similarly, from (1.78) the near-zone expansion for $\mathbf{A}$ is

$$A_i(\mathbf{R}, t) = \frac{\mu_0}{4\pi} \left[ \frac{1}{R} \dot{p}_i(t') + \frac{R_j}{R^3} \left( \frac{1}{2} \dot{q}_{ij}(t') - \varepsilon_{ijk} m_k(t') \right) \right.$$
$$\left. + \frac{3R_j R_k - R^2 \delta_{jk}}{2R^5} \left( \frac{1}{3} \dot{q}_{ijk}(t') - \varepsilon_{ijl} m_{lk}(t') \right) + \cdots \right]. \tag{1.87}$$

The contributions of the magnetic multipoles to (1.87) are identical to those in the magnetostatic potential (1.47).

## 1.12  Macroscopic media

So far we have been considering a distribution of charge in vacuum, and we now extend the discussion to bulk matter. In particular, we show how the electromagnetic potentials, obtained in the previous section for the near zone, can be used to probe certain electromagnetic properties for a macroscopic medium (see (1.99)–(1.107)). The medium is understood to consist of nuclei with bound electrons ("molecules") and perhaps also of unbound point charges.

The steps are as follows. First consider a volume element $dV$ in the medium, choose an external origin located at the field point $P$, and let $\mathbf{R}$ denote the position vector of the element $dV$ relative to $P$. We assume that $P$ is in the near zone of the fields due to $dV$ and use (1.86) and (1.87) to obtain the contributions $d\Phi^P$ and $d\mathbf{A}^P$ to the potentials at $P$. In writing these down we should let $\mathbf{R} \to -\mathbf{R}$ due to our choice of origin at $P$. Then

$$d\Phi^P(t) = \frac{1}{4\pi\varepsilon_0}\left[\frac{1}{R}\rho_f(\mathbf{R},t') - \frac{R_i}{R^3}P_i(\mathbf{R},t') + \frac{3R_iR_j - R^2\delta_{ij}}{2R^5}Q_{ij}(\mathbf{R},t')\right.$$
$$\left. - \frac{5R_iR_jR_k - R^2(R_i\delta_{jk} + R_j\delta_{ki} + R_k\delta_{ij})}{2R^7}Q_{ijk}(\mathbf{R},t') + \cdots\right]dV$$

$$(1.88)$$

and [19]

$$dA_i^P(t) = \frac{\mu_0}{4\pi}\left[\frac{1}{R}\dot{P}_i(\mathbf{R},t') - \frac{R_j}{R^3}\left(\frac{1}{2}\dot{Q}_{ij}(\mathbf{R},t') - \varepsilon_{ijk}M_k(\mathbf{R},t')\right)\right.$$
$$\left. + \frac{3R_jR_k - R^2\delta_{jk}}{2R^5}\left(\frac{1}{3}\dot{Q}_{ijk}(\mathbf{R},t') - \varepsilon_{ijl}M_{lk}(\mathbf{R},t')\right) + \cdots\right]dV. \quad (1.89)$$

Here $\rho_f$ is the macroscopic "free" or "true" charge density [21,22] defined by

$$\rho_f(\mathbf{R},t') = \left\langle\sum_\alpha q^{(\alpha)}\delta(\mathbf{R} - \mathbf{R}^{(\alpha)})\right\rangle + \left\langle\sum_\alpha q_c^{(\alpha)}\delta(\mathbf{R} - \mathbf{R}_c^{(\alpha)})\right\rangle, \quad (1.90)$$

where $\mathbf{R}^{(\alpha)}$ is the position vector of the multipole origin of molecule $\alpha$, $\mathbf{R}_c^{(\alpha)}$ is the position vector of conduction charge $\alpha$, and the summations are over all molecules and conduction charges in the medium. Similarly, $P_i$ is the macroscopic electric dipole moment density defined by

$$P_i(\mathbf{R},t') = \left\langle\sum_\alpha p_i^{(\alpha)}\delta(\mathbf{R} - \mathbf{R}^{(\alpha)})\right\rangle. \quad (1.91)$$

In (1.90) and (1.91) the angular brackets denote a spatial average at time $t'$, and $q^{(\alpha)}$ and $p_i^{(\alpha)}$ are the electric charge and electric dipole moment of the $\alpha$th molecule as defined in (1.9) and (1.10). An essential feature of the averages in (1.90) and (1.91) is that they replace a discrete distribution of multipoles with a continuous distribution which varies on a scale much larger than molecular dimensions [21–23]. The other multipole moment densities in (1.88) and (1.89) (the electric quadrupole $Q_{ij}$, the electric octopole $Q_{ijk}$, the magnetic dipole $M_i$, and magnetic quadrupole $M_{ij}$) are defined in an analogous manner in terms of the corresponding moments in (1.11), (1.12), (1.50), and (1.51). These macroscopic moment densities are well-behaved functions of $\mathbf{R}$, and they are primitive moment densities because they are averages of primitive molecular moments (Sections 1.1 and 1.6).

Next we consider a cluster of adjacent volume elements $dV$ having a total volume $\mathcal{V} = \int dV$, and a field point $P$ which is outside $\mathcal{V}$. It is assumed that the dimensions of $\mathcal{V}$ are sufficiently small that $P$ is in the near zone of all the elements comprising $\mathcal{V}$. Then the contributions to the potentials at $P$ from the multipoles in $\mathcal{V}$ are obtained by integrating (1.88) and (1.89) over $\mathcal{V}$. In doing so it is helpful to first express the explicit functions of $\mathbf{R}$ in (1.88) and (1.89) in terms of gradients, using the relations

$$\nabla_i \frac{1}{R} = -\frac{R_i}{R^3} \tag{1.92}$$

$$\nabla_j \nabla_i \frac{1}{R} = \frac{3R_i R_j - R^2 \delta_{ij}}{R^5} \tag{1.93}$$

$$\nabla_k \nabla_j \nabla_i \frac{1}{R} = -3 \frac{5R_i R_j R_k - R^2(R_i \delta_{jk} + R_j \delta_{ki} + R_k \delta_{ij})}{R^7}. \tag{1.94}$$

Using (1.92)–(1.94) in (1.88) and (1.89) yields the scalar and vector potentials at the field point $P$ and at time $t$, due to the multipoles in $\mathcal{V}$

$$\Phi^P(t) = \frac{1}{4\pi\varepsilon_0} \left[ \int_{\mathcal{V}} \frac{1}{R}\rho_f \, dV + \int_{\mathcal{V}} P_i \nabla_i \frac{1}{R} \, dV + \frac{1}{2} \int_{\mathcal{V}} Q_{ij} \nabla_j \nabla_i \frac{1}{R} \, dV \right.$$
$$\left. + \frac{1}{6} \int_{\mathcal{V}} Q_{ijk} \nabla_k \nabla_j \nabla_i \frac{1}{R} \, dV + \cdots \right] \tag{1.95}$$

and [19]

$$A_i^P(t) = \frac{\mu_0}{4\pi} \left[ \int_{\mathcal{V}} \frac{1}{R} \dot{P}_i \, dV + \int_{\mathcal{V}} T_{ij} \nabla_j \frac{1}{R} \, dV + \int_{\mathcal{V}} T_{ijk} \nabla_k \nabla_j \frac{1}{R} \, dV + \cdots \right], \tag{1.96}$$

where

$$T_{ij} = \frac{1}{2}\dot{Q}_{ij} - \varepsilon_{ijk} M_k \tag{1.97}$$

$$T_{ijk} = \frac{1}{6}\dot{Q}_{ijk} - \frac{1}{2}\varepsilon_{ijl} M_{lk}. \tag{1.98}$$

In (1.95) and (1.96) the multipole moment densities are those at the position $\mathbf{R}$ in the medium at time $t' = t - R/c$.

The first term in (1.95) has the $R^{-1}$ dependence associated with charge density in (1.68), while the first term in (1.96) has the $R^{-1}$ dependence characteristic of current density in (1.69). To interpret the other terms in (1.95) and (1.96), which are associated successively with increasing powers of $R^{-1}$, we use Gauss' theorem to transform these terms (Appendix E). Then

$$\Phi^P = \frac{1}{4\pi\varepsilon_0}\left[\int_{\mathcal{V}}\frac{1}{R}\left\{\rho_f - \nabla_i P_i + \frac{1}{2}\nabla_j\nabla_i Q_{ij} - \frac{1}{6}\nabla_k\nabla_j\nabla_i Q_{ijk}+\cdots\right\}dV\right.$$

$$+ \oint_S \frac{1}{R}\left\{P_i - \frac{1}{2}\nabla_j Q_{ij} + \frac{1}{6}\nabla_k\nabla_j Q_{ijk} + \cdots\right\}n_i\,dS$$

$$- \oint_S \frac{R_j}{R^3}\left\{\frac{1}{2}Q_{ij} - \frac{1}{6}\nabla_k Q_{ijk} + \cdots\right\}n_i\,dS$$

$$\left.+ \oint_S \frac{3R_jR_k - R^2\delta_{jk}}{R^5}\left\{\frac{1}{6}Q_{ijk} + \cdots\right\}n_i\,dS + \cdots\right] \tag{1.99}$$

and

$$A_i^P = \frac{\mu_0}{4\pi}\left[\int_{\mathcal{V}}\frac{1}{R}\left\{\dot{P}_i - \nabla_j\left(\frac{1}{2}\dot{Q}_{ij} - \varepsilon_{ijk}M_k\right) + \nabla_k\nabla_j\left(\frac{1}{6}\dot{Q}_{ijk} - \frac{1}{2}\varepsilon_{ijl}M_{lk}\right) + \cdots\right\}dV\right.$$

$$+ \oint_S\frac{1}{R}\left\{\left(\frac{1}{2}\dot{Q}_{ij} - \varepsilon_{ijk}M_k\right)n_j - \nabla_j\left(\frac{1}{6}\dot{Q}_{ijk} - \frac{1}{2}\varepsilon_{ijl}M_{lk}\right)n_k + \cdots\right\}dS$$

$$\left.- \oint_S\frac{R_k}{R^3}\left\{\left(\frac{1}{6}\dot{Q}_{ijk} - \frac{1}{2}\varepsilon_{ijl}M_{lk}\right)n_j + \cdots\right\}dS + \cdots\right]. \tag{1.100}$$

Here $dS$ is an element of area on the surface $S$ enclosing the volume $\mathcal{V}$ and $\mathbf{n}$ is the outward normal to $dS$. Equations (1.99) and (1.100) are the desired forms of the scalar and vector potentials due to matter in $\mathcal{V}$, expressed in terms of the multipole moment densities. They lead to the following interpretations.

Because of the characteristic dependence in (1.99) of $\Phi$ on $R^{-1}$ for charge, $R_i/R^3$ for electric dipole moment, $(3R_iR_j - R^2\delta_{ij})/2R^5$ for electric quadrupole moment (Section 1.1), we interpret the successive terms in curly brackets in (1.99) as:

(i) Total charge density $\rho_f + \rho_b$ at a volume element $dV$, where $\rho_b$ is the bound charge density

$$\rho_b = -\nabla_i P_i + \frac{1}{2}\nabla_i\nabla_j Q_{ij} - \frac{1}{6}\nabla_k\nabla_j\nabla_i Q_{ijk} + \cdots. \tag{1.101}$$

(ii) Bound surface charge density at a surface element $dS$

$$\sigma_b = \left(P_i - \frac{1}{2}\nabla_j Q_{ij} + \frac{1}{6}\nabla_k\nabla_j Q_{ijk} + \cdots\right)n_i. \tag{1.102}$$

(iii) Bound surface electric dipole moment density at $dS$

$$\mathcal{P}_{bi} = -\left(\frac{1}{2}Q_{ij} - \frac{1}{6}\nabla_k Q_{ijk} + \cdots\right)n_j. \tag{1.103}$$

(iv) Bound surface electric quadrupole moment density at $dS$

$$\mathcal{Q}_{bij} = \left(\frac{1}{3}Q_{ijk} + \cdots\right) n_k. \tag{1.104}$$

Now $\mathcal{V}$ is a small but otherwise arbitrary volume within the medium. Thus the charge density $\rho_b(\mathbf{R}, t)$ given by (1.101) applies at any point $\mathbf{R}$ of the medium. With regard to the surface densities (1.102)–(1.104), one may ask to what surface do they apply? As derived above, they apply to the surface $S$ of the volume $\mathcal{V}$. However, $\mathcal{V}$ is an imagined construction within the macroscopic medium, and there cannot be any "surface" density associated with it. This becomes clear when we take account of all the volumes $\mathcal{V}$ comprising the medium: the "surface" terms (1.102)–(1.104) due to adjacent volumes inside the medium cancel because $P_i$, $Q_{ij}$, etc. are continuous and because the unit vector $n_i$ of adjacent volumes points in opposite directions. However, no such cancellation occurs on any part of a surface $S$ which is coincident with the surface of the macroscopic medium. Thus the densities (1.102)–(1.104) apply at the surface of the medium.

Similarly, because of the characteristic dependence in (1.100) of $\mathbf{A}$ on $R^{-1}$ for current density, $R_i/R^3$ for magnetic dipole moment (Section 1.6), we interpret the successive terms in curly brackets in (1.100) as:

(i) Bound current density at a volume element $dV$

$$J_{bi} = \dot{P}_i - \nabla_j \left(\frac{1}{2}\dot{Q}_{ij} - \varepsilon_{ijk}M_k\right) + \nabla_k\nabla_j \left(\frac{1}{6}\dot{Q}_{ijk} - \frac{1}{2}\varepsilon_{ijl}M_{lk}\right) + \cdots . \tag{1.105}$$

(ii) Bound surface current density at a surface element $dS$

$$K_{bi} = \left\{ \left(\frac{1}{2}\dot{Q}_{ij} - \varepsilon_{ijk}M_k\right) - \nabla_k \left(\frac{1}{6}\dot{Q}_{ijk} - \frac{1}{2}\varepsilon_{ikl}M_{lj}\right) + \cdots \right\} n_j. \tag{1.106}$$

(iii) Bound surface magnetic dipole moment current density at $dS$

$$L_{bij} = - \left(\frac{1}{6}\dot{Q}_{ijk} - \frac{1}{2}\varepsilon_{ikl}M_{lj}\right) n_k + \cdots . \tag{1.107}$$

As in the above discussion of (1.101)–(1.104) we conclude that $\mathbf{J}_b(\mathbf{R}, t)$ is the bound current density at a point $\mathbf{R}$ of the medium, and $K_{bi}$ and $L_{bij}$ are densities on the surface of the medium. Note that in deducing (1.100) and (1.105) we have omitted contributions due to motion of free charge and of molecules as units (motion of the centre of mass of a molecule). Taking these into account one obtains first a "free" current density $\mathbf{J}_f$ associated with motion of the charges that determine the free charge density (1.90); thus the total current density is $\mathbf{J}_f + \mathbf{J}_b$. Second, there are additional, rather complicated contributions to $\mathbf{J}_b$ in (1.105) involving the molecular moments and velocities. In stationary media these velocities are generally small and fluctuating; we therefore neglect the additional terms in $\mathbf{J}_b$ in what follows. For further discussion of this point, see [21–23] and references therein.

In the next section we use equations (1.101) and (1.105) for the bound charge and current densities in a macroscopic medium to obtain multipole forms for the response fields $\mathbf{D}$ and $\mathbf{H}$ in the inhomogeneous macroscopic Maxwell equations. The various surface densities obtained above play a role in the matching conditions on the fields, and some examples are discussed in Chapter 6.

## 1.13  Maxwell's macroscopic equations: multipole forms for D and H

Maxwell's microscopic equations are

$$\boldsymbol{\nabla} \cdot \mathbf{e} = \varepsilon_0^{-1}\eta \tag{1.108}$$

$$\boldsymbol{\nabla} \times \mathbf{e} = -\dot{\mathbf{b}} \tag{1.109}$$

$$\boldsymbol{\nabla} \cdot \mathbf{b} = 0 \tag{1.110}$$

$$\boldsymbol{\nabla} \times \mathbf{b} = \mu_0(\mathbf{j} + \varepsilon_0\dot{\mathbf{e}}), \tag{1.111}$$

where $\mathbf{e}(\mathbf{r}, t)$ and $\mathbf{b}(\mathbf{r}, t)$ are the electric and magnetic fields and $\eta(\mathbf{r}, t)$ and $\mathbf{j}(\mathbf{r}, t)$ are the microscopic charge and current densities.

For a macroscopic system these fields and source densities are averaged in the manner mentioned above [21–23]. When this is done, bound charge and current arise, as is evident in (1.101) and (1.105), and these supplement the averaged free source densities $\rho_f$ and $\mathbf{J}_f$. Thus Maxwell's macroscopic equations may be written

$$\boldsymbol{\nabla} \cdot \mathbf{E} = \varepsilon_0^{-1}(\rho_f + \rho_b) \tag{1.112}$$

$$\boldsymbol{\nabla} \times \mathbf{E} = -\dot{\mathbf{B}} \tag{1.113}$$

$$\boldsymbol{\nabla} \cdot \mathbf{B} = 0 \tag{1.114}$$

$$\boldsymbol{\nabla} \times \mathbf{B} = \mu_0(\mathbf{J}_f + \mathbf{J}_b + \varepsilon_0\dot{\mathbf{E}}), \tag{1.115}$$

in which the fields and source densities are now macroscopic.

With $\rho_b$ and $\mathbf{J}_b$ given by (1.101) and (1.105), the inhomogeneous Maxwell equations (1.112) and (1.115) may be rearranged to obtain, in tensor form and to electric octopole–magnetic quadrupole order,

$$\nabla_i\left(\varepsilon_0 E_i + P_i - \frac{1}{2}\nabla_j Q_{ij} + \frac{1}{6}\nabla_k\nabla_j Q_{ijk} + \cdots\right) = \rho_f \tag{1.116}$$

and

$$\varepsilon_{ijk}\nabla_j\left(\mu_0^{-1}B_k - M_k + \frac{1}{2}\nabla_l M_{kl} + \cdots\right)$$

$$= J_{fi} + \varepsilon_0\dot{E}_i + \dot{P}_i - \frac{1}{2}\nabla_j\dot{Q}_{ij} + \frac{1}{6}\nabla_k\nabla_j\dot{Q}_{ijk} + \cdots. \tag{1.117}$$

On the basis of these we define new macroscopic fields $\mathbf{D}$ and $\mathbf{H}$ by

$$D_i = \varepsilon_0 E_i + P_i - \frac{1}{2}\nabla_j Q_{ij} + \frac{1}{6}\nabla_k \nabla_j Q_{ijk} + \cdots \qquad (1.118)$$

$$H_i = \mu_0^{-1} B_i - M_i + \frac{1}{2}\nabla_j M_{ij} + \cdots. \qquad (1.119)$$

The terms in (1.118) and (1.119) are consistent to the order of electric octopole–magnetic quadrupole. Because $P_i, Q_{ij}, \ldots$ and $M_i, \ldots$ include induced contributions, which result from the response of matter to applied fields (see Section 1.3), $\mathbf{D}$ and $\mathbf{H}$ are termed response fields. Note that, as emphasized in Section 1.12, the moment densities in (1.118) and (1.119) are based on primitive moments.

According to (1.116)–(1.119), the response fields satisfy the inhomogeneous macroscopic Maxwell equations

$$\mathbf{\nabla} \cdot \mathbf{D} = \rho_f \qquad (1.120)$$

$$\mathbf{\nabla} \times \mathbf{H} = \mathbf{J}_f + \dot{\mathbf{D}}. \qquad (1.121)$$

Equations (1.118) and (1.119) can also be obtained using series expansions of the averaged microscopic charge and current densities, $\langle \eta \rangle$ and $\langle \mathbf{j} \rangle$, in coordinate space. Explicit calculations to electric quadrupole–magnetic dipole order are given in [21–23].

The averaging techniques referred to in this section are an essential aspect of the transition from microscopic to macroscopic electromagnetism. Detailed accounts of this topic are available in the literature [21–23], and here we make some brief comments. The averaging is a mathematical procedure designed to smooth out spatial variations which occur on a scale of order molecular dimensions. The resulting averaged multipole moment densities and fields are well-behaved quantities that vary on a scale $l$ which is large compared to molecular dimensions $l_o$. It is the condition that the parameter $l_o/l$ is small which makes possible the desired multipole expansions (1.101) and (1.105). One may say that the considerable simplification achieved in transforming microscopic electromagnetism into a macroscopic, continuum theory involves a certain cost, namely the loss of information regarding changes in physical quantities in microscopic space intervals. It is to physical phenomena where this loss is immaterial that the macroscopic theory should best apply. A detailed discussion of this point, and its elucidation in terms of truncated Fourier analysis, is given by Robinson [22].

For a wide variety of phenomena the multipole expansions are a good approximation. For example, in experiments in molecular physics using lasers the macroscopic fields are essentially harmonic, and the parameter $l_o/(\text{wavelength})$ is of order $10^{-3}$. There are, however, systems for which the multipole expansions are not valid, such as thermal plasmas [24], and these are not considered in this book.

Multipole theory provides a simplified yet powerful method of analyzing a variety of electromagnetic many-body problems by treating the system as a continuum, and much of the present book is concerned with properties and applications

of this approach. We will be particularly interested in discussing the theory and
its applications beyond electric dipole order. In certain instances we will treat
an effect from both microscopic and macroscopic viewpoints, and demonstrate
the equivalence of the results (see Sections 5.9–5.12).

## 1.14   Discussion

(i) In applications involving the response fields $\mathbf{D}$ and $\mathbf{H}$ given by (1.118) and
(1.119), it is necessary to include multipole contributions of comparable
magnitude. The expressions (1.78), (1.87), and (1.96) show that these are
ordered according to the following hierarchy

$$\text{electric dipole} \gg \begin{cases} \text{electric quadrupole} \\ \text{magnetic dipole} \end{cases} \gg \begin{cases} \text{electric octopole} \\ \text{magnetic quadrupole} \end{cases} \gg \cdots .$$

(1.122)

Thus in multipole theory, effects may be classified according to the lowest
multipole order in the expansions of $\mathbf{D}$ and $\mathbf{H}$ that is necessary to explain
the effect, based on the above hierarchy. For example, the dynamic magnetoelectric effect is described to lowest order by terms of electric quadrupole–
magnetic dipole order (Chapter 4). For other examples see Section 2.12 and
Chapter 5. Note that the higher-order effects are not necessarily "small":
thus it is an interesting theoretical and experimental challenge to find ways
of enhancing the magnetoelectric effect.

(ii) The most frequently used approximations to (1.118) and (1.119) are [7,23]

$$D_i = \varepsilon_0 E_i + P_i \tag{1.123}$$

$$H_i = \mu_0^{-1} B_i - M_i. \tag{1.124}$$

These are based on retaining in (1.105) only terms of electric dipole–
magnetic dipole order, that is,

$$J_{bi} = \dot{P}_i + \varepsilon_{ijk}\nabla_j M_k. \tag{1.125}$$

This cannot be correct in general because it omits a term involving the
electric quadrupole moment density $Q_{ij}$ in (1.105) which, as shown by
(1.96) and (1.97), enters the vector potential to the same order in $R^{-1}$
as does $\mathbf{M}$. Taking account of $Q_{ij}$ (i.e. working to electric quadrupole–
magnetic dipole order of the hierarchy (1.122)) means that (1.123) and
(1.125) become

$$D_i = \varepsilon_0 E_i + P_i - \frac{1}{2}\nabla_j Q_{ij} \tag{1.126}$$

$$J_{bi} = \dot{P}_i - \nabla_j \left( \frac{1}{2}\dot{Q}_{ij} - \varepsilon_{ijk} M_k \right) \tag{1.127}$$

(see (1.105) and (1.118)), while (1.124) is unchanged.

(iii) It is sometimes argued that neglect of the electric quadrupole term in (1.126) and (1.127) is justified by the smallness of this term [23]. However, we will show later that in order to obtain physically meaningful results from multipole theory when applied to, for example, optical activity or the magnetoelectric effect, it is essential to adhere to the hierarchy (1.122) (see Chapters 5 and 8). Furthermore, the amplitudes of harmonic response fields are origin dependent in the electric dipole–magnetic dipole approximation (see Section 4.6).

(iv) It is clear that the response fields $\mathbf{D}$ and $\mathbf{H}$ in (1.120) and (1.121) are not uniquely defined (see also Section 7.2). The non-uniqueness of $\mathbf{D}$ and $\mathbf{H}$ is an important feature of multipole theory which will be discussed in detail in later chapters (see Chapters 7 and 8). Here we simply give an example which illustrates this non-uniqueness and we comment briefly on various formulations of $\mathbf{D}$ and $\mathbf{H}$ which appear in the literature.

For time-harmonic fields varying as $\exp(-i\omega t)$, (1.117) can be written as

$$\varepsilon_{ijk}\nabla_j(\mu_0^{-1}B_k) = J_{fi} + \frac{\partial}{\partial t}\left(\varepsilon_0 E_i + P_i - \frac{1}{2}\nabla_j Q_{ij} + \frac{1}{6}\nabla_k\nabla_j Q_{ijk}\right.$$
$$\left. + \frac{i}{\omega}\varepsilon_{ijk}\nabla_j\left(M_k - \frac{1}{2}\nabla_l M_{kl}\right) + \cdots\right). \quad (1.128)$$

This has the form of (1.121) with

$$D_i = \varepsilon_0 E_i + P_i - \frac{1}{2}\nabla_j Q_{ij} + \frac{1}{6}\nabla_k\nabla_j Q_{ijk} + \frac{i}{\omega}\varepsilon_{ijk}\nabla_j\left(M_k - \frac{1}{2}\nabla_l M_{kl}\right) + \cdots$$
$$(1.129)$$

and

$$B_i = \mu_0 H_i. \quad (1.130)$$

Note that $D_i$ given by (1.129) also satisfies (1.120) because the additional terms involving $M_k$ and $M_{kl}$ in (1.129) have zero divergence.

For a harmonic plane wave the multipole moment densities in (1.129) can be expressed in terms of the fields and their derivatives that induce these densities (see Sections 2.7 and 2.11). If a complex form is used for the electric field of the wave, then all terms in the multipole moment densities are expressible in terms of $\mathbf{E}$, and (1.129) takes the form

$$D_i = \tilde{\varepsilon}_{ij} E_j, \quad (1.131)$$

where $\tilde{\varepsilon}_{ij}$ is a complex dynamic permittivity. Constitutive relations of the type (1.130) and (1.131) have been used by a number of authors in theories of various optical effects, usually without regard to their multipole basis or translational properties [25–27].

## 1.15    Primitive moments versus traceless moments

In Sections 1.1 and 1.6 we described how, in electrostatics and magnetostatics of a charge distribution, the electric and magnetic fields can be expressed in terms of symmetric, traceless multipole moments rather than primitive moments. We now consider this topic in electrodynamics, first for a charge distribution in vacuum and then for macroscopic media.

### 1.15.1    *A charge distribution*

In the calculations leading to multipole expansions of the dynamic scalar and vector potentials (1.76) and (1.78), it was convenient to use the primitive moments $p_i$, $q_{ij}$, $q_{ijk}$, ... and $m_i$, $m_{ij}$, ... . We now consider whether these potentials, and the resulting electromagnetic fields, can be expressed solely in terms of symmetric, traceless multipole moments. We discuss the contributions of electric quadrupole, electric octopole, and magnetic quadrupole. For simplicity, we restrict ourselves to the far- and near-zone limits (Section 1.11). Thus, we use the relations (1.14), (1.15), and (1.52) to eliminate the primitive moments $q_{ij}$, $q_{ijk}$, and $m_{ij}$ in favour of the traceless moments $\Theta_{ij}$, $\Omega_{ijk}$, and $\Gamma_{ij}$ in the expansions (1.79), (1.80), (1.86), and (1.87). Then the question is: do the traces of the primitive moments $q_{kk}$, $q_{ikk}$, and $m_{kk}$ contribute?

(i) Consider first the expansions in the far zone. For the electric quadrupole contribution the answers to the above question are:

| $\Phi$ | $\mathbf{A}$ | $\mathbf{E}$ | $\mathbf{B}$ |
|---|---|---|---|
| Yes | Yes | No | No, |

where for $\mathbf{E}$ and $\mathbf{B}$ we have used (1.84) and (1.85). That is, for the electric quadrupole contribution in the far-zone, the electric and magnetic fields can be expressed solely in terms of the traceless moment $\Theta_{ij}$, although the potentials $\Phi$ and $\mathbf{A}$ cannot. For the electric octopole–magnetic quadrupole contribution we have:

|  | $\Phi$ | $\mathbf{A}$ | $\mathbf{E}$ | $\mathbf{B}$ |
|---|---|---|---|---|
| $\Omega_{ijk}$ : | Yes | Yes | Yes | Yes |
| $\Gamma_{ij}$ : | — | Yes | Yes | Yes, |

so that neither the potentials nor the fields can be expressed only in terms of the traceless moments $\Omega_{ijk}$ and $\Gamma_{ij}$ [28].

(ii) In the near-zone one has, for the electric quadrupole order,

| $\Phi$ | $\mathbf{A}$ | $\mathbf{E}$ | $\mathbf{B}$ |
|---|---|---|---|
| No | Yes | No | No. |

Thus the dynamic vector potential cannot be expressed solely in terms of the traceless moment $\Theta_{ij}$. For the electric octopole–magnetic quadrupole contribution:

$$\begin{array}{cccc}
\Phi & \mathbf{A} & \mathbf{E} & \mathbf{B} \\
\end{array}$$

$$\Omega_{ijk}: \quad \text{No} \quad \text{Yes} \quad \text{No} \quad \text{No}$$
$$\Gamma_{ij}: \quad - \quad \text{Yes} \quad - \quad \text{Yes.}$$

These results indicate that for consistency in dynamic multipole expansions for a charge distribution, it is generally necessary to use primitive moments.

### 1.15.2   Macroscopic media

The electromagnetic response fields $\mathbf{D}$ and $\mathbf{H}$ in (1.118) and (1.119) are expressed in terms of the primitive macroscopic multipole moment densities $P_i$, $Q_{ij}$, $Q_{ijk}$, ..., $M_i$, $M_{ij}$, .... If one attempts to express them in terms of traceless macroscopic moments in the manner described above then, in general, there is no reason why contributions from traces of the primitive moments $Q_{kk}, Q_{ikk}$, etc. should vanish. Consider, for example, the contribution of the primitive electric quadrupole moment $Q_{ij}$ to $\mathbf{D}$ in (1.118). The traceless moment is $\Theta_{ij} = \frac{1}{2}(3Q_{ij} - Q_{kk}\delta_{ij})$ (see (1.14): to avoid complicating the notation we use the same symbol $\Theta$ for the macroscopic moment). Then in terms of the traceless moment, and to electric quadrupole order, (1.118) is

$$D_i = \epsilon_o E_j + P_i - \frac{1}{3}\nabla_j\Theta_{ij} - \frac{1}{6}\nabla_i Q_{kk}, \qquad (1.132)$$

and the contribution from $Q_{kk}$ does not vanish. This suggests that in macroscopic electrodynamics one should work in terms of primitive moments, as is done by several authors [19], [22], [29–31].

An alternative approach, designed to bring the theory into conformity with electrostatics (Section 1.1), redefines the response fields $\mathbf{D}$ and $\mathbf{H}$ in such a way that they depend only on traceless moments. Thus, in the example (1.132), one modifies $\mathbf{D}$ to [23]

$$D_i' = \epsilon_o E_i + P_i - \frac{1}{3}\nabla_j\Theta_{ij} \qquad (1.133)$$

by transferring the term involving the trace $Q_{kk}$ into the free charge density $\rho_f$ in (1.120), that is,

$$\rho_f \to \rho_f + \frac{1}{6}\nabla^2 Q_{kk}. \qquad (1.134)$$

To electric quadrupole order the field $H_i = \mu_o^{-1}B_i - M_i$ is unchanged. Thus in the Maxwell equation (1.121) the free current density changes according to

$$J_{fi} \to J_{fi} - \frac{1}{6}\nabla_i\dot{Q}_{kk}. \qquad (1.135)$$

The electric octopole–magnetic quadrupole contributions to $\mathbf{D}$ and $\mathbf{H}$ can be redefined in a similar way, and they result in complicated modifications to the sources $\rho_f$ and $\mathbf{J}_f$. If one uses response fields based on traceless moments then, in general, these modifications to $\rho_f$ and $\mathbf{J}_f$ are essential. For example, if the trace terms in (1.134) and (1.135) are omitted, the macroscopic Maxwell equations with $\mathbf{D}$ given by (1.133) lose their origin independence [19].

In this monograph the presentation is based on primitive multipole moments. One reason for this is convenience: use of traceless moments can result in the appearance of "mixed" (traceless and primitive) quantities, as in (1.132), which complicates the analysis (such as in the multipole theory of reflection, Chapter 6, or in origin independence of the macroscopic Maxwell equations). Often results are unaffected by the choice of moments, as in the proof of origin dependence of the constitutive tensor in multipole theory (Chapter 4), the theory of optical activity in an anisotropic medium (Section 5.3 and Ref. 32), and the theory of birefringence induced by an electric field gradient (Section 5.11).

However, the physical consequences of using traceless moments are not always acceptable. For example, it has been argued that in general it is the primitive moment $Q_{ij}$ and not the traceless moment $\Theta_{ij}$ which contributes to the dielectric response of a fluid [31]. Difficulties associated with the use of traceless moments have been discussed in Ref. 33, and the need to take account of the traces of multipole moments has also been emphasized by Melrose and McPhedran [24]. They point out that even in the simple example of multipole radiation, it is only for transverse radiated fields that traceless moments may be used. In media where the radiated fields have longitudinal components, it is necessary to use primitive moments [24]. Another reason for preferring primitive moments is associated with the response nature of $\mathbf{D}$ and $\mathbf{H}$ to the fields $\mathbf{E}$ and $\mathbf{B}$ (Section 1.13). Because the multipole moments, including their traces, are functions of $\mathbf{E}, \mathbf{B}$, and their space and time derivatives (Section 2.11), it therefore seems desirable to retain these traces in the response fields rather than incorporating them in the free source densities.

## References

[1] Buckingham, A. D. (1959). Molecular quadrupole moments. *Quarterly Reviews*, **13**, 183–214. (This influential review also treats higher multipoles.)
[2] Barron, L. D. (1982). *Molecular light scattering and optical activity.* Cambridge University Press, Cambridge
[3] Landau, L. D. and Lifshitz, E. M. (1975). *The classical theory of fields*, Sect. 41. Pergamon, Oxford.
[4] Fowler, P. W. and Steiner, E. (1990). Electric moments and field gradients in the van der Waals $H_2$ system. *Molecular Physics*, **70**, 377–390.
[5] Low, F. E. (1977). *Classical field theory.* Wiley, New York.
[6] Jackson, J. D. (1999). *Classical electrodynamics.* Wiley, New York.
[7] Schwinger, J., DeRaad, L. L., Milton, K. A., and Tsai, W. Y. (1998). *Classical electrodynamics*, pp. 41–42. Perseus, Reading, MA.
[8] Buckingham, A. D. and Pople, J. A. (1955). Theoretical studies of the Kerr effect I: deviations from a linear polarization law. *Proceedings of the Physical Society* A, **68**, 905–909.
[9] Buckingham, A. D. (1979). Polarizability and hyperpolarizability. *Philosophical Transactions of the Royal Society London* A, **293**, 239–248, and references therein.

[10] Buckingham A. D. (1959). Direct method of measuring molecular quadrupole moments. *Journal of Chemical Physics*, **30**, 1580–1585.

[11] Buckingham, A. D. and Longuet-Higgins, H. C. (1968). The quadrupole moments of dipolar molecules. *Molecular Physics*, **14**, 63–72.

[12] Buckingham, A. D. and Disch, R. L. (1963). The quadrupole moment of the carbon dioxide molecule. *Proceedings of the Royal Society London* A, **273**, 275–289.

[13] Buckingham, A. D., Graham, C., and Williams, J. H. (1983). Electric field-gradient-induced birefringence in $N_2$, $C_2H_5$, $C_3H_6$, $Cl_2$, $N_2O$ and $CH_3F$. *Molecular Physics*, **49**, 703–710.

[14] Watson, J. N., Craven, I. E., and Ritchie, G. L. D. (1997). Temperature dependence of electric field-gradient-induced birefringence in carbon dioxide and carbon disulfide. *Chemical Physics Letters*, **274**, 1–6.

[15] Graham, C., Imrie, D. A., and Raab, R. E. (1998). Measurement of the electric quadrupole moments of $CO_2$, CO, $N_2$, $Cl_2$ and $BF_3$. *Molecular Physics*, **93**, 49–56.

[16] Buckingham, A. D. and Stiles, P. J. (1972). Magnetic multipoles and the "pseudo-contact" chemical shift. *Molecular Physics*, **24**, 99–108.

[17] See Ref. 6, Sect. 6.5.

[18] Jackson, J. D. and Okun, L. B. (2001). Historical roots of gauge invariance. *Reviews of Modern Physics*, **73**, 663–680, for the association of this gauge with L. V. Lorenz.

[19] Graham, E. B., Pierrus, J., and Raab, R. E. (1992). Multipole moments and Maxwell's equations. *Journal of Physics* B: *Atomic, Molecular, and Optical Physics*, **25**, 4673–4684.

[20] See Ref. 6, Sect. 9.1.

[21] Russakoff, G.(1970). A derivation of the macroscopic Maxwell equations. *American Journal of Physics*, **38**, 1188–1195. See also Scaife, B. K. P. (1989). *Principles of dielectrics*, Ch. 10. Clarendon, Oxford.

[22] Robinson, F. N. H. (1973). *Macroscopic electromagnetism*. Pergamon, Oxford.

[23] See Ref. 6, Sect. 6.6.

[24] Melrose, D. B. and McPhedran, R. C. (1991). *Electromagnetic processes in dispersive media*. Cambridge University Press, Cambridge.

[25] Landau, L. D. and Lifshitz, E. M. (1960). *Electrodynamics of continuous media*, p. 249. Pergamon, Oxford.

[26] Hornreich, R. M. and Shtrikman, S. (1968). Theory of gyrotropic birefringence. *Physical Review*, **171**, 1065–1074.

[27] Agranovich, V. M. and Ginzburg, V. L. (1984). *Crystal optics with spatial dispersion, and excitons*. Springer, Berlin.

[28] Raab, R. E. (1975). Magnetic multipole moments. *Molecular Physics*, **29**, 1323–1331.

[29] Rosenfeld, L. (1951). *Theory of electrons*, Ch. 2. North-Holland, Amsterdam.

multipole decomposition not valide –

[30] Nakano, H. and Kimura, H. (1969). Quantum statistical-mechanical theory of optical activity. *Journal of the Physical Society of Japan*, **27**, 519–535.

[31] Logan, D. E. (1982). On the dielectric theory of fluids II. Gases in non-uniform fields. *Molecular Physics*, **46**, 271–285.

[32] Buckingham, A. D. and Dunn, M. B. (1971). Optical activity of oriented molecules. *Journal of the Chemical Society* A, 1988–1991.

[33] Gunning, M. J. and Raab, R. E. (1997). Physical implications of the use of primitive and traceless electric quadrupole moments. *Molecular Physics*, **91**, 589–595.

# 2

## QUANTUM THEORY OF MULTIPOLE MOMENTS
## AND POLARIZABILITIES

*Awake, O north wind; and come thou south; blow*
*upon my garden, that the spices thereof may flow out.*
Song of Solomon

In this chapter, starting with the semi-classical Hamiltonian for a charge distribution in an electromagnetic field (Section 2.1), we derive quantum-mechanical expressions for the multipole polarizability-type (or distortion) tensors which describe the induction of multipole moments by an electromagnetic field. This is done for electrostatic fields (Section 2.2), magnetostatic fields (Section 2.4), and the fields of a harmonic plane electromagnetic wave and its various derivative fields (Sections 2.5 and 2.7). We also use the semi-classical Hamiltonian for time-dependent fields to obtain multipole moment operators, and this is done in a manner which yields consistent results for static and dynamic fields, and which has the correct classical limit (Section 2.6).

For time-dependent fields we present two calculations of the multipole polarizability tensors. The first is based on a standard choice of gauge (Section 2.5) and the second on a less familiar but more convenient choice, the Barron–Gray gauge (Section 2.6). We do not discuss calculations based on canonical transformations and which yield non-covariant Hamiltonians containing integrals of the multipole moment densities. For these the reader is referred to the literature [1,2].

There are a number of advantages in having quantum-mechanical expressions for multipole polarizability tensors: they reveal on inspection any intrinsic symmetries which for electromagnetic tensors may not be otherwise accessible; they show relationships between certain tensors; their contribution to a particular effect may be calculated; the dependence of the tensor on choice of origin can be determined; and by including a complex line–shape function in the polarizability tensor, one may describe dispersion and resonance features.

## 2.1 Semi-classical quantum mechanics

Consider a system of $N$ particles in vacuum, subject to an external electromagnetic field. The particles have mass $m^{(\alpha)}$ and charge $q^{(\alpha)}$ where $\alpha = 1, 2, \ldots, N$. We neglect the small corrections and effects that arise from a quantum treatment of the electromagnetic field. Accordingly, the interaction between the field and charge distribution is described by the semi-classical Hamiltonian

$$H = \sum_{\alpha=1}^{N} \left\{ (2m^{(\alpha)})^{-1} \left[ \mathbf{\Pi}^{(\alpha)} - q^{(\alpha)} \mathbf{A}(\mathbf{r}^{(\alpha)}, t) \right]^2 + q^{(\alpha)} \Phi(\mathbf{r}^{(\alpha)}, t) \right\} + V. \quad (2.1)$$

In this, $\mathbf{\Pi}^{(\alpha)}$ is the generalized momentum operator of particle $\alpha$, $\mathbf{r}^{(\alpha)}$ is its position operator for an arbitrary choice of origin in the distribution, $\mathbf{A}$ and $\Phi$ are the vector and scalar potentials of the external field, and $V$ is the potential energy operator of the system in the absence of the field.

We assume that the interaction energy with the field is much smaller than the field-free energy. This permits use of perturbation theory. Accordingly, $H$ is written in powers of the field terms in (2.1) as

$$H = H^{(0)} + H^{(1)} + H^{(2)}, \quad (2.2)$$

where $H^{(0)}$, $H^{(1)}$, and $H^{(2)}$ are respectively the zeroth-, first-, and second-order perturbation Hamiltonians given by

$$H^{(0)} = \sum_{\alpha=1}^{N} (2m^{(\alpha)})^{-1} \left( \mathbf{\Pi}^{(\alpha)} \right)^2 + V \quad (2.3)$$

$$H^{(1)} = \sum_{\alpha=1}^{N} \left\{ -q^{(\alpha)} (2m^{(\alpha)})^{-1} \left[ 2\mathbf{A}(\mathbf{r}^{(\alpha)}, t) \cdot \mathbf{\Pi}^{(\alpha)} - i\hbar \mathbf{\nabla}^{(\alpha)} \cdot \mathbf{A}(\mathbf{r}^{(\alpha)}, t) \right] \right.$$
$$\left. + q^{(\alpha)} \Phi(\mathbf{r}^{(\alpha)}, t) \right\} \quad (2.4)$$

$$H^{(2)} = \sum_{\alpha=1}^{N} (q^{(\alpha)})^2 (2m^{(\alpha)})^{-1} \mathbf{A}^2(\mathbf{r}^{(\alpha)}, t). \quad (2.5)$$

In obtaining (2.4) from (2.1) we used the commutator

$$\mathbf{\Pi} \cdot \mathbf{A} - \mathbf{A} \cdot \mathbf{\Pi} = -i\hbar \mathbf{\nabla} \cdot \mathbf{A}. \quad (2.6)$$

We now consider in turn a perturbation due to an electrostatic field (Section 2.2), a magnetostatic field (Section 2.4), and a time-dependent electromagnetic field (Sections 2.5 and 2.6).

## 2.2 Electrostatic perturbation

For an electrostatic field

$$\mathbf{A}(\mathbf{r}^{(\alpha)}, t) = 0, \quad \Phi(\mathbf{r}^{(\alpha)}, t) = \Phi(\mathbf{r}^{(\alpha)}). \quad (2.7)$$

Then from (2.4) and (2.5)

$$H^{(1)} = \sum_{\alpha=1}^{N} q^{(\alpha)} \Phi(\mathbf{r}^{(\alpha)}) \quad (2.8)$$

$$H^{(2)} = 0. \quad (2.9)$$

To express $H^{(1)}$ in multipole form we assume that $\Phi$ varies slowly inside the charge distribution and make a Taylor expansion of $\Phi$ about the origin. Using $\mathbf{E} = -\boldsymbol{\nabla}\Phi$, we have

$$H^{(1)} = q\Phi - p_i E_i - \frac{1}{2} q_{ij}\nabla_j E_i - \frac{1}{6} q_{ijk}\nabla_k\nabla_j E_i + \cdots . \qquad (2.10)$$

Here the multipole moment operators $p_i$, $q_{ij}$, ... are based on the definitions (1.9)–(1.12), and the potential, the field, and its gradients have their values at the origin.

Prior to application of the perturbation, the system is in the unperturbed energy eigenstate $|n^{(0)}\rangle$. We assume that this state, but not necessarily the others, is non-degenerate. The energy of the system is expanded in powers of the perturbation

$$W_n = W_n^{(0)} + W_n^{(1)} + W_n^{(2)} + \cdots , \qquad (2.11)$$

where from time-independent non-degenerate perturbation theory [3]

$$W_n^{(0)} = \langle H^{(0)}\rangle_{nn} \qquad (2.12)$$

$$W_n^{(1)} = \langle H^{(1)}\rangle_{nn} \qquad (2.13)$$

$$W_n^{(2)} = \langle H^{(2)}\rangle_{nn} + \sum_{s\neq n}(\hbar\omega_{ns})^{-1}\langle H^{(1)}\rangle_{ns}\langle H^{(1)}\rangle_{sn}. \qquad (2.14)$$

Here, and in what follows, we use the notation

$$\langle \Omega\rangle_{sn} = \langle s^{(0)}|\Omega|n^{(0)}\rangle \qquad (2.15)$$

for the matrix elements of an operator $\Omega$ on the vector space describing the system, where $|s^{(0)}\rangle$ and $|n^{(0)}\rangle$ denote unperturbed orthonormal eigenvectors of the system.

In (2.14)

$$\hbar\omega_{ns} = W_n^{(0)} - W_s^{(0)}, \qquad (2.16)$$

where $\omega_{ns}$ is the angular frequency of the transition between the unperturbed states $n$ and $s$. The summation in (2.14) is over all unperturbed energy eigenstates of the system, other than the initial state, and is understood to include an integration over states that are continuous. In addition to (2.12)–(2.14) we require the perturbed eigenvector $|n\rangle$ to first order in the perturbation [3]

$$|n\rangle = |n^{(0)}\rangle + \sum_{s\neq n}c_s|s^{(0)}\rangle, \qquad (2.17)$$

where

$$c_s = (\hbar\omega_{ns})^{-1}\langle H^{(1)}\rangle_{sn}. \qquad (2.18)$$

The expectation value of the electric dipole moment operator $p_i$ in the perturbed state $|n\rangle$ is then

$$\langle n|p_i|n \rangle = \left\langle n^{(0)} + \sum_{s \neq n} c_s s^{(0)} | p_i | n^{(0)} + \sum_{t \neq n} c_t t^{(0)} \right\rangle$$

$$= \langle p_i \rangle_{nn} + \sum_{s \neq n} (\hbar \omega_{ns})^{-1} (\langle p_i \rangle_{ns} \langle H^{(1)} \rangle_{sn} + \langle p_i \rangle_{ns}^* \langle H^{(1)} \rangle_{sn}^*)$$

$$= \langle p_i \rangle_{nn} + 2 \sum_{s \neq n} (\hbar \omega_{ns})^{-1} \mathcal{R}e \{ \langle p_i \rangle_{ns} \langle H^{(1)} \rangle_{sn} \}. \qquad (2.19)$$

Here we have used the notation (2.15) and the Hermitian property of $p_i$ (from their definitions (cf. Section 1.1) the electric multipole moment operators are Hermitian because $\mathbf{r}$ is and also because $r_i r_j - r_j r_i = 0$). Note that in $\langle H^{(1)} \rangle_{sn}$ the term in $q$ in (2.10) does not contribute because $q$ is a constant and the eigenvectors are orthogonal. Thus, from (2.10) and (2.19)

$$\langle n|p_i|n \rangle = \langle p_i \rangle_{nn} + 2 \sum_{s \neq n} (\hbar \omega_{ns})^{-1} \mathcal{R}e \Big\{ \langle p_i \rangle_{ns} \Big( \langle p_j \rangle_{sn} E_j$$

$$+ \frac{1}{2} \langle q_{jk} \rangle_{sn} \nabla_k E_j + \frac{1}{6} \langle q_{jkl} \rangle_{sn} \nabla_l \nabla_k E_j + \cdots \Big) \Big\}. \qquad (2.20)$$

We write this as

$$p_i = p_i^{(0)} + \alpha_{ij} E_j + \frac{1}{2} a_{ijk} \nabla_k E_j + \frac{1}{6} b_{ijkl} \nabla_l \nabla_k E_j + \cdots . \qquad (2.21)$$

Here $p_i^{(0)} = \langle p_i \rangle_{nn}$ is the permanent (or unperturbed) electric dipole moment in the state $|n^{(0)}\rangle$, and $\alpha_{ij}$, $a_{ijk}$, $b_{ijkl}$, ... are known collectively as static or *dc* polarizabilities (in the state $|n^{(0)}\rangle$). They are given by

$$\alpha_{ij} = 2 \sum_{s \neq n} (\hbar \omega_{sn})^{-1} \mathcal{R}e \{ \langle p_i \rangle_{ns} \langle p_j \rangle_{sn} \} = \alpha_{ji} \qquad (2.22)$$

$$a_{ijk} = 2 \sum_{s \neq n} (\hbar \omega_{sn})^{-1} \mathcal{R}e \{ \langle p_i \rangle_{ns} \langle q_{jk} \rangle_{sn} \} \qquad (2.23)$$

$$b_{ijkl} = 2 \sum_{s \neq n} (\hbar \omega_{sn})^{-1} \mathcal{R}e \{ \langle p_i \rangle_{ns} \langle q_{jkl} \rangle_{sn} \}. \qquad (2.24)$$

In (2.22) we have indicated the symmetry which follows from the Hermitian property of $p_i$.

A similar approach for the expectation values of the electric quadrupole moment and electric octopole moment operators in the perturbed state $|n\rangle$ gives

$$q_{ij} = q_{ij}^{(0)} + \mathfrak{a}_{ijk} E_k + \frac{1}{2} d_{ijkl} \nabla_l E_k + \cdots \qquad (2.25)$$

$$q_{ijk} = q_{ijk}^{(0)} + \mathfrak{b}_{ijkl} E_l + \cdots , \qquad (2.26)$$

where

$$\mathfrak{a}_{ijk} = 2 \sum_{s \neq n} (\hbar \omega_{sn})^{-1} \mathcal{R}e\{\langle q_{ij}\rangle_{ns}\langle p_k\rangle_{sn}\} = \mathfrak{a}_{kij} \tag{2.27}$$

$$\mathfrak{b}_{ijkl} = 2 \sum_{s \neq n} (\hbar \omega_{sn})^{-1} \mathcal{R}e\{\langle q_{ijk}\rangle_{ns}\langle p_l\rangle_{sn}\} = \mathfrak{b}_{lijk} \tag{2.28}$$

$$d_{ijkl} = 2 \sum_{s \neq n} (\hbar \omega_{sn})^{-1} \mathcal{R}e\{\langle q_{ij}\rangle_{ns}\langle q_{kl}\rangle_{sn}\} = d_{klij}. \tag{2.29}$$

Thus to electric octopole order the independent polarizability tensors for an electrostatic field are

$$\alpha_{ij}, \quad a_{ijk}, \quad b_{ijkl}, \quad d_{ijkl}. \tag{2.30}$$

The classical (phenomenological) counterparts of (2.21), (2.25), and (2.26) are (1.24)–(1.26).

It is also of interest to obtain the above results from an expansion of the energy to second order in the perturbation. From (2.10) and (2.13) we have

$$W_n^{(1)} = q\Phi - \langle p_i\rangle_{nn} E_i - \frac{1}{2}\langle q_{ij}\rangle_{nn}\nabla_j E_i - \frac{1}{6}\langle q_{ijk}\rangle_{nn}\nabla_k\nabla_j E_i + \cdots, \tag{2.31}$$

since $q$ is a constant and $|n^{(0)}\rangle$ is normalized. Adopting the notation used in (2.21), we write

$$W_n^{(1)} = q\Phi - p_i^{(0)} E_i - \frac{1}{2} q_{ij}^{(0)} \nabla_j E_i - \frac{1}{6} q_{ijk}^{(0)} \nabla_k \nabla_j E_i + \cdots, \tag{2.32}$$

which is the quantum-mechanical counterpart of the classical result (1.37) for a rigid distribution. Similarly, from (2.9), (2.10), and (2.14) one can show that

$$W_n^{(2)} = -\frac{1}{2}\alpha_{ij} E_i E_j - \frac{1}{2}a_{ijk} E_i(\nabla_k E_j) - \frac{1}{6}b_{ijkl} E_i(\nabla_l\nabla_k E_j)$$
$$- \frac{1}{8}d_{ijkl}(\nabla_j E_i)(\nabla_l E_k) + \cdots, \tag{2.33}$$

where the polarizabilities are given by (2.22)–(2.24) and (2.29). Combining (2.32) and (2.33) in (2.11) we have

$$W = W^{(0)} + q\Phi - p_i^{(0)} E_i - \frac{1}{2}\alpha_{ij} E_i E_j - \frac{1}{2}a_{ijk} E_i(\nabla_k E_j)$$
$$- \frac{1}{6}b_{ijkl} E_i(\nabla_l\nabla_k E_j) - \frac{1}{2}q_{ij}^{(0)}(\nabla_j E_i) - \frac{1}{8}d_{ijkl}(\nabla_j E_i)(\nabla_l E_k)$$
$$- \frac{1}{6}q_{ijk}^{(0)}(\nabla_k\nabla_j E_i) + \cdots. \tag{2.34}$$

Because each order of field gradient is independent of any other, we have from (2.34)

$$-\frac{\partial W}{\partial E_i} = p_i^{(0)} + \alpha_{ij}E_j + \frac{1}{2}\,a_{ijk}\nabla_k E_j + \frac{1}{6}\,b_{ijkl}\nabla_l\nabla_k E_j + \cdots \qquad (2.35)$$

$$-2\frac{\partial W}{\partial(\nabla_j E_i)} = q_{ij}^{(0)} + a_{kij}E_k + \frac{1}{2}\,d_{ijkl}\nabla_l E_k + \cdots \qquad (2.36)$$

$$-6\frac{\partial W}{\partial(\nabla_k\nabla_j E_i)} = q_{ijk}^{(0)} + b_{lijk}E_l + \cdots . \qquad (2.37)$$

In obtaining (2.35) and (2.36) we have used the symmetries in (2.22) and (2.29). Comparing (2.35)–(2.37) with (2.21), (2.25), and (2.26) we see that the total (permanent plus induced) multipole moments in an electrostatic field can be obtained from the energy (2.34) according to

$$p_i = -\frac{\partial W}{\partial E_i}, \qquad q_{ij} = -2\frac{\partial W}{\partial(\nabla_j E_i)}, \qquad q_{ijk} = -6\frac{\partial W}{\partial(\nabla_k\nabla_j E_i)}, \qquad \text{etc.} \quad (2.38)$$

Note that (2.34) is the extension to a polarizable distribution of the result (2.32) for a rigid distribution. The following calculations show that the relations (2.38) are not restricted to moments which are linear in the field and its derivatives, as in (2.21), (2.25), and (2.26).

## 2.3   Buckingham's derivation of electrostatic multipole moments

We present an elegant derivation due to Buckingham [4] who has shown that the results (2.38) are generally valid in electrostatics. The Hamiltonian is $H = H^{(0)} + H^{(1)}$ where for electrostatic fields $H^{(1)}$ is given by (2.10). Then

$$\frac{\partial H}{\partial E_i} = -p_i, \qquad \frac{\partial H}{\partial(\nabla_j E_i)} = -\frac{1}{2}\,q_{ij}, \qquad \text{etc.} \qquad (2.39)$$

Let $|\Psi\rangle$ denote a normalized eigenket of $H$ with eigenvalue $W = \langle\Psi|H|\Psi\rangle$. Then using the first equation in (2.39) we have

$$\begin{aligned}
\frac{\partial W}{\partial E_i} &= -\langle\Psi|p_i|\Psi\rangle + \left\langle\frac{\partial\Psi}{\partial E_i}\Big|H\Big|\Psi\right\rangle + \left\langle\Psi\Big|H\Big|\frac{\partial\Psi}{\partial E_i}\right\rangle \\
&= -\langle\Psi|p_i|\Psi\rangle + W\frac{\partial}{\partial E_i}\langle\Psi|\Psi\rangle \\
&= -\langle\Psi|p_i|\Psi\rangle,
\end{aligned} \qquad (2.40)$$

which is the first equation in (2.38). Similarly

$$\begin{aligned}
\frac{\partial W}{\partial(\nabla_j E_i)} &= -\frac{1}{2}\langle\Psi|q_{ij}|\Psi\rangle + \left\langle\frac{\partial\Psi}{\partial(\nabla_j E_i)}\Big|H\Big|\Psi\right\rangle + \left\langle\Psi\Big|H\Big|\frac{\partial\Psi}{\partial(\nabla_j E_i)}\right\rangle \\
&= -\frac{1}{2}\langle\Psi|q_{ij}|\Psi\rangle,
\end{aligned} \qquad (2.41)$$

which is the second equation in (2.38). And so on for higher moments.

The results (2.40), (2.41), etc. can be applied to systems whose electrostatic energy is represented by a power series in the field and its derivatives, and they yield electric multipole moments which may be non-linear functions of the field and its derivatives [4]. The above approach is essentially an application of the Hellmann–Feynman theorem [5] in which the electric field and its derivatives at the origin are treated as parameters in the Hamiltonian (cf. (2.10)).

## 2.4  Magnetostatic perturbation

For a magnetostatic field

$$\mathbf{A} = \mathbf{A}(\mathbf{r}) \quad \text{and} \quad \Phi = 0. \tag{2.42}$$

We see from (2.4), (2.5), and (2.42) that a magnetostatic field has a second-order perturbation Hamiltonian, unlike the electrostatic case. By making a Taylor expansion of $\mathbf{A}$ one can determine the contributions of the magnetic field and its derivatives to the first- and second-order perturbation energies. These contributions allow identification of quantum expressions for various magnetic polarizability (or more correctly, magnetizability) tensors. However, this procedure based on a Taylor expansion of $\mathbf{A}$ is intricate, and because the same results can be obtained in the *dc* limit of a more direct theory using a harmonic plane electromagnetic wave (Section 2.7), we consider here only uniform magnetostatic fields. Non-uniform fields are discussed in Section 2.9.

The calculations are simplified by working in the Coulomb gauge

$$\boldsymbol{\nabla} \cdot \mathbf{A}(\mathbf{r}) = 0. \tag{2.43}$$

For a uniform magnetostatic field $\mathbf{B}$, the choice of vector potential

$$\mathbf{A} = \frac{1}{2}\mathbf{B} \times \mathbf{r} \tag{2.44}$$

satisfies the gauge condition (2.43). Using (2.42)–(2.44) we can express (2.4) and (2.5) as

$$H^{(1)} = -\sum_{\alpha=1}^{N} \frac{q^{(\alpha)}}{2m^{(\alpha)}} \left(\mathbf{B} \times \mathbf{r}^{(\alpha)}\right) \cdot \boldsymbol{\Pi}^{(\alpha)} = -\mathbf{m} \cdot \mathbf{B} \tag{2.45}$$

and

$$H^{(2)} = -\sum_{\alpha=1}^{N} \frac{(q^{(\alpha)})^2}{8m^{(\alpha)}} \left[ r_i^{(\alpha)} r_j^{(\alpha)} - (r^{(\alpha)})^2 \delta_{ij} \right] B_i B_j. \tag{2.46}$$

Here

$$\mathbf{m} = \sum_{\alpha=1}^{N} \frac{q^{(\alpha)}}{2m^{(\alpha)}} \mathbf{r}^{(\alpha)} \times \boldsymbol{\Pi}^{(\alpha)} \tag{2.47}$$

is the magnetic dipole moment operator of the distribution (cf. (1.50)). Spin angular momentum may be included in $\mathbf{m}$ in an *ad hoc* way by combining the spin angular momentum operator $\mathbf{s}$ with the orbital angular momentum operator $\mathbf{l} = \mathbf{r} \times \mathbf{\Pi}$. Thus

$$\mathbf{m} = \sum_{\alpha=1}^{N} \frac{q^{(\alpha)}}{2m^{(\alpha)}} \left( \mathbf{l}^{(\alpha)} + g^{(\alpha)} \mathbf{s}^{(\alpha)} \right), \tag{2.48}$$

where $g^{(\alpha)}$ is the $g$-factor of particle $\alpha$. Because $\mathbf{l}$ and $\mathbf{s}$ are Hermitian, so too is $\mathbf{m}$. From (2.11)–(2.14), (2.45), and (2.46) we have

$$W_n = W_n^{(0)} - \langle m_i \rangle_{nn} B_i - \left\{ \sum_{s \neq n} (\hbar \omega_{sn})^{-1} \mathcal{R}e\{ \langle m_i \rangle_{ns} \langle m_j \rangle_{sn} \} \right.$$

$$\left. + \sum_{\alpha=1}^{N} \frac{(q^{(\alpha)})^2}{8m^{(\alpha)}} \left\langle r_i^{(\alpha)} r_j^{(\alpha)} - (r^{(\alpha)})^2 \delta_{ij} \right\rangle_{nn} \right\} B_i B_j + \cdots . \tag{2.49}$$

Here the term linear in $B_i$ comes from the first-order contribution $W_n^{(1)}$, and the term quadratic in $B_i$ from the second-order contribution $W_n^{(2)}$. We write (2.49) as

$$W = W^{(0)} - m_i^{(0)} B_i - \frac{1}{2} \chi_{ij} B_i B_j + \cdots , \tag{2.50}$$

where

$$m_i^{(0)} = \langle m_i \rangle_{nn} = \sum_{\alpha=1}^{N} \frac{q^{(\alpha)}}{2m^{(\alpha)}} \left\langle l_i^{(\alpha)} + g^{(\alpha)} s_i^{(\alpha)} \right\rangle_{nn} \tag{2.51}$$

is the permanent magnetic dipole moment in the unperturbed quantum state $|n^{(0)}\rangle$ and

$$\chi_{ij} = 2 \sum_{s \neq n} (\hbar \omega_{sn})^{-1} \mathcal{R}e\left\{ \langle m_i \rangle_{ns} \langle m_j \rangle_{sn} \right\} + \sum_{\alpha=1}^{N} \frac{(q^{(\alpha)})^2}{4m^{(\alpha)}} \left\langle r_i^{(\alpha)} r_j^{(\alpha)} - (r^{(\alpha)})^2 \delta_{ij} \right\rangle_{nn} \tag{2.52}$$

is the permanent $dc$ magnetizability, also in the state $|n^{(0)}\rangle$. In obtaining this formula we have used the Hermitian property of $\mathbf{m}$. The intrinsic symmetry $\chi_{ij} = \chi_{ji}$ is evident from (2.52).

It is of interest to comment on the origin dependence of the $dc$ magnetizability. This question was first considered by Van Vleck [6] who showed that the expression in (2.52) is independent of the choice of coordinate origin used for the position operators $\mathbf{r}^{(\alpha)}$. Specifically, the origin dependence of the one summation term in (2.52) cancels that in the other (Section 3.7). Such origin independence of the quantum-mechanical expression for an observable is to be expected when the measured value of the observable does not depend on the choice of origin inside a molecule or other distribution of particles. Note that if the contribution of $H^{(2)}$ is neglected relative to that of $H^{(1)}$ in (2.49), then the terms quadratic in $\mathbf{r}^{(\alpha)}$ in (2.52) would fall away, and $\chi_{ij}$ would be origin dependent.

One can also evaluate the expectation value of a magnetic multipole moment operator using the perturbed energy eigenvectors. The result is expressed in terms of the field and its various gradients (cf. the analogous electrostatic treatment leading to (2.20)). In this calculation it is not only the eigenvector that is changed by the magnetic field but also the moment operator itself, since this contains the momentum operator $\mathbf{\Pi}$ and therefore requires the replacement

$$\mathbf{\Pi} \to \mathbf{\Pi} - q\mathbf{A}. \tag{2.53}$$

Thus, for example, the expectation value of the perturbed magnetic dipole moment operator, denoted $\mathbf{m}'$, in the perturbed state $|n\rangle$ is

$$\langle n|m_i'|n\rangle = \left\langle n^{(0)} + \sum_{s \neq n} c_s s^{(0)} \left| \sum_{\alpha=1}^{N} \frac{q^{(\alpha)}}{2m^{(\alpha)}} \left[ \mathbf{r}^{(\alpha)} \times \left( \mathbf{\Pi}^{(\alpha)} - q^{(\alpha)} \mathbf{A}^{(\alpha)} \right) \right]_i \right.$$
$$\left. + g^{(\alpha)} s_i^{(\alpha)} \right| n^{(0)} + \sum_{t \neq n} c_t t^{(0)} \right\rangle. \tag{2.54}$$

By means of (2.18), (2.44), and (2.45) this can be expressed as

$$m_i' = m_i^{(0)} + \chi_{ij} B_j + \cdots , \tag{2.55}$$

where $m_i'$ denotes $\langle n|m_i'|n\rangle$, and $m_i^{(0)}$ and $\chi_{ij}$ are given by (2.51) and (2.52), respectively. From (2.50) and (2.55) we have

$$m_i' = -\frac{\partial W}{\partial B_i}, \tag{2.56}$$

which is the magnetostatic analogue of the electrostatic result for $p_i$ in (2.38).

## 2.5   Time-dependent fields: standard gauge

The procedure used above for obtaining total multipole moments from the energy (see (2.38) and (2.56)) is not applicable to time-dependent fields. For these the quantum-mechanical expressions for electrodynamic polarizability tensors are derived instead from the expectation value of the appropriate multipole moment operator. In this section we restrict ourselves to a linear dependence of induced moment on field and its various derivatives, and hence we neglect the contribution of $H^{(2)}$ in (2.2).

The evaluation of $H^{(1)}$ in (2.4) is simplified if we work in the gauge

$$\mathbf{\nabla} \cdot \mathbf{A}(\mathbf{r}, t) = 0 \tag{2.57}$$

with

$$\Phi(\mathbf{r}, t) = 0 \tag{2.58}$$

at each particle. (The condition (2.58) can be imposed because the sources of the external field are outside the system of particles.) Then (2.4) becomes

$$H^{(1)} = -\sum_{\alpha=1}^{N} \frac{q^{(\alpha)}}{m^{(\alpha)}} \mathbf{A}(\mathbf{r}^{(\alpha)}, t) \cdot \mathbf{\Pi}^{(\alpha)}. \tag{2.59}$$

With the aid of a Taylor expansion of $\mathbf{A}$ about the origin, this can be written

$$H^{(1)} = -\sum_{\alpha=1}^{N} \frac{q^{(\alpha)}}{m^{(\alpha)}} \left\{ A_i(0, t) + \left[ \nabla_j^{(\alpha)} A_i(\mathbf{r}^{(\alpha)}, t) \right]_0 r_j^{(\alpha)} + \cdots \right\} \Pi_i^{(\alpha)}. \tag{2.60}$$

Note that in the approximation associated with the leading term $A_i(0, t)$ there is no magnetic field although there is an electric field

$$\mathbf{E} = -\dot{\mathbf{A}}. \tag{2.61}$$

We now consider in turn the leading two terms in (2.60).

The first-order perturbation eigenvector to be used in the expectation value of a multipole moment operator is

$$|n(t)\rangle = e^{-iW_n^{(0)}t/\hbar} |n^{(0)}\rangle + \sum_{s\neq n} c_s(t) e^{-iW_s^{(0)}t/\hbar} |s^{(0)}\rangle. \tag{2.62}$$

Here the first-order mixing coefficients are given in time-dependent perturbation theory by [3]

$$c_s(t) = -\frac{i}{\hbar} \int_0^t e^{-i\omega_{ns}t} \langle s^{(0)} | H^{(1)} | n^{(0)} \rangle \, dt, \tag{2.63}$$

where $\omega_{ns}$ is defined in (2.16). The contribution to (2.63) of the leading term in (2.60) is then

$$c_s(t) = \frac{i}{\hbar} \left\langle s^{(0)} \left| \sum_{\alpha=1}^{N} \frac{q^{(\alpha)}}{m^{(\alpha)}} \Pi_i^{(\alpha)} \right| n^{(0)} \right\rangle \int_0^t A_i(0, t) e^{-i\omega_{ns}t} \, dt. \tag{2.64}$$

The matrix element in (2.64) is evaluated by using the commutator

$$\left[ H^{(0)}, p_i \right] = -i\hbar \sum_{\alpha=1}^{N} \frac{q^{(\alpha)}}{m^{(\alpha)}} \Pi_i^{(\alpha)}, \tag{2.65}$$

where $p_i$ is the electric dipole moment operator, and we have assumed a velocity-independent potential in (2.3). Then

$$\left\langle s^{(0)} \left| \sum_{\alpha=1}^{N} \frac{q^{(\alpha)}}{m^{(\alpha)}} \Pi_i^{(\alpha)} \right| n^{(0)} \right\rangle = i\omega_{sn} \langle p_i \rangle_{sn}. \tag{2.66}$$

To proceed further we require an explicit form for the vector potential, which we take to be that for a harmonic plane wave

$$\mathbf{A}(\mathbf{r}, t) = \mathbf{A}_0 \cos(\mathbf{k} \cdot \mathbf{r} - \omega t). \tag{2.67}$$

Integrating (2.64) by parts and using (2.61) and (2.66) yields

$$c_s(t) = \frac{\omega_{sn}}{\hbar(\omega_{sn}^2 - \omega^2)} e^{i\omega_{sn}t} \langle p_i \rangle_{sn} \left( E_i + i\frac{\omega_{sn}}{\omega^2} \dot{E}_i \right), \tag{2.68}$$

where $E_i$ and $\dot{E}_i$ are the values at the origin. In (2.68) we have omitted a $dc$ term coming from the lower limit of integration.

We note that (2.68) displays an interaction-type term between electric dipole and electric field, which was not evident in the leading term of the Hamiltonian (2.60). Consequently this first contribution to the interaction between a harmonic plane wave and a charge distribution is termed the electric dipole approximation. Equation (2.68) also contains a term in the time derivative of the electric field. The significance of the contribution of this term to the induced electric dipole moment is discussed in Section 2.12.

Using (2.68) in (2.62), we find that the expectation value of the electric dipole moment operator can be expressed, to first order in the field, as

$$\langle n(t) | p_i | n(t) \rangle = \langle p_i \rangle_{nn} + \frac{2}{\hbar} \sum_{s \neq n} \frac{\omega_{sn}}{\omega_{sn}^2 - \omega^2} \mathcal{R}e \left\{ \langle p_i \rangle_{ns} \langle p_j \rangle_{sn} \right\} E_j$$

$$- \frac{2}{\hbar} \sum_{s \neq n} \frac{\omega_{sn}^2}{\omega^2(\omega_{sn}^2 - \omega^2)} \mathcal{I}m \left\{ \langle p_i \rangle_{ns} \langle p_j \rangle_{sn} \right\} \dot{E}_j. \tag{2.69}$$

Note that the restriction $s \neq n$ in the summation in (2.62) may be removed in (2.69) because from (2.16) $\omega_{sn} = 0$ when $s = n$. We write (2.69) as

$$p_i = p_i^{(0)} + \alpha_{ij} E_j + \frac{1}{\omega} \alpha_{ij}' \dot{E}_j, \tag{2.70}$$

where $p_i^{(0)} = \langle p_i \rangle_{nn}$ is the permanent electric dipole moment in the state $|n^{(0)}\rangle$, and

$$\alpha_{ij} = \frac{2}{\hbar} \sum_s \frac{\omega_{sn}}{\omega_{sn}^2 - \omega^2} \mathcal{R}e \left\{ \langle p_i \rangle_{ns} \langle p_j \rangle_{sn} \right\} \tag{2.71}$$

$$\alpha_{ij}' = -\frac{2}{\hbar} \sum_s \frac{\omega_{sn}^2}{\omega(\omega_{sn}^2 - \omega^2)} \mathcal{I}m \left\{ \langle p_i \rangle_{ns} \langle p_j \rangle_{sn} \right\}. \tag{2.72}$$

Because $\mathbf{p}$ is Hermitian, these frequency-dependent polarizabilities possess the intrinsic symmetries

$$\alpha_{ij} = \alpha_{ji} \tag{2.73}$$

$$\alpha_{ij}' = -\alpha_{ji}'. \tag{2.74}$$

Here, and in what follows, we use the same symbol to denote both a frequency-dependent polarizability and the corresponding static limit. The context will indicate which of the two meanings applies.

Equation (2.72) can be transformed by means of the identity

$$\frac{\omega_{sn}^2}{\omega^2(\omega_{sn}^2 - \omega^2)} = \frac{1}{\omega_{sn}^2 - \omega^2} + \frac{1}{\omega^2}. \tag{2.75}$$

This yields

$$\alpha_{ij}' = -\frac{2}{\hbar}\sum_s \frac{\omega}{\omega_{sn}^2 - \omega^2}\mathcal{I}m\{\langle p_i\rangle_{ns}\langle p_j\rangle_{sn}\} - \frac{2}{\hbar\omega}\mathcal{I}m\left\{\sum_s \langle p_i\rangle_{ns}\langle p_j\rangle_{sn}\right\}. \tag{2.76}$$

From the closure relation

$$\sum_s \langle p_i\rangle_{ns}\langle p_j\rangle_{sn} = \langle p_i p_j\rangle_{nn}, \tag{2.77}$$

which is real because $\mathbf{p}$ is Hermitian and $p_i p_j = p_j p_i$. Thus

$$\alpha_{ij}' = -\frac{2}{\hbar}\sum_s \frac{\omega}{\omega_{sn}^2 - \omega^2}\mathcal{I}m\{\langle p_i\rangle_{ns}\langle p_j\rangle_{sn}\}. \tag{2.78}$$

Note that $\alpha_{ij}'$ vanishes at $\omega = 0$.

Next we consider the second term in (2.60). We write

$$[\nabla_j A_i]_0\, r_j\Pi_i = \frac{1}{2}[\nabla_j A_i + \nabla_i A_j]_0\, r_j\Pi_i + \frac{1}{2}[\nabla_j A_i - \nabla_i A_j]_0\, r_j\Pi_i$$

$$= \frac{1}{2}[\nabla_j A_i]_0\,(r_i\Pi_j + r_j\Pi_i) - \frac{1}{2}\varepsilon_{ijk}[\nabla\times\mathbf{A}]_{0k}\, r_j\Pi_i. \tag{2.79}$$

Using (2.79) and the commutator

$$\left[H^{(0)}, q_{ij}\right] = -i\hbar\sum_{\alpha=1}^N \frac{q^{(\alpha)}}{m^{(\alpha)}}\left(r_i^{(\alpha)}\Pi_j^{(\alpha)} + r_j^{(\alpha)}\Pi_i^{(\alpha)} - i\hbar\delta_{ij}\right), \tag{2.80}$$

we find that the contribution of the second term in (2.60) to the mixing coefficient (2.63) is

$$c_s(t) = \frac{\omega_{sn}}{\hbar(\omega_{sn}^2 - \omega^2)}\,e^{i\omega_{sn}t}\left\{\frac{1}{2}\langle q_{ij}\rangle_{sn}\left(\nabla_j E_i + i\frac{\omega_{sn}}{\omega^2}\nabla_j \dot{E}_i\right)\right.$$

$$\left. + \langle m_i\rangle_{sn}\left(B_i + i\frac{1}{\omega_{sn}}\dot{B}_i\right)\right\}. \tag{2.81}$$

Here the fields and their derivatives are evaluated at the origin.

It is evident from (2.81) that the same term in the Taylor expansion of the vector potential of a harmonic plane wave leads to both electric quadrupole and magnetic dipole interactions. The leading two terms in (2.60) constitute the electric quadrupole–magnetic dipole approximation of the interaction between the wave and a charge distribution. As with the electric dipole result in (2.70), each multipole interaction-type term in (2.81) involves both a field quantity and its time derivative.

By also including (2.81) in (2.62) we obtain the extension of (2.70),

$$
p_i = p_i^{(0)} + \alpha_{ij} E_j + \frac{1}{\omega} \alpha'_{ij} \dot{E}_j + \frac{1}{2} a_{ijk} \nabla_k E_j + \frac{1}{2\omega} a'_{ijk} \nabla_k \dot{E}_j
$$
$$
+ G_{ij} B_j + \frac{1}{\omega} G'_{ij} \dot{B}_j. \tag{2.82}
$$

Here the quantum-mechanical expressions for the polarizabilities are given by (2.71), (2.78), and

$$
a_{ijk} = \frac{2}{\hbar} \sum_s \frac{\omega_{sn}}{\omega_{sn}^2 - \omega^2} \mathcal{R}e\left\{\langle p_i \rangle_{ns} \langle q_{jk} \rangle_{sn}\right\} = a_{ikj} \tag{2.83}
$$

$$
a'_{ijk} = -\frac{2}{\hbar} \sum_s \frac{\omega}{\omega_{sn}^2 - \omega^2} \mathcal{I}m\left\{\langle p_i \rangle_{ns} \langle q_{jk} \rangle_{sn}\right\} = a'_{ikj} \tag{2.84}
$$

$$
G_{ij} = \frac{2}{\hbar} \sum_s \frac{\omega_{sn}}{\omega_{sn}^2 - \omega^2} \mathcal{R}e\left\{\langle p_i \rangle_{ns} \langle m_j \rangle_{sn}\right\} \tag{2.85}
$$

$$
G'_{ij} = -\frac{2}{\hbar} \sum_s \frac{\omega}{\omega_{sn}^2 - \omega^2} \mathcal{I}m\left\{\langle p_i \rangle_{ns} \langle m_j \rangle_{sn}\right\}. \tag{2.86}
$$

In regard to these results we note the following: (i) In writing down (2.83)–(2.86) we have used the property $\omega_{sn} = 0$ when $s = n$, and also that the quantities in curly brackets in (2.84) and (2.86) are real when $s = n$ (because the expectation values of Hermitian operators are real). These enable us to omit in the summations in (2.83)–(2.86) the restriction $s \neq n$ coming from (2.62). (ii) The intrinsic symmetry of $a_{ijk}$ and $a'_{ijk}$ derives from that of $q_{jk}$. (iii) Just as $\alpha'_{ij}$ in (2.78) vanishes at zero frequency, so too do the polarizabilities associated with the time-derivative fields in (2.82), namely $a'_{ijk}$ and $G'_{ij}$. (iv) The mixing coefficients in (2.68) and (2.81) may also be used to calculate expectation values of higher multipole moment operators and from these to obtain expressions for yet other multipole polarizabilities.

In concluding this section we point out that the gauge choice in (2.57), despite its appeal for the immediate simplification it effects in the perturbation Hamiltonian, possesses a number of disadvantages. (i) The multipole moments, that enter the various interaction terms arising from the Taylor expansion of $\mathbf{A}(\mathbf{r}, t)$, become evident as such only after several manipulations, such as those in (2.66), (2.79), and (2.80), and then only in the form of matrix elements. (ii) Similarly, the fields that partner the multipole moments in these interactions are recognizable only after a time integration with an assumed harmonic plane wave form for $\mathbf{A}(\mathbf{r}, t)$. (iii) In most instances one or other of the polarizability tensors associated with a given field or its time derivative also requires manipulation, here by means of (2.75), in order to reduce it to a simpler form. Altogether, the procedure is cumbersome, in marked contrast to the direct approach offered by the gauge choice made in the next section.

## 2.6   Time-dependent fields: the Barron–Gray gauge

In applying the perturbation Hamiltonians (2.4) and (2.5) to multipole theory, it would be clearly advantageous to make power-series expansions of the electromagnetic potentials about the origin. Furthermore, the coefficients in these expressions are fixed by the requirement that the potentials should yield the fields in the usual manner (see below). This approach was initiated by Barron and Gray [7] who proposed the following potentials

$$A_i(\mathbf{r},t) = \varepsilon_{ijk} \left\{ \frac{1}{2} B_j(0,t) r_k + \frac{1}{3} [\nabla_l B_j(\mathbf{r},t)]_o r_k r_l \right.$$

$$\left. + \frac{1}{8} [\nabla_m \nabla_l B_j(\mathbf{r},t)]_o r_k r_l r_m + \cdots \right\} \quad (2.87)$$

$$\Phi(\mathbf{r},t) = \Phi(0,t) - E_i(0,t) r_i - \frac{1}{2} [\nabla_j E_i(\mathbf{r},t)]_o r_i r_j$$

$$- \frac{1}{6} [\nabla_k \nabla_j E_i(\mathbf{r},t)]_o r_i r_j r_k + \cdots . \quad (2.88)$$

It is readily shown that these potentials do not satisfy either the Coulomb gauge $\nabla \cdot \mathbf{A} = 0$ or the Lorenz gauge $\nabla \cdot \mathbf{A} + c^{-2}\dot{\Phi} = 0$. Thus they are in a gauge of their own, which we refer to as the Barron–Gray gauge. Note that (2.87) and (2.88) are extensions of the potentials (2.42) and (2.44) for uniform magnetostatic fields. The electric field $\mathbf{E} = -\nabla\Phi - \dot{\mathbf{A}}$ corresponding to (2.87) and (2.88) is

$$E_i(\mathbf{r},t) = E_i(0,t) + [\nabla_j E_i(\mathbf{r},t)]_o r_j + \frac{1}{2}[\nabla_k \nabla_j E_i(\mathbf{r},t)]_o r_j r_k + \cdots . \quad (2.89)$$

In obtaining (2.89) we have used the Maxwell equation $\nabla \times \mathbf{E} = -\dot{\mathbf{B}}$. The magnetic field $\mathbf{B} = \nabla \times \mathbf{A}$ corresponding to (2.87) is

$$B_i(\mathbf{r},t) = B_i(0,t) + [\nabla_j B_i(\mathbf{r},t)]_o r_j + \frac{1}{2}[\nabla_k \nabla_j B_i(\mathbf{r},t)]_o r_j r_k + \cdots , \quad (2.90)$$

where we have used the Maxwell equation $\nabla \cdot \mathbf{B} = 0$. Thus the potentials (2.87) and (2.88) imply Taylor expansions about an origin for arbitrary time-dependent electric and magnetic fields. It is therefore clear that the choices (2.87) and (2.88) can be made only if the expansions (2.89) and (2.90) are valid.

We now use the potentials (2.87) and (2.88) to express the first- and second-order perturbation Hamiltonians (2.4) and (2.5) in terms of the electric and magnetic fields. This is a lengthy process but the end justifies the means, as one obtains for $H^{(1)}$ an explicit multipole form and for $H^{(2)}$ a series of magnetic susceptibility terms [7,8]

$$H^{(1)} = q\Phi(t) - p_i E_i(t) - \frac{1}{2} q_{ij} E_{ij}(t) - \frac{1}{6} q_{ijk} E_{ijk}(t) - \cdots$$
$$- m_i B_i(t) - \frac{1}{2} m_{ij} B_{ij}(t) - \frac{1}{6} m_{ijk} B_{ijk}(t) - \cdots \tag{2.91}$$

$$H^{(2)} = -\frac{1}{2} \chi_{ij} B_i(t) B_j(t) - \frac{1}{2} \chi_{ijk} B_i(t) B_{jk}(t) - \frac{1}{6} \chi_{ijkl} B_i(t) B_{jkl}(t)$$
$$- \cdots - \frac{4}{27} \chi_{ijkl} B_{ik}(t) B_{jl}(t) - \cdots . \tag{2.92}$$

Here we have introduced the convenient notation for fields and gradients at the origin

$$E_i(t) = E_i(0,t), \quad E_{ij}(t) = [\nabla_j E_i(\mathbf{r},t)]_o, \quad E_{ijk}(t) = [\nabla_k \nabla_j E_i(\mathbf{r},t)]_o, \quad \text{etc.} \tag{2.93}$$

The electric and magnetic multipole moment operators and magnetic susceptibility operators for a charge distribution in (2.91) and (2.92) are given by [8]

$$q = \sum_\alpha q^{(\alpha)}, \quad p_i = \sum_\alpha q^{(\alpha)} r_i^{(\alpha)}, \quad q_{ij} = \sum_\alpha q^{(\alpha)} r_i^{(\alpha)} r_j^{(\alpha)}, \quad \text{etc.} \tag{2.94}$$

$$m_i = \sum_\alpha \frac{q^{(\alpha)}}{2m^{(\alpha)}} l_i^{(\alpha)} \tag{2.95}$$

$$m_{ij} = \sum_\alpha \frac{q^{(\alpha)}}{3m^{(\alpha)}} \left( r_j^{(\alpha)} l_i^{(\alpha)} + l_i^{(\alpha)} r_j^{(\alpha)} \right) \tag{2.96}$$

$$m_{ijk} = \sum_\alpha \frac{3q^{(\alpha)}}{8m^{(\alpha)}} \left( r_k^{(\alpha)} r_j^{(\alpha)} l_i^{(\alpha)} + l_i^{(\alpha)} r_j^{(\alpha)} r_k^{(\alpha)} \right) = m_{ikj} \tag{2.97}$$

$$\chi_{ij} = \sum_\alpha \frac{(q^{(\alpha)})^2}{4m^{(\alpha)}} \left( r_i^{(\alpha)} r_j^{(\alpha)} - (r^{(\alpha)})^2 \delta_{ij} \right) = \chi_{ji} \tag{2.98}$$

$$\chi_{ijk} = \sum_\alpha \frac{(q^{(\alpha)})^2}{3m^{(\alpha)}} \left( r_i^{(\alpha)} r_j^{(\alpha)} - (r^{(\alpha)})^2 \delta_{ij} \right) r_k^{(\alpha)} = \chi_{jik} \tag{2.99}$$

$$\chi_{ijkl} = \sum_\alpha \frac{(q^{(\alpha)})^2}{8m^{(\alpha)}} \left( r_i^{(\alpha)} r_j^{(\alpha)} - (r^{(\alpha)})^2 \delta_{ij} \right) r_k^{(\alpha)} r_l^{(\alpha)} = \chi_{jikl} = \chi_{ijlk}, \tag{2.100}$$

where $\mathbf{l} = \mathbf{r} \times \mathbf{\Pi}$ is the angular momentum operator. The above operators are
Hermitian and they reduce to their classical forms (Chapter 1) in the classical
limit. Other magnetic quadrupole and octopole moment operators have been
published [9] which, although Hermitian, differ from those in (2.96) and (2.97) by
exhibiting full permutation symmetry of their tensor subscripts and not reducing
to their classical forms.

If spin is included then $H^{(1)}$ in (2.4) contains the additional term

$$- \sum_\alpha \frac{q^{(\alpha)}}{2m^{(\alpha)}} \, g^{(\alpha)} \mathbf{s}^{(\alpha)} \cdot \left[ \boldsymbol{\nabla}^{(\alpha)} \times \mathbf{A}(\mathbf{r}^{(\alpha)}, t) \right]. \tag{2.101}$$

By including this term, with $\mathbf{A}$ given by (2.87), we obtain for the general expression of a magnetic $2^n$-pole moment operator

$$m_{ijk\cdots z} = \sum_\alpha n \frac{q^{(\alpha)}}{4m^{(\alpha)}} \left[ r_j^{(\alpha)} r_k^{(\alpha)} \cdots r_z^{(\alpha)} \left( \frac{2}{n+1} l_i^{(\alpha)} + g^{(\alpha)} s_i^{(\alpha)} \right) \right.$$
$$\left. + \left( \frac{2}{n+1} l_i^{(\alpha)} + g^{(\alpha)} s_i^{(\alpha)} \right) r_j^{(\alpha)} r_k^{(\alpha)} \cdots r_z^{(\alpha)} \right]. \tag{2.102}$$

In the presence of a magnetic field the magnetic moment operators in (2.95)–(2.97) and (2.102) are modified because of the replacement (2.53) in $\mathbf{l} = \mathbf{r} \times \boldsymbol{\Pi}$. Using (2.53) and (2.87) in (2.102), one can show that the perturbed moment operators are

$$m_i' = m_i + \chi_{ij} B_j(t) + \frac{1}{2} \chi_{ijk} B_{jk}(t) + \frac{1}{6} \chi_{ijkl} B_{jkl}(t) + \cdots \tag{2.103}$$

$$m_{ij}' = m_{ij} + \chi_{kij} B_k(t) + \frac{16}{27} \chi_{kijl} B_{kl}(t) + \cdots \tag{2.104}$$

$$m_{ijk}' = m_{ijk} + \chi_{lijk} B_l(t) + \cdots = m_{ikj}', \tag{2.105}$$

where the field and its gradients are evaluated at the origin. We remark that the techniques in this section, which have been applied to a linear dependence of the polarizability tensors on a field or its derivatives, can also be used for non-linear phenomena.

## 2.7  Polarizabilities for harmonic plane wave fields

As mentioned in Section 2.5, polarizabilities associated with time-dependent fields are to be found from the expectation value of the appropriate multipole moment operator. Denoting this operator by $\Omega'$, we write

$$\Omega' = \Omega^{(0)} + \Omega^{(1)}, \tag{2.106}$$

where $\Omega^{(0)}$ is its unperturbed form and $\Omega^{(1)}$ applies for a magnetic moment which, as in (2.103)–(2.105), is perturbed to first order by a magnetic field and its various gradients. Then from (2.62) and (2.106)

$$\langle n(t) | \Omega' | n(t) \rangle = \langle \Omega^{(0)} \rangle_{nn} + \langle \Omega^{(1)} \rangle_{nn} + 2 \sum_{s \neq n} \mathcal{R}e\{ c_s \, e^{i\omega_{ns} t} \langle \Omega^{(0)} \rangle_{ns} \}, \tag{2.107}$$

where the notation (2.15) and (2.16) is used. We now assume that the external electromagnetic field is represented by harmonic plane waves

$$\mathbf{E} = \mathbf{E}_0 \cos(\mathbf{k} \cdot \mathbf{r} - \omega t) \tag{2.108}$$

and similarly for $\mathbf{B}$. Then by substituting $H^{(1)}$ from (2.91) into (2.63) we can determine the mixing coefficients $c_s(t)$. This yields to electric octopole–magnetic quadrupole order [8]

$$
\begin{aligned}
c_s(t) = \frac{e^{-i\omega_{ns}t}}{\hbar(\omega^2 - \omega_{ns}^2)} \Big\{ & \langle p_i \rangle_{sn} \left[ \omega_{ns} E_i(t) - i\dot{E}_i(t) \right] + \frac{1}{2} \langle q_{ij} \rangle_{sn} \left[ \omega_{ns} E_{ij}(t) - i\dot{E}_{ij}(t) \right] \\
& + \frac{1}{6} \langle q_{ijk} \rangle_{sn} \left[ \omega_{ns} E_{ijk}(t) - i\dot{E}_{ijk}(t) \right] + \cdots + \langle m_i \rangle_{sn} \left[ \omega_{ns} B_i(t) - i\dot{B}_i(t) \right] \\
& + \frac{1}{2} \langle m_{ij} \rangle_{sn} \left[ \omega_{ns} B_{ij}(t) - i\dot{B}_{ij}(t) \right] + \cdots \Big\}.
\end{aligned}
\tag{2.109}
$$

From (2.107) and (2.109) we obtain the expectation values of the total electric and magnetic multipole moments for a charge distribution to the order of electric octopole and magnetic quadrupole. Thus

$$
\begin{aligned}
\langle n(t)|p_i|n(t)\rangle = {}& p_i^{(0)} + \alpha_{ij} E_j(t) + \frac{1}{\omega} \alpha'_{ij} \dot{E}_j(t) + \frac{1}{2} a_{ijk} E_{jk}(t) + \frac{1}{2\omega} a'_{ijk} \dot{E}_{jk}(t) \\
& + \frac{1}{6} b_{ijkl} E_{jkl}(t) + \frac{1}{6\omega} b'_{ijkl} \dot{E}_{jkl}(t) + \cdots + G_{ij} B_j(t) \\
& + \frac{1}{\omega} G'_{ij} \dot{B}_j(t) + \frac{1}{2} H_{ijk} B_{jk}(t) + \frac{1}{2\omega} H'_{ijk} \dot{B}_{jk}(t) + \cdots \tag{2.110}
\end{aligned}
$$

$$
\begin{aligned}
\langle n(t)|q_{ij}|n(t)\rangle = {}& q_{ij}^{(0)} + \mathsf{a}_{ijk} E_k(t) + \frac{1}{\omega} \mathsf{a}'_{ijk} \dot{E}_k(t) + \frac{1}{2} d_{ijkl} E_{kl}(t) \\
& + \frac{1}{2\omega} d'_{ijkl} \dot{E}_{kl}(t) + \cdots + L_{ijk} B_k(t) + \frac{1}{\omega} L'_{ijk} \dot{B}_k(t) + \cdots
\end{aligned}
\tag{2.111}
$$

$$
\langle n(t)|q_{ijk}|n(t)\rangle = q_{ijk}^{(0)} + \mathsf{b}_{ijkl} E_l(t) + \frac{1}{\omega} \mathsf{b}'_{ijkl} \dot{E}_l(t) + \cdots \tag{2.112}
$$

$$
\begin{aligned}
\langle n(t)|m_i|n(t)\rangle = {}& m_i^{(0)} + \mathcal{G}_{ij} E_j(t) + \frac{1}{\omega} \mathcal{G}'_{ij} \dot{E}_j(t) + \frac{1}{2} \mathcal{L}_{ijk} E_{jk}(t) \\
& + \frac{1}{2\omega} \mathcal{L}'_{ijk} \dot{E}_{jk}(t) + \cdots + \chi_{ij} B_j(t) + \frac{1}{\omega} \chi'_{ij} \dot{B}_j(t) + \cdots \tag{2.113}
\end{aligned}
$$

$$
\langle n(t)|m_{ij}|n(t)\rangle = m_{ij}^{(0)} + \mathcal{H}_{ijk} E_k(t) + \frac{1}{\omega} \mathcal{H}'_{ijk} \dot{E}_k(t) + \cdots . \tag{2.114}
$$

Note that the only contribution of the term involving $\Omega^{(1)}$ in (2.107) to the above results is the last two terms in (2.113).

The multipole polarizabilities in (2.110)–(2.114) are for a charge distribution and they are given by

$$\alpha_{ij} = \frac{2}{\hbar} \sum_s \omega_{sn} Z_{sn} \, \mathcal{R}e\{\langle p_i \rangle_{ns} \langle p_j \rangle_{sn}\} = \alpha_{ji} \tag{2.115}$$

$$\alpha'_{ij} = -\frac{2}{\hbar} \sum_s \omega Z_{sn} \, \mathcal{I}m\{\langle p_i \rangle_{ns} \langle p_j \rangle_{sn}\} = -\alpha'_{ji} \tag{2.116}$$

$$a_{ijk} = \frac{2}{\hbar} \sum_s \omega_{sn} Z_{sn} \, \mathcal{R}e\{\langle p_i \rangle_{ns} \langle q_{jk} \rangle_{sn}\} = \mathfrak{a}_{jki} \tag{2.117}$$

$$a'_{ijk} = -\frac{2}{\hbar} \sum_s \omega Z_{sn} \, \mathcal{I}m\{\langle p_i \rangle_{ns} \langle q_{jk} \rangle_{sn}\} = -\mathfrak{a}'_{jki} \tag{2.118}$$

$$G_{ij} = \frac{2}{\hbar} \sum_s \omega_{sn} Z_{sn} \, \mathcal{R}e\{\langle p_i \rangle_{ns} \langle m_j \rangle_{sn}\} = \mathcal{G}_{ji} \tag{2.119}$$

$$G'_{ij} = -\frac{2}{\hbar} \sum_s \omega Z_{sn} \, \mathcal{I}m\{\langle p_i \rangle_{ns} \langle m_j \rangle_{sn}\} = -\mathcal{G}'_{ji} \tag{2.120}$$

$$b_{ijkl} = \frac{2}{\hbar} \sum_s \omega_{sn} Z_{sn} \, \mathcal{R}e\{\langle p_i \rangle_{ns} \langle q_{jkl} \rangle_{sn}\} = \mathfrak{b}_{jkli} \tag{2.121}$$

$$b'_{ijkl} = -\frac{2}{\hbar} \sum_s \omega Z_{sn} \, \mathcal{I}m\{\langle p_i \rangle_{ns} \langle q_{jkl} \rangle_{sn}\} = -\mathfrak{b}'_{jkli} \tag{2.122}$$

$$d_{ijkl} = \frac{2}{\hbar} \sum_s \omega_{sn} Z_{sn} \, \mathcal{R}e\{\langle q_{ij} \rangle_{ns} \langle q_{kl} \rangle_{sn}\} = d_{klij} \tag{2.123}$$

$$d'_{ijkl} = -\frac{2}{\hbar} \sum_s \omega Z_{sn} \, \mathcal{I}m\{\langle q_{ij} \rangle_{ns} \langle q_{kl} \rangle_{sn}\} = -d'_{klij} \tag{2.124}$$

$$H_{ijk} = \frac{2}{\hbar} \sum_s \omega_{sn} Z_{sn} \, \mathcal{R}e\{\langle p_i \rangle_{ns} \langle m_{jk} \rangle_{sn}\} = \mathcal{H}_{jki} \tag{2.125}$$

$$H'_{ijk} = -\frac{2}{\hbar} \sum_s \omega Z_{sn} \, \mathcal{I}m\{\langle p_i \rangle_{ns} \langle m_{jk} \rangle_{sn}\} = -\mathcal{H}'_{jki} \tag{2.126}$$

$$L_{ijk} = \frac{2}{\hbar} \sum_s \omega_{sn} Z_{sn} \, \mathcal{R}e\{\langle q_{ij} \rangle_{ns} \langle m_k \rangle_{sn}\} = L_{jik} = \mathcal{L}_{kij} \tag{2.127}$$

$$L'_{ijk} = -\frac{2}{\hbar} \sum_s \omega Z_{sn} \, \mathcal{I}m\{\langle q_{ij} \rangle_{ns} \langle m_k \rangle_{sn}\} = L'_{jik} = -\mathcal{L}'_{kij} \tag{2.128}$$

$$\chi_{ij} = \frac{2}{\hbar} \sum_s \omega_{sn} Z_{sn} \, \mathcal{R}e\{\langle m_i \rangle_{ns} \langle m_j \rangle_{sn}\}$$

$$+ \sum_{\alpha=1}^{N} \frac{(q^{(\alpha)})^2}{4m^{(\alpha)}} \left\langle r_i^{(\alpha)} r_j^{(\alpha)} - (r^{(\alpha)})^2 \delta_{ij} \right\rangle_{nn} = \chi_{ji} \tag{2.129}$$

$$\chi'_{ij} = -\frac{2}{\hbar} \sum_s \omega Z_{sn} \, \mathcal{I}m\{\langle m_i \rangle_{ns} \langle m_j \rangle_{sn}\} = -\chi'_{ji}, \tag{2.130}$$

where

$$Z_{sn} = (\omega_{sn}^2 - \omega^2)^{-1}. \tag{2.131}$$

The following points are noted concerning the polarizabilities in (2.115)–(2.130):

(i) Despite the restriction $s \neq n$ in (2.107), the summation over $s$ in (2.115)–(2.130) may include $s = n$ for the reasons given in Section 2.5.

(ii) The permutation symmetry of subscripts of the electric moment operators $q_{ij}$ and $q_{ijk}$ carries through to the polarizabilities that contain matrix elements of these operators (see Section 2.10).

(iii) The various polarizability tensors introduced above can be classified according to their multipole order. Some of the multipole orders in (2.115)–(2.130) are obvious: thus $\alpha_{ij}$ is of electric dipole order, $a_{ijk}$ is of electric quadrupole order, $G_{ij}$ is of magnetic dipole order, $b_{ijkl}$ is of electric octopole order, etc. Others are less obvious: for example in $d_{ijkl}$ the product of two electric quadrupole moments implies a 4th power of the position vectors $r_i$, which is the same as in $b_{ijkl}$. Thus $d_{ijkl}$ is of electric octopole order. A classification of these tensors in terms of their multipole order and space–time properties is given in Table 3.2.

(iv) Expressions for the static polarizabilities can be obtained from (2.115)–(2.130) by setting $\omega = 0$, including in $Z_{sn}$ in (2.131). Thus all those designated by a prime vanish. These tensors describe the induction of a multipole moment by the time derivative of a field or the time derivative of one of its gradients.

(v) The expression for $Z_{sn}$ in (2.131) does not apply when the charge distribution absorbs radiation. A modification of $Z_{sn}$ to allow for absorption is described in the next section.

(vi) The relationships (2.115)–(2.130) between polarizability tensors follow from the Hermitian property of the multipole moment operators, and they apply also in absorption, as will be apparent in Section 2.8.

## 2.8 Absorption of radiation

The polarizability expressions in (2.115)–(2.130) were derived without regard to possible absorption of radiation by the charge distribution. When the radiation frequency $\omega$ equals the frequency $\omega_{sn}$ of a transition between an upper level $s$ and the ground state $n$, the function $Z_{sn}$ in (2.131) becomes infinite and so do all the polarizabilities. Because this is unphysical, the theory requires modification. The reader is referred to a clear account by Barron [10] of the physical basis and theory of such a modification. Here we give a brief description of the main results.

The form of $Z_{sn}$ that is derived in this theory, namely [11, 12]

$$Z_{sn} = f + ig, \tag{2.132}$$

comprises the dispersion and absorption line-shape functions $f(\omega, \omega_{sn}, \Gamma_{sn})$ and $g(\omega, \omega_{sn}, \Gamma_{sn})$, where $\Gamma_{sn}$, a measure of line width of the $sn$ transition, is independent of $\omega$. Far from an absorption band (i.e. when $|\omega - \omega_{sn}| \gg \Gamma_{sn}$), the contribution of $g$ may be neglected relative to that of $f$, and the function $Z_{sn}$

tends to the expression (2.131), which is thus purely dispersive. In the vicinity of an absorption (or resonance) frequency $\omega_{sn}$

$$f = \frac{\omega_{sn}^2 - \omega^2}{(\omega_{sn}^2 - \omega^2)^2 + \omega^2 \Gamma_{sn}^2} \qquad (2.133)$$

$$g = \frac{\omega \Gamma_{sn}}{(\omega_{sn}^2 - \omega^2)^2 + \omega^2 \Gamma_{sn}^2}. \qquad (2.134)$$

Thus the absorption (or resonance) curve (2.134) is Lorentzian and $\Gamma_{sn}$ is approximately its width at half maximum height. Provided the frequency separation between any two neighbouring absorption bands is much greater than their respective half-widths, then $Z_{sn}$ in (2.131) should be replaced by (2.132) in all the polarizability expressions (2.115)–(2.130).

The symmetries which a polarizability tensor possesses are not affected by the particular form of $Z_{sn}$ in its expression. This is because these symmetries arise in various ways from properties of the remaining factors of the expression, namely the effect of taking a real or imaginary part; the Hermitian property of the multipole moment operators; the permutation symmetry of their subscripts, and their space–time transformations (see Chapter 3). This feature of the intrinsic symmetries also follows from the property that the dispersive and absorptive parts of polarizabilities of the type considered here are connected by Kramers–Kronig relations [10]. This means that the absorptive part of the polarizabilities possesses the same intrinsic symmetry as the dispersive part.

## 2.9   Additional static magnetic polarizabilities

In this section we give an example of how the Barron–Gray gauge of Section 2.6 may be used to obtain static magnetic polarizabilities by direct calculation (rather than as the *dc* limit of the dynamic polarizabilities in Section 2.7). We remind the reader that in Section 2.4 we deferred deriving magnetic polarizabilities associated with static field gradients because of the intricate nature of the gauge used there; we are now in a position to do so much more directly.

The Barron–Gray gauge (2.87) and (2.88) applies also to electrostatic and magnetostatic fields. For the latter, $\Phi(\mathbf{r}) = 0$ and

$$A_i(\mathbf{r}) = \varepsilon_{ijk} \left\{ \frac{1}{2} B_j r_k + \frac{1}{3} \left( \nabla_l B_j \right) r_k r_l + \frac{1}{8} \left( \nabla_m \nabla_l B_j \right) r_k r_l r_m + \cdots \right\}, \qquad (2.135)$$

where the field and its gradients are evaluated at the origin. From (2.135), (2.91), and (2.63) we obtain the mixing coefficients for magnetostatic fields given by the Barron–Gray gauge. To magnetic quadrupole order

$$c_s = (\hbar \omega_{sn})^{-1} \left\{ \langle m_i \rangle_{sn} B_i + \frac{1}{2} \langle m_{ij} \rangle_{sn} B_{ij} + \cdots \right\}. \qquad (2.136)$$

Equation (2.136) is the magnetostatic limit of (2.109). Also, for static fields (2.107) is

$$\langle n|\Omega'|n\rangle = \langle \Omega^{(0)}\rangle_{nn} + \langle \Omega^{(1)}\rangle_{nn} + 2\sum_{s\neq n} \mathcal{R}e\{c_s\langle \Omega^{(0)}\rangle_{ns}\}. \tag{2.137}$$

From (2.136), (2.137), (2.103), and (2.104) one readily obtains to magnetic quadrupole order

$$\langle n|m'_i|n\rangle = m_i^{(0)} + \chi_{ij}\,B_j + \frac{1}{2}\chi_{ijk}\,B_{jk} + \cdots \tag{2.138}$$

$$\langle n|m'_{ij}|n\rangle = m_{ij}^{(0)} + \xi_{ijk}\,B_k + \cdots . \tag{2.139}$$

Here $\chi_{ij}$ is given by (2.52) and

$$\chi_{ijk} = \sum_{s\neq n} \frac{2}{\hbar\omega_{sn}}\mathcal{R}e\{\langle m_i\rangle_{ns}\langle m_{jk}\rangle_{sn}\}$$

$$+ \sum_{\alpha=1}^{N} \frac{(q^{(\alpha)})^2}{3m^{(\alpha)}}\left\langle \left(r_i^{(\alpha)}r_j^{(\alpha)} - \left(r^{(\alpha)}\right)^2\delta_{ij}\right)r_k^{(\alpha)}\right\rangle_{nn} = \xi_{jki}. \tag{2.140}$$

Physical effects due to static magnetic field gradients are known: for example, the lowering of the energy of closed shells in an atom [13].

## 2.10   Symmetries

We comment briefly on various symmetries possessed by polarizability tensors and related quantities.

(i) We have already noted symmetries such as those in (2.22), (2.27)–(2.29), (2.115), and (2.116) which follow from the Hermitian property of the operators in these equations. We remark that for *dc* polarizabilities these symmetries can also be obtained from classical energy considerations.

(ii) We have also noted additional symmetries that follow from the permutation symmetry of electric quadrupole and higher moment operators (Section 2.7). For example,

$$a_{ijk} = a_{ikj} \tag{2.141}$$

$$b_{ijkl} = b_{ijlk} = b_{ikjl} \tag{2.142}$$

$$d_{ijkl} = d_{klij} = d_{jikl}. \tag{2.143}$$

(iii) Symmetries may also be deduced from quantum-mechanical expressions for the frequency-dependent hyperpolarizabilities that account for a range of non-linear effects. For instance, in

$$D_i = \varepsilon_0 E_i + \alpha_{ij}\,E_j + \frac{1}{2}\beta_{ijk}\,E_j E_k + \cdots , \tag{2.144}$$

where **E** is the electric field of a harmonic plane wave, the phenomenon of second-harmonic generation is due to the leading hyperpolarizability

$\beta_{ijk}$. From its quantum-mechanical expression this tensor can be shown to possess the symmetry [14]

$$\beta_{ijk} = \beta_{jik}, \tag{2.145}$$

in addition to the permutation symmetry of subscripts $j$ and $k$ that is evident from (2.144). Knowledge of the expressions for such tensors avoids the use of Kleinman's conjecture, which is effectively the assumption that a symmetry such as (2.145) of a frequency-dependent polarizability is that of its *dc* limit, as determined from thermodynamic considerations [15].

(iv) The symmetries (i)–(iii) above are known as intrinsic symmetries, and they are distinct from the particular point-group symmetry of a polarizability tensor. The latter symmetry depends on the geometric and time nature of the molecule, or other object, to which the tensor belongs (Chapter 3).

(v) All tensors, including polarizabilities, may also possess a space–time symmetry independently of any other symmetry. Thus they may be either polar or axial and time even or time odd (see Section 3.5).

(vi) In multipole theory, material constants for constitutive relations are constructed as linear combinations of polarizability tensors (Chapter 4). For non-dissipative media these material constants possess symmetries which are associated with the Lagrangian structure of the theory and which are usually distinct from the intrinsic symmetries of the polarizability tensors (Chapter 4).

(vii) Similarly, in the wave equation obtained for propagation of electromagnetic waves in a dielectric, certain multipole tensor coefficients possess symmetries which can be distinct from intrinsic symmetries (Section 5.1).

## 2.11 Macroscopic multipole moment and polarizability densities

The results derived so far in this chapter apply to a charge distribution. We now extend the theory to describe multipole moments induced in bulk matter by macroscopic harmonic fields $\mathbf{E}(\mathbf{r}, t)$ and $\mathbf{B}(\mathbf{r}, t)$. This is done by averaging the expectation values (2.110)–(2.114) for multipole moments using the technique referred to in Section 1.12. This averaging introduces the macroscopic multipole moment densities $P_i$, $Q_{ij}$, $M_i$, … of Section 1.12.

Consider (2.110) as an example: after performing the spatial average of (1.91) we have for the macroscopic electric dipole moment density [16]

$$P_i = P_i^{(0)} + \alpha_{ij} E_j + \frac{1}{\omega} \alpha'_{ij} \dot{E}_j + \frac{1}{2} a_{ijk} \nabla_k E_j + \frac{1}{2\omega} a'_{ijk} \nabla_k \dot{E}_j$$
$$+ \frac{1}{6} b_{ijkl} \nabla_l \nabla_k E_j + \frac{1}{6\omega} b'_{ijkl} \nabla_l \nabla_k \dot{E}_j + \cdots + G_{ij} B_j$$
$$+ \frac{1}{\omega} G'_{ij} \dot{B}_j + \frac{1}{2} H_{ijk} \nabla_k B_j + \frac{1}{2\omega} H'_{ijk} \nabla_k \dot{B}_j + \cdots . \tag{2.146}$$

Here we have used $\mathbf{r}$, rather than $\mathbf{R}$, for the position of a field point. Also, in (2.146) the tensors $\alpha_{ij}$, $\alpha'_{ij}$, $a_{ijk}$, etc. are polarizability densities, for which we continue to use the same symbols as those for the polarizabilities in (2.110)–(2.114). The context will indicate which of the meanings applies. The quantum-mechanical expressions for polarizability densities are obtained by performing a spatial average of the corresponding polarizability tensors (2.115)–(2.130). Approximate forms are sometimes used, based on a particular normalized weight function in the spatial average, namely one which is constant inside a macroscopic volume element $\Delta V$ and zero outside. Thus, for example, from (2.115)

$$\alpha_{ij} = 2(\hbar\Delta V)^{-1}\sum_s \omega_{sn}Z_{sn}\mathcal{R}e\{\langle p_i\rangle_{ns}\langle p_j\rangle_{sn}\} = \alpha_{ji} \qquad (2.147)$$

is a density at the position $\mathbf{r}$ of the element $\Delta V$. Here the dipole moment operators $p_i$ are given in (2.94), with summation over all charged particles $q^{(\alpha)}$ in $\Delta V$.

Other macroscopic multipole moment densities are obtained in a similar way. Thus, from (2.111) one has the macroscopic electric quadrupole moment density

$$Q_{ij} = Q_{ij}^{(0)} + \mathfrak{a}_{ijk}E_k + \frac{1}{\omega}\mathfrak{a}'_{ijk}\dot{E}_k + \frac{1}{2}d_{ijkl}\nabla_l E_k$$
$$+ \frac{1}{2\omega}d'_{ijkl}\nabla_l\dot{E}_k + \cdots + L_{ijk}B_k + \frac{1}{\omega}L'_{ijk}\dot{B}_k + \cdots . \qquad (2.148)$$

Expressions for the moment densities of electric octopole ($Q_{ijk}$), magnetic dipole ($M_i$), and magnetic quadrupole ($M_{ij}$) follow from (2.112)–(2.114). These multipole moment and polarizability densities are important for the further development and applications of the theory (Chapters 4–9).

## 2.12  Phenomenology of the wave–matter interaction

In interacting with matter, a harmonic plane electromagnetic wave does so not only through its electric and magnetic fields, but also through the space and time derivatives of these fields. For example, the electric field of a wave, because of its finite wavelength, is not uniform over an object in its path, such as a molecule in a dilute gas or a unit cell in a crystal. Thus for the interaction of the wave with such an object, not just the fields but also the various field gradients and their time derivatives should play a role. With regard to the latter, because of the harmonic nature of the waves, $\ddot{E}_i = -\omega^2 E_i$, $\dddot{E}_i = -\omega^2\dot{E}_i$, etc., so that a field and its various gradients have only two linearly independent time derivatives, which we take to be the zeroth and the first. Altogether we expect that the following attributes of the wave may be relevant for its interaction with matter

$$E_i,\ \dot{E}_i,\ \nabla_j E_i,\ \nabla_j\dot{E}_i,\ \nabla_k\nabla_j E_i,\ \nabla_k\nabla_j\dot{E}_i,\ \ldots; B_i,\ \dot{B}_i,\ \nabla_j B_i,\ \nabla_j\dot{B}_i,\ldots .$$
$$(2.149)$$

In fact, we have already seen in (2.110)–(2.114) the role played by these quantities in inducing multipole moments.

The action of a field gradient of an electromagnetic wave in inducing an effect in matter is termed spatial dispersion [17]. H. A. Lorentz in 1878 was probably the first to invoke this concept in a theory of linear birefringence in certain cubic crystals [18], followed a few years later by J. W. Gibbs in an account of optical activity [19]. These two effects, and a selection of others, which are induced by the fields and derivatives in (2.149), acting either singly or jointly, are listed below, together with their multipole order. (Effects marked with an asterisk are considered in detail in Chapters 5 and 9.)

1. Electric dipole

  (i) $E_i$:

       Linear birefringence in uniaxial and biaxial crystals [20].
       Depolarization of scattered light [21].

  (ii) $\dot{E}_i$:

       *Intrinsic Faraday effect in ferromagnetic crystals [22, 23].
       Microscopic effects associated with ultrafast pulses [24].

2. Electric quadrupole–magnetic dipole

  (i) $\nabla_j E_i$, $\dot{B}_i$:

       *Optical activity in anisotropic media [25, 26].
       Differential scattering of left and right circularly polarized light [27].

  (ii) $\dot{B}_i$:

       *Optical activity in fluids and cubic crystals [25, 26].

  (iii) $\nabla_j \dot{E}_i$, $B_i$:

       *Gyrotropic birefringence in magnetic crystals [28, 29].
       *Reflection effects in paramagnetic $Cr_2O_3$ [30, 31].
       Dynamic magnetoelectric effect [32].

  (iv) $\nabla_j \dot{E}_i$:

       *Non-reciprocal linear birefringence in magnetic cubic crystals [29].

3. Electric octopole–magnetic quadrupole

  (i) $\nabla_k \nabla_j E_i$, $\nabla_j \dot{B}_i$:

       *Lorentz birefringence in cubic crystals [18, 33, 34].
       *Jones birefringence in crystals [35, 36].
       Differential scattering of orthogonally polarized light [37].

  (ii) $\nabla_k \nabla_j \dot{E}_i$, $\nabla_j B_i$:

       *Intrinsic rotation in antiferromagnetic cubic crystals [16].

The above list is not exhaustive and does not include optical effects in static fields such as those of Faraday, Pockels, Kerr, Buckingham, etc.

Because a visible wavelength ($\approx$500 nm) is very much greater than a linear dimension of a molecule or unit cell, the reader might suppose that the contribution of a field-gradient term to an effect would be negligible. However, where this contribution is the leading one, for example, in an optically active crystal like quartz (see 2(i) above), such a supposition is not correct. The rotation of linearly polarized light along 1 cm of optic axis in quartz is about 300° for a wavelength of 500 nm [38].

## References

[1] Atkins, P. W. and Woolley, R. G. (1970). The interaction of molecular multipoles with the electromagnetic field in the canonical formulation of non-covariant quantum electrodynamics. *Proceedings of the Royal Society London* A, **319**, 549–563.

[2] Felderhof, B. U. and Ayu-Gyamfi, D. (1974). The multipole Hamiltonian for molecules. *Physica*, **71**, 399–414.

[3] Atkins, P. W. and Friedman, R. S. (1997) *Molecular quantum mechanics*, Ch. 6. Oxford University Press, Oxford.

[4] Buckingham, A. D. (1959). Molecular quadrupole moments. *Quaterly Reviews*, **13**, 183–214.

[5] de Lange, O. L. and Raab, R. E. (1991). *Operator methods in quantum mechanics*, p. 25. Clarendon, Oxford.

[6] Van Vleck, J. H. (1932). *The theory of electric and magnetic susceptibilities*, p. 276. Clarendon, Oxford.

[7] Barron, L. D. and Gray, C. G. (1973). The multipole interaction Hamiltonian for time dependent fields. *Journal of Physics* A: *Mathematical, Nuclear, and General*, **6**, 59–61.

[8] Raab, R. E. (1975). Magnetic multipole moments. *Molecular Physics*, **29**, 1323–1331.

[9] Buckinghan, A. D. and Stiles, P. J. (1972). Magnetic multipoles and the "pseudo-contact" chemical shift. *Molecular Physics*, **24**, 99–108.

[10] Barron, L. D. (1982). *Molecular light scattering and optical activity*. Cambridge University Press, Cambridge.

[11] Buckingham, A. D. (1967). Permanent and induced molecular moments and long-range intermolecular forces. In *Intermolecular forces* (ed. Hirschfelder, J. O.), Advances in Chemical Physics, **12**, 107–142. Interscience, New York.

[12] Buckingham, A. D. and Raab, R. E. (1975). Electric-field-induced differential scattering of right and left circularly polarized light. *Proccedings of the Royal Society London* A, **345**, 365–377.

[13] Mattis, D. C. (1968). Magnetic polarizability of closed shells. *Physical Review Letters*, **20**, 792–794.

[14] Buckingham, A. D. and Longuet-Higgins, H. C. (1968). The quadrupole moments of dipolar molecules. *Molecular Physics*, **14**, 63–72.

[15] Franken, P. A. and Ward, J. F. (1963). Optical harmonics and nonlinear phenomena. *Reviews of Modern Physics*, **35**, 23–39.

[16] Graham, E. B. and Raab, R. E. (1991). Non-reciprocal optical rotation in cubic antiferromagnets. *Philosophical Magazine* B, **64**, 267–274.

[17] Agranovich, V. M. and Ginzburg, V. L. (1984). *Crystal optics with spatial dispersion, and excitons.* Springer, Berlin.

[18] Lorentz, H. A. (1878). Concerning the relation between the velocity of propagation of light and the density and composition of media. In *H. A. Lorentz, collected papers* (ed. Zeeman, P. and Fokker, A. D.), Vol. 2, pp. 1–119. Nijhoff, The Hague.

[19] Gibbs, J. W. (1882). Notes on the electromagnetic theory of light. No.II–on refraction in perfectly transparent media which exhibit the phenomena of circular birefringence. *American Journal of Science*, **23**, 460–476.

[20] Gunning, M. J. and Raab, R. E. (1998). Systematic eigenvalue approach to crystal optics: an analytic alternative to the geometric ellipsoid model. *Journal of the Optical Society of America* A, **15**, 2199–2207.

[21] Bogaard, M. P., Buckingham, A. D., Pierens, R. K., and White, A. H. (1978). Rayleigh scattering depolarization ratio and the molecular polarizability anisotropy for gases. *Journal of the Chemical Society Faraday Transactions* I, **74**, 3008–3015.

[22] Argyres, P. N. (1955). Theory of the Faraday and Kerr effects in ferromagnetics. *Physical Review*, **97**, 334–345.

[23] Graham, E. B. and Raab, R. E. (1991). Electric dipole effects in magnetic crystals. *Journal of Applied Physics*, **69**, 2549–2551.

[24] Ziolkowski, R. W., Arnold, J. M., and Gogny, D. M. (1995). Ultrafast pulse interactions with two-level atoms. *Physical Review* A, **52**, 3082–3094.

[25] Buckingham, A. D. and Dunn, M. B. (1971). Optical activity of oriented molecules. *Journal of the Chemical Society* A, 1988–1991.

[26] Raab, R. E. and Cloete, J. H. (1994). An eigenvalue theory of circular birefringence and dichroism in a non-magnetic chiral medium. *Journal of Electromagnetic Waves and Applications*, **8**, 1073–1089.

[27] Barron, L. D. and Buckingham, A. D. (1971). Rayleigh and Raman scattering from optically active molecules. *Molecular Physics*, **20**, 1111–1119.

[28] Hornreich, R. M. and Shtrikman, S. (1968). Theory of gyrotropic birefringence. *Physical Review*, **171**, 1065–1074.

[29] Graham, E. B. and Raab, R. E. (1992). Magnetic effects in antiferromagnetic crystals in the electric quadrupole-magnetic dipole approximation. *Philosophical Magazine*, **66**, 269–284.

[30] Krichevtsov, B. B., Pavlov, V. V., Pisarev, R. V., and Gridnev, V. N. (1993). Spontaneous non-reciprocal reflection of light from antiferromagnetic $Cr_2O_3$. *Journal of Physics: Condensed Matter*, **5**, 8233–8244.

[31] Graham, E. B. and Raab, R. E. (1997). Macroscopic theory of reflection from antiferromagnetic $Cr_2O_3$. *Journal of Physics: Condensed Matter*, **9**, 1863–1869.

[32] Raab, R. E. and Sihvola, A. H. (1997). On the existence of linear non-reciprocal bi-isotropic (NRBI) media. *Journal of Physics* A: *Mathematical and General*, **30**, 1335–1344.

[33] Condon, E. U. and Seitz, F. (1932). Lorentz double refraction in the regular system. *Journal of the Optical Society of America*, **22**, 393–401.

[34] Graham, E. B. and Raab, R. E. (1990). Light propagation in cubic and other anisotropic crystals. *Proceedings of the Royal Society London* A, **430**, 593–614.

[35] Graham, E. B. and Raab, R. E. (1983). On the Jones birefringence. *Proceedings of the Royal Society London* A, **390**, 73–90.

[36] Graham, C. and Raab, R. E. (1994). Eigenvector approach to the evaluation of the Jones $N$ matrices of nonabsorbing crystalline media. *Journal of the Optical Society of America* A, **11**, 2137–2144.

[37] de Figueiredo, I. M. B. and Raab, R. E. (1981). A molecular theory of new differential light-scattering effects in a fluid. *Proceedings of the Royal Society London* A, **375**, 425–441.

[38] Jenkins, F. A. and White, H. E. (1976). *Fundamentals of Optics*, p. 583. McGraw–Hill, New York.

# 3

## SPACE AND TIME PROPERTIES

*Or madly squeeze a right-hand foot*
*Into a left-hand shoe.*
Lewis Carrol
(*Through the Looking-Glass*)

This chapter presents a collection of useful results for the space and time properties of certain physical quantities. The presentation is considerably condensed, and the reader who requires more details is referred to the extensive literature on the topics in Sections 3.1–3.6. In Sections 3.1–3.3 we give an account of vectors and Cartesian tensors with emphasis on their behaviour under coordinate transformations. An introduction to time reversal in Section 3.4 is used to derive the formula for transformation of the matrix element of an operator under time reversal. The formalism of Sections 3.1–3.4 is applied in Section 3.5 to obtain a classification of various physical quantities represented by Cartesian tensors according to their space and time behaviour: we consider mechanical and electromagnetic quantities, as well as multipole moments and polarizabilities. A discussion of the symmetry of an object and its effect on property tensors is given in Section 3.6. We study the effect of space translation on multipole polarizabilities in Section 3.7; the results obtained there are essential to our further discussion of multipole theory. In Section 3.8 we discuss a pictorial method of determining symmetry conditions for effects in macroscopic media.

## 3.1 Coordinate transformations

Consider two systems of Cartesian axes with a common origin $O$. Their respective axes are denoted by $x_1$, $x_2$, $x_3$ (or $x_i$, $i = 1$, 2, 3) and $x'_1$, $x'_2$, $x'_3$ (or $x'_i$). In general, the two sets of axes are not coincident. One may be obtained from the other through a transformation of axes, such as a rotation about an axis through $O$, reflection in a plane containing $O$, inversion through $O$, and a combination of these.

Cartesian axes are either right hand or left hand in terms of the right-hand screw convention applied to the order $x_1$, $x_2$, $x_3$. A transformation which leaves the hand of axes unchanged is a proper transformation and one that changes the hand is termed improper. Clearly, a rotation is a proper transformation, whereas reflection and inversion are both improper.

Any improper transformation is equivalent to a combination of an inversion and a rotation. For example, reflection of axes in the $x_1x_2$ plane means

$$x_1 \to x'_1 = x_1, \qquad x_2 \to x'_2 = x_2, \qquad x_3 \to x'_3 = -x_3. \qquad (3.1)$$

Inverting the original axes gives

$$x_1 \to x_1' = -x_1, \qquad x_2 \to x_2' = -x_2, \qquad x_3 \to x_3' = -x_3. \qquad (3.2)$$

When this is followed by a rotation of $\pi$ about the axis $x_3'$, one obtains from (3.2)

$$x_1' \to x_1'' = x_1, \qquad x_2' \to x_2'' = x_2, \qquad x_3' \to x_3'' = -x_3, \qquad (3.3)$$

which is (3.1).

## 3.2   Vectors

It is convenient to preface our discussion of tensors by considering a special case of a tensor, namely a vector. Thus we consider a quantity $\mathbf{V}$ which possesses both magnitude and direction and therefore has components in any system of axes. The component of $\mathbf{V}$ on the Cartesian axis $x_j$ is denoted by $V_j$ and on $x_i'$ by $V_i'$. Each component of $\mathbf{V}$ on the axes $x_j$ may be projected onto the axis $x_i'$ and then added to give

$$V_i' = V_1 \cos\theta_{i1} + V_2 \cos\theta_{i2} + V_3 \cos\theta_{i3} = l_{ij} V_j, \qquad (3.4)$$

where $\theta_{ij}$ is the angle between $x_i'$ and $x_j$, and

$$l_{ij} = \cos\theta_{ij} \qquad (3.5)$$

is the direction cosine of $x_i'$ relative to $x_j$. In (3.4) we continue to use the summation convention introduced in Section 1.1: thus a repeated subscript implies summation from 1 to 3. The explicit form of the transformation matrix $l_{ij}$ depends on the type of transformation involved.

(i) *Rotations*

   We define a vector $\mathbf{V}$ as any ordered set $(V_1, V_2, V_3)$ which transforms according to (3.4) under rotations. The explicit form of the rotation matrix $l_{ij}$ depends on the axis of rotation and examples are given in most textbooks [1]. Within the broad category of vectors so defined, it is useful to specify subcategories in terms of the behaviour under improper transformations and also translations of the coordinate system.

(ii) *Improper transformations*

   Not all vectors transform according to (3.4) when the transformation is improper. We illustrate this by applying an inversion of axes (since this is involved in all improper transformations) to the angular momentum vector $\mathbf{L} = m\mathbf{r} \times \mathbf{u}$. Here $\mathbf{r} = (x_1, x_2, x_3)$ is the position vector of a particle of mass $m$, and $\mathbf{u} = \dot{\mathbf{r}}$ is its velocity. Equation (3.2) gives the effect of inversion on the components of $\mathbf{r}$. It follows that in (3.5)

$$\theta_{11} = \theta_{22} = \theta_{33} = \pi, \quad |\theta_{ij}| = \pi/2 \text{ if } i \neq j. \qquad (3.6)$$

   Thus (3.4) requires $\mathbf{L}$ to transform according to

$$L_1' = -L_1, \qquad L_2' = -L_2, \qquad L_3' = -L_3. \tag{3.7}$$

That is, in the same manner as $\mathbf{r}$. However, $\mathbf{L}$ must transform according to its defining expression, for example, its component

$$L_1 = m(x_2 u_3 - x_3 u_2). \tag{3.8}$$

We assume that $m$ is invariant (under both proper and improper transformations of axes, Section 3.2). Then, because $\mathbf{u}$ transforms like $\mathbf{r}$, (3.8) yields

$$L_1' = L_1; \qquad \text{similarly } L_2' = L_2, \quad L_3' = L_3. \tag{3.9}$$

We conclude from (3.7) and (3.9) that (3.4) does not apply to the definition of $\mathbf{L}$ for an inversion of axes. It is left to the reader to show that (3.4) does correctly describe the transformation of $\mathbf{L}$ for any rotation of axes. Vectors are classified by whether, in terms of their definition, they do or do not change sign under inversion of axes. Those that change sign are called polar vectors, those that do not are axial vectors (or pseudovectors). Thus a vector to which (3.4) applies for both proper and improper transformations is defined to be a polar vector, and one whose transformations are described by

$$V_i' = \pm l_{ij} V_j, \tag{3.10}$$

where the upper (lower) sign applies for a proper (improper) transformation, satisfies the definition of an axial vector. Examples of polar and axial vectors are given in Section 3.5.

(iii) *Translations*

In addition to the coordinate transformations mentioned above, we also consider translations of the coordinate system for an arbitrary shift in the origin from $O$ to $\bar{O}$. The corresponding changes in the components of $\mathbf{V}$ are denoted by $\Delta V_i$. If $\Delta V_i = 0$ for $i = 1$, 2, 3 the vector $\mathbf{V}$ is referred to as origin independent (translationally invariant); if $\Delta V_i \neq 0$ for one or more values of $i$, then $\mathbf{V}$ is origin dependent. Examples of origin-independent vectors are velocity, acceleration, and the angular momentum of a system of particles about their centre of mass. Examples of origin-dependent vectors are the position vector, torque on a particle, and the angular momentum of a particle in classical mechanics.

The behaviour of vectors under coordinate transformations is, of course, an important property of the vector equations of physics. These vector equations (e.g. Newton's law of motion and Faraday's law) are linear relations between the components of two or more vectors. As such they contain a "relativity principle". Because vectors are transformed under rotation, reflection, and inversion according to the linear homogeneous equations (3.4) and (3.10), the validity of a vector equation is unaffected by these coordinate transformations: the components of the vectors in these equations change, but not the relations between them.

Vector equations should also be invariant under translations. For example, in the classical equation of motion for a system of $N$ particles interacting via two-body central potentials

$$m^{(\alpha)}\ddot{\mathbf{r}}^{(\alpha)} = -\sum_{\beta=1}^{N} \nabla V_{\alpha\beta}(|\mathbf{r}^{(\alpha)} - \mathbf{r}^{(\beta)}|) \quad (\alpha = 1, 2, \ldots, N \text{ and } \beta \neq \alpha), \quad (3.11)$$

the translational invariance is obvious because the vectors on each side of (3.11) are origin independent. In the equation $\dot{\mathbf{L}} = \mathbf{N}$ relating the rate of change of angular momentum to torque, the vectors are origin dependent but the equation is translationally invariant because each side changes by the same amount due to a shift of origin. We have already mentioned an example involving the macroscopic Maxwell equations and constitutive relations, where the origin independence is less obvious (Section 1.15). In later chapters we will show that origin independence is a powerful condition in multipole theory, assisting one in distinguishing a physically acceptable theory from versions which are unphysical (Chapters 6–8).

## 3.3   Cartesian tensors

The above discussion of a vector, which is a quantity possessing one subscript, can readily be generalized to a quantity possessing an arbitrary number of subscripts. Such a quantity is called a tensor, and the number of distinct subscripts is the rank of the tensor. Thus a vector is a tensor of rank one.

To motivate our discussion of Cartesian tensors, we consider products of vectors. (i) Let $V_i$ and $W_j$ be two polar vectors. Then each obeys (3.4) for both proper and improper transformations, so that their product transforms according to

$$V_i' W_j' = l_{ir}\, l_{js} V_r W_s. \quad (3.12)$$

(ii) If $V_i$ and $W_j$ are both axial vectors then from (3.10) their product transforms according to (3.12) for both proper and improper transformations. (iii) If $V_i$ is polar and $W_j$ is axial, then from (3.4) and (3.10)

$$V_i' W_j' = \pm l_{ir}\, l_{js} V_r W_s, \quad (3.13)$$

where the upper (lower) sign is for a proper (improper) transformation.

We now define a second-rank Cartesian tensor $T_{ij}$ as a quantity with nine components which transform according to

$$T_{ij}' = l_{ir}\, l_{js}\, T_{rs} \quad (3.14)$$

for a polar tensor, and

$$T_{ij}' = \pm l_{ir}\, l_{js}\, T_{rs} \quad (3.15)$$

for an axial tensor. In (3.15) the upper (lower) sign applies for a proper (improper) transformation. In terms of these definitions, the above discussion shows that the product of two polar or of two axial vectors is a second-rank polar tensor, while the product of a polar and an axial vector is a second-rank axial tensor. In a similar way, tensors of rank 3 are defined by

$$T'_{ijk} = l_{ir}\, l_{js}\, l_{kt}\, T_{rst} \quad \text{(polar)} \tag{3.16}$$

$$T'_{ijk} = \pm l_{ir}\, l_{js}\, l_{kt}\, T_{rst} \quad \text{(axial),} \tag{3.17}$$

and so on for tensors of higher rank. A tensor of rank zero is called a scalar and the definitions are

$$S' = S \quad \text{(polar)} \tag{3.18}$$

$$S' = \pm S \quad \text{(axial),} \tag{3.19}$$

where the upper (lower) sign applies to proper (improper) transformations.

It is important to note that the direction cosines making up the nine elements of the transformation matrix $l_{ij}$ in (3.5) are not all independent. This follows directly from the property that the direction cosines are components of the unit vectors $\mathbf{n}'_i$ of the primed coordinate system relative to the unprimed system

$$\mathbf{n}'_i = (l_{i1}, l_{i2}, l_{i3}). \tag{3.20}$$

Orthonormality of these unit vectors requires

$$l_{ir}\, l_{jr} = \delta_{ij}. \tag{3.21}$$

These relations enable us to invert any tensor equation: for example, from (3.4) and (3.21) we have

$$V_j = l_{ij}\, V'_i. \tag{3.22}$$

From (3.22) and (3.4) it follows that

$$l_{ri}\, l_{rj} = \delta_{ij}. \tag{3.23}$$

Equations (3.21) and (3.23) are the orthogonality relations for the transformation matrix.

From the theory outlined above one can obtain the following useful results:

(i) From (3.4), (3.12), and (3.23) the product $V_i W_i$ of two vectors, both either polar or axial, transforms as

$$V'_i W'_i = l_{ir}\, l_{is}\, V_r W_s = \delta_{rs}\, V_r W_s = V_r W_r. \tag{3.24}$$

Thus $V_i W_i$ is a polar scalar. If $V_i$ and $W_i$ are unlike vectors, then $V_i W_i$ is an axial scalar.

(ii) In general, two subscripts of an individual tensor of rank 2 or higher may be the same, in which case the tensor is said to be contracted with respect to the repeated subscripts. Consider, for example, the transformation of the axial tensor $T_{ijkj}$, which is contracted with respect to $j$.

$$T'_{ijkj} = \pm\, l_{ir}\, l_{js}\, l_{kt}\, l_{ju}\, T_{rstu} = \pm\, l_{ir}\, l_{kt}\, \delta_{su}\, T_{rstu} = \pm\, l_{ir}\, l_{kt}\, T_{rsts}. \qquad (3.25)$$

Thus $T_{ijkj}$ is a second-rank axial tensor. In general, the rank of a tensor is the number of its unrepeated subscripts.

(iii) The product of two tensors of like type is polar, whereas the product of a polar and an axial tensor is axial; see the examples (3.12) and (3.13). In addition, the product of a tensor of rank $n$ with a tensor of rank $m$ is a tensor of rank $n + m$, provided none of the subscripts in the set of $n$ is repeated in the set of $m$.

(iv) Consider the tensor equation

$$U_{i\ldots} = \mathfrak{a}_{\ldots}\, T_{r\ldots}, \qquad (3.26)$$

where $T$ is a tensor of rank $n$ and $U$ is a tensor of rank $m$ with different subscripts from those of $T$. It is clear that $\mathfrak{a}$ is a tensor of rank $n+m$ which has all the subscripts of $T$ and $U$, so that those of $T$ are contracted with the corresponding ones in $\mathfrak{a}$ to yield the tensor $U$ with its $m$ subscripts. An example is the expression for an electric quadrupole moment induced by an electric field

$$q_{ij} = \mathfrak{a}_{ijk}E_k. \qquad (3.27)$$

Equations like (3.27) define a tensor as the constant of proportionality between a "cause" tensor (here $E_k$) and an "effect" tensor (here $q_{ij}$). The various polarizabilities in Chapter 2 are of this type, being defined by the multipole moments in (2.110)–(2.114) that are induced by the fields $\mathbf{E}$ and $\mathbf{B}$ and their various derivatives.

(v) A polar tensor of rank $n$ transforms like the product of $n$ polar vectors, each of which changes sign under inversion of axes. An axial tensor of rank $n$ transforms like the product of $n - 1$ polar vectors with an axial vector, which like $\mathbf{L}$ in (3.9) does not change sign when axes are inverted. Accordingly, we conclude that under inversion an even-rank tensor changes sign only if it is axial, whereas an odd-rank tensor does so only if polar.

(vi) A tensor is said to be isotropic if each of its components retains the same value under an arbitrary rotation of axes. Because isotropic tensors play a role in averaging a molecular property over all orientations in a field-free gas and in expressing the tensors of a spherically symmetric object, we consider the commonly used ones [2]: There is no isotropic tensor of rank 1. Of rank 2 the Kronecker delta $\delta_{ij}$ is the only one, while the Levi–Civita tensor $\varepsilon_{ijk}$ defined in (1.32) is the only one of rank 3. That these tensors are isotropic follows from their transformations under rotation:

(a) From (3.14), (3.15), and (3.21)

$$\delta'_{ij} = l_{ir}\, l_{js}\, \delta_{rs} = l_{ir}\, l_{jr} = \delta_{ij}. \qquad (3.28)$$

Hence $\delta_{ij}$ is isotropic.

(b) From (3.16), (3.17), and (1.32)

$$\varepsilon'_{ijk} = l_{ir}\,l_{js}\,l_{kt}\,\varepsilon_{rst}$$

$$= l_{i1}\,l_{js}\,l_{kt}\,\varepsilon_{1st} + l_{i2}\,l_{js}\,l_{kt}\,\varepsilon_{2st} + l_{i3}\,l_{js}\,l_{kt}\,\varepsilon_{3st}$$

$$= l_{i1}(l_{j2}\,l_{k3} - l_{j3}\,l_{k2}) + l_{i2}(l_{j3}\,l_{k1} - l_{j1}\,l_{k3}) + l_{i3}(l_{j1}\,l_{k2} - l_{j2}\,l_{k1}).$$
$$(3.29)$$

If any two of $i, j, k$ are equal, it follows from (3.29) that $\varepsilon'_{ijk} = 0$. If $i, j, k$ is in cyclic order then use of (3.21) in (3.29) shows that $\varepsilon'_{ijk} = 1$; similarly, if $i, j, k$ is in anti-cyclic order then $\varepsilon'_{ijk} = -1$. Thus $\varepsilon'_{ijk} = \varepsilon_{ijk}$.

Isotropic tensors of higher rank are multiples of $\delta_{ij}$ and $\varepsilon_{ijk}$, such as $\delta_{ij}\,\delta_{kl}$ and $\varepsilon_{ijk}\,\delta_{lm}$.

(vii) It is useful to determine the manner in which the vector operator

$$\nabla = \left( \frac{\partial}{\partial x_1}, \frac{\partial}{\partial x_2}, \frac{\partial}{\partial x_3} \right) \qquad (3.30)$$

transforms. Now

$$\frac{\partial}{\partial x'_i} = \frac{\partial x_j}{\partial x'_i}\,\frac{\partial}{\partial x_j} = l_{ij}\,\frac{\partial}{\partial x_j}, \qquad (3.31)$$

where we have used (3.22). Thus $\nabla$ is a polar vector.

(viii) A tensor is said to be origin independent (translationally invariant) if none of its components is changed by an arbitrary shift of the origin of coordinates. If one, or more, of its components is changed by this shift, the tensor is origin dependent. We have already seen in Sections 1.2 and 1.7 that multipole moments are examples of tensors which may be either origin dependent or origin independent.

## 3.4   Time reversal

The purpose of this section is to give a brief discussion of time reversal, and then to obtain the formula (3.44) for the manner in which the matrix element of an operator transforms under time reversal. In classical mechanics, time reversal changes the sign of the time coordinate

$$t \rightarrow t' = -t. \qquad (3.32)$$

Thus the trajectory $\mathbf{r}(t)$ of a particle changes to $\mathbf{r}(-t)$ while the velocity $\mathbf{u} = \dot{\mathbf{r}}$ and the momentum $\boldsymbol{\Pi} = m\mathbf{u}$ change sign. If the classical Hamiltonian is invariant under $t \rightarrow -t$, $\mathbf{r} \rightarrow \mathbf{r}$, and $\boldsymbol{\Pi} \rightarrow -\boldsymbol{\Pi}$ (e.g. for motion in a conservative field), then the time-reversed trajectory is also a possible motion of the particle. Tensors which are unchanged by time reversal are called time even; those which change sign under time reversal are time odd.

In quantum mechanics, the time-reversal operator $T$ is required to have the properties

$$T\mathbf{r} = \mathbf{r}T, \qquad T\mathbf{\Pi} = -\mathbf{\Pi}T, \qquad T\mathbf{L} = -\mathbf{L}T, \qquad (3.33)$$

where $\mathbf{r}$, $\mathbf{\Pi}$, and $\mathbf{L}$ are the position, momentum, and angular momentum operators. By requiring that $T$ leaves the canonical commutation relations $x_i\Pi_j - \Pi_j x_i = i\hbar\delta_{ij}$ unchanged, and using the first two equations in (3.33), we have $Ti = -iT$. Thus $T$ must involve complex conjugation and we write [3]

$$T = UK, \qquad (3.34)$$

where $K$ denotes complex conjugation and $U$ is a unitary operator. For a linear space with kets

$$|u\rangle, \quad |v\rangle, \quad |w\rangle, \quad \cdots \qquad (3.35)$$

the time-reversed space has kets

$$T|u\rangle = |u'\rangle, \qquad T|v\rangle = |v'\rangle, \qquad T|w\rangle = |w'\rangle, \qquad \cdots . \qquad (3.36)$$

It can be shown from (3.34) and (3.36) that a scalar product transforms under time reversal according to [3]

$$\langle u|v\rangle \xrightarrow{T} \langle u'|v'\rangle = \langle u|v\rangle^*. \qquad (3.37)$$

The time-reversal operator (3.34) is said to be anti-unitary: for a unitary operator one would delete the complex conjugation in (3.37). A rigorous treatment of the above can be given in terms of Wigner's theorem [4, 5]. The explicit form of the operator $U$ in (3.34) depends on the representation one uses and also on the nature of the system [3, 6]: for a non-relativistic, spinless particle in the coordinate representation $U = I$. In what follows we do not require an explicit form for $U$.

If $\Omega$ is an operator on the space of kets, the time-reversed operator corresponding to $\Omega$ is

$$\Omega' = T\Omega T^{-1}. \qquad (3.38)$$

Now if

$$\Omega|w\rangle = |v\rangle \qquad (3.39)$$

then

$$T\Omega T^{-1} T|w\rangle = T|v\rangle. \qquad (3.40)$$

That is,

$$\Omega'|w'\rangle = |v'\rangle. \qquad (3.41)$$

An operator $\Omega$ is time even if, according to (3.38), $\Omega' = \Omega$ and time odd if $\Omega' = -\Omega$. From (3.33), $\mathbf{r}$ is a time-even operator whereas $\mathbf{\Pi}$ and $\mathbf{L}$ are time odd.

Under time reversal a matrix element transforms according to

$$\langle u|\Omega|w\rangle \xrightarrow{T} \langle u'|\Omega|w'\rangle$$
$$= \pm \langle u'|\Omega'|w'\rangle \tag{3.42}$$
$$= \pm \langle u'|v'\rangle, \tag{3.43}$$

where in (3.42) we have assumed time-even or time-odd operators ($\Omega' = \pm\Omega$) and in (3.43) we have used (3.41). From (3.43), (3.37), and (3.39) we obtain the rule for transformation of matrix elements of an operator under time reversal

$$\langle u|\Omega|w\rangle \xrightarrow{T} \pm \langle u|\Omega|w\rangle^*, \tag{3.44}$$

where the upper (lower) sign applies for time-even (time-odd) operators.

For example, by working with eigenfunctions $\exp(i\mathbf{k}\cdot\mathbf{r}/\hbar)$ of the momentum operator $\mathbf{\Pi} = -i\hbar\nabla$ in the coordinate representation, one sees that according to (3.44) the momentum eigenvalues $\mathbf{k}$ change sign under $T$, as required. The application of (3.44) to polarizability tensors is considered in the next section.

## 3.5   The space and time nature of various tensors

We use the theory outlined in previous sections to classify certain mechanical and electromagnetic quantities, and also some multipole moments and polarizabilities, in terms of their behaviour under time reversal $T$ and coordinate transformations. For the latter we are concerned with inversion of axes as described in Section 3.1 (also termed space inversion, the parity operation, or simply parity, and denoted by $P$). The dependence of multipole properties on space translation is considered separately in Section 3.7.

At the outset one is faced with a certain arbitrariness associated with space and time properties of the fundamental scalars, mass $m$ and charge $q$. These are fixed by convention: we take $m$ and $q$ to be time-even polar scalars. (At the end of this section we discuss an example of an axial scalar. There is no indication that such a property should be assigned to mass or charge.) Next, we recall that the position vector $\mathbf{r}$ is a time-even polar vector, and note that the product of two tensors with the same behaviour under $T$ is a time-even tensor, otherwise the product is time odd. The polar or axial nature of higher rank tensors, such as products of vectors, has been discussed in Section 3.3.

Based on the above, one can determine the space and time properties of various tensors. The procedure is straightforward, employing, for example, definitions of kinematical, dynamical, and electromagnetic quantities, and fundamental equations of physics (Newton's law of motion, Maxwell's equations).

(i) *Mechanical quantities*

From their definitions, and using the time reversal (3.32) of classical mechanics, velocity $\mathbf{u} = d\mathbf{r}/dt$ is a time-odd polar vector and acceleration $\mathbf{a} = d^2\mathbf{r}/dt^2$ is a time-even polar vector. The law of motion $\mathbf{F} = m\mathbf{a}$, together with our convention for $m$, shows that $\mathbf{F}$ is a time-even polar vector.

From the components of angular momentum $\mathbf{L}$, such as (3.8), we see that $\mathbf{L}$ is a time-odd axial vector. From its definition, torque $\mathbf{N} = \mathbf{r} \times \mathbf{F}$ is a time-even axial vector. These properties of $\mathbf{L}$ and $\mathbf{N}$ are consistent with respect to the relation $d\mathbf{L}/dt = \mathbf{N}$. Kinetic energy is a time-even polar scalar, as is potential energy and hence the mechanical energy $W$. With regard to the isotropic tensors $\delta_{ij}$ and $\varepsilon_{ijk}$, relations such as $r^2 = x_i x_j \delta_{ij}$ and $L_i = m\varepsilon_{ijk}\, x_j u_k$ show that $\delta_{ij}$ is a time-even polar tensor and $\varepsilon_{ijk}$ is a time-even axial tensor. The vector operator $\nabla$ is a time-even polar vector (see (3.31) and (3.32)).

(ii) *Electromagnetic quantities*

The defining relations for the electric and magnetic fields $\mathbf{E}$ and $\mathbf{B}$

$$\mathbf{F} = q\mathbf{E} \quad \text{and} \quad \mathbf{F} = q\mathbf{u} \times \mathbf{B}, \qquad (3.45)$$

and our convention for $q$, show that $\mathbf{E}$ is a time-even polar vector and $\mathbf{B}$ is a time-odd axial vector. Charge density $\rho = dq/dV$ is a time-even polar scalar, and current density $\mathbf{J} = \rho\mathbf{u}$ is a time-odd polar vector. These properties of $\rho$, $\mathbf{J}$, $\mathbf{E}$, and $\mathbf{B}$ are consistent with respect to the Maxwell equations (1.112)–(1.115). From the inhomogeneous Maxwell equations (1.120) and (1.121) we conclude that $\mathbf{D}$ is a time-even polar vector and $\mathbf{H}$ is a time-odd axial vector. We mention that our convention for charge (time-even polar scalar) is consistent with common practice [7,8]. The arbitrary and inconsequential nature of this convention has been discussed by Nye [7].

(iii) *Multipole moments and polarizabilities*

The transformations of the polarizability tensors in (2.115)–(2.130) can be found from their defining relationships, in which they are the constants of proportionality between an induced multipole moment and the field, or one of its derivatives, that induces the moment. These relationships are contained in (2.110)–(2.114). To use them one requires the space and time behaviour of the multipole moments: from the definitions (1.10)–(1.12), (1.50), and (1.51) these are seen to be

$$p_i, \ q_{ij}, \ q_{ijk} \text{ are time-even polar tensors,}$$
$$m_i, \ m_{ij}, \text{ are time-odd axial tensors.}$$

In Table 3.1 we list the space and time character of the various mechanical, electromagnetic, and isotropic tensors considered above. In Table 3.2 this is done for the polarizability tensors of different multipole order whose expressions appear in (2.115)–(2.130). We note that the time-odd polarizability tensors in Table 3.2 are all zero for a non-magnetic medium (see Section 3.6). The above method of determining the time behaviour of polarizability tensors is based on the use of time reversal in classical physics. It is also of interest to use the results of the previous section and apply the time-reversal operator in quantum mechanics directly to (2.115)–(2.130), that is by using the rule (3.44). We consider

**Table 3.1** *Space and time properties of some mechanical, electromagnetic, and isotropic tensors.*

|      | Time-even | | Time-odd | |
|------|-----------|-------|----------|-------|
| Rank | Polar | Axial | Polar | Axial |
| 0 | $m, q, W, \rho$ | $\phi$ | | |
| 1 | $r_i, a_i, F_i, \nabla_i, E_i, D_i, p_i$ | $N_i$ | $u_i, J_i$ | $L_i, B_i, H_i, m_i$ |
| 2 | $\delta_{ij}, q_{ij}$ | | | $m_{ij}$ |
| 3 | $q_{ijk}$ | $\varepsilon_{ijk}$ | | |

**Table 3.2** *Space and time properties and multipole classification of the independent polarizability tensors in (2.115)–(2.130).*

| Multipole order | Time-even | | Time-odd | |
|-----------------|-----------|-------|----------|-------|
|  | Polar | Axial | Polar | Axial |
| Electric dipole | $\alpha_{ij}$ | | $\alpha'_{ij}$ | |
| Electric quadrupole –magnetic dipole | $a_{ijk}$ | $G'_{ij}$ | $a'_{ijk}$ | $G_{ij}$ |
| Electric octopole –magnetic quadrupole | $b_{ijkl}, d_{ijkl}, \chi_{ij}$ | $H'_{ijk}, L'_{ijk}$ | $b'_{ijkl}, d'_{ijkl}, \chi'_{ij}$ | $H_{ijk}, L_{ijk}$ |

two examples. In these we work only with the relevant parts of (2.115)–(2.130), namely $\mathcal{Re}\{\}$ or $\mathcal{Im}\{\}$. Thus in (2.115) and (2.120)

$$\text{(i)} \quad \mathcal{Re}\{\langle p_i \rangle_{ns} \langle p_j \rangle_{sn}\} \xrightarrow{T} \mathcal{Re}\{\langle p_i \rangle^*_{ns} \langle p_j \rangle^*_{sn}\} = \mathcal{Re}\{\langle p_i \rangle_{ns} \langle p_j \rangle_{sn}\}$$

since from Table 3.1, $p_i$ is time even.

$$\text{(ii)} \quad \mathcal{Im}\{\langle p_i \rangle_{ns} \langle m_j \rangle_{sn}\} \xrightarrow{T} \mathcal{Im}\{\langle p_i \rangle^*_{ns} \langle -m_j \rangle^*_{sn}\} = \mathcal{Im}\{\langle p_i \rangle_{ns} \langle m_j \rangle_{sn}\}$$

since $m_j$ is time odd (Table 3.1). Thus it follows that $\alpha_{ij}$ and $G'_{ij}$ in (2.115) and (2.120) are time-even tensors, in agreement with the entries in Table 3.2. Further applications of time reversal in molecular physics are discussed by Barron and Buckingham [9].

To conclude this section we consider an interesting example of an axial scalar. This is the angle $\phi$ through which the plane of linearly polarized light is rotated by an optically active medium. (The sign of $\phi$ depends on the convention used for it.) Reflection of the light beam back along its original path reverses the rotation according to the same convention. Thus $\phi$ is an axial scalar; see (3.19). In addition to the behaviour of $\phi$ under $P$ we also require for use in Section 3.8 its transformation under $T$. Imagine time reversing an experiment with linearly polarized light passing through an optically active fluid of time-even molecules. The light velocity is reversed, the tip of the electric vector retraces the same helical path as in the original experiment, and the random distribution of molecules remains unchanged. In terms of the convention for the sense of $\phi$, its sign is the same. Hence $\phi$ is a time-even axial scalar.

## 3.6   Symmetry and property tensors

In this section we give a condensed account of the space–time symmetry of an object, and also the effect of this symmetry on the property tensors of that object. For a detailed treatment of this subject the reader is referred to the monographs by Birss [10] and Bhagavantam [11]. We list below the main concepts required in this section.

(i) The symmetry of an object may change during its motion or the motion of its parts, for instance the asymmetrical vibrations of the $CO_2$ molecule. Therefore we follow the usual practice of considering only stationary rigid bodies. On these one performs notional symmetry operations at an instant of time. Such an operation restores an object to its original appearance, that is, brings it into coincidence with itself.

(ii) Symmetry operations may be purely spatial, like those mentioned in Section 3.1, or a combination of spatial operations and time reversal, or time reversal on its own. We consider initially only spatial symmetry operations.

(iii) A spatial symmetry operation on an object may always be referred to a system of Cartesian axes with its origin at a point in the object. The totality of all the independent spatial symmetry operations on a time-symmetric object, relative to the same origin of axes, comprises the point-group symmetry of that object.

(iv) Some examples of symmetry operations that make up an object's point group are: rotation about an axis through the origin $O$ by an angle $2\pi/n$ that restores coincidence, where $n = \infty$ for a sphere or axially symmetric object (e. g. a dumb-bell or the $N_2$ molecule), otherwise $n$ is an integer; reflection in a plane through $O$; inversion through $O$, etc.

(v) All objects possess at least one symmetry operation, namely a rotation of $2\pi$ about any axis. This is the identity operation. Except for objects that possess only the identity operation, all others have in addition one or more of the possible symmetry operations. These might include several different rotation axes, and even the same axis with different integral $n$ values, and also different reflection planes.

(vi) For a crystal there are 32 different point groups whose symmetry operations are purely spatial. These are sometimes referred to as the non-magnetic point groups. In respect of an $n$-fold rotation axis (the term derives from the angle $2\pi/n$ mentioned in (iv)), the only values of $n$ that are allowed for an infinite three-dimensional crystal are 1, 2, 3, 4, and 6 [11]. Molecular symmetries are described by the same 32 non-magnetic point groups as those of crystals, together with additional ones such as a five-fold rotation axis, the two types of axial symmetry exemplified by the $H_2$ and HCl molecules, and spherical symmetry (as in the He atom).

(vii) Objects possessing spontaneous magnetization (such as ferromagnetic and ferrimagnetic crystals) which reverses direction under time reversal are termed time antisymmetric. This notion also applies to the unit cell of a crystal having aligned spins and molecules with resultant angular momentum. Time-symmetric objects, such as diamagnetic and paramagnetic crystals and diamagnetic molecules do not possess such properties. Time-antisymmetric objects are usually referred to as magnetic, as are their time-odd properties.

(viii) It is possible that a spatial operation combined with time reversal is a symmetry operation for a magnetic object. The inclusion of time reversal in this way increases the number of point groups to a total of 90 for magnetic crystals. These include the 32 that are applicable to non-magnetic crystals, and do not involve time reversal; the symmetry operations of the remaining 58 comprise spatial operations as well as combinations of a spatial operation and time reversal. Also, each magnetic point group reverts to the associated non-magnetic point group, from which it is derived, when its time-odd properties are disregarded (see below).

(ix) Mainly two different notations are used for designating the 32 non-magnetic symmetry classes of crystals, namely the Schönflies and International conventions. The former is also used for molecules, while the latter lends itself more readily to specifying the 90 classes for magnetic crystals. (A listing of both conventions is given in [10], pp. 36–38.)

(x) The 32 point-group classes for non-magnetic crystals are divided into 7 systems. The name and number in each are: triclinic 2; monoclinic 3; orthorhombic 3; tetragonal 7; trigonal 5; hexagonal 7; cubic 5 (see Ref. 10 for the basis of this classification). Because the spatial symmetries on which these seven systems are based are common to magnetic crystals, the latter have the same named systems but with more members in each. In crystal optics a broader grouping of the non-cubic systems is encountered, which depends on the number of optic axes: biaxial systems (triclinic, monoclinic, orthorhombic) and uniaxial systems (tetragonal, trigonal, hexagonal).

(xi) The Cartesian axes used in this book to specify symmetry operations and tensor components for crystals are those of Birss [10, 12]. These crystallographic axes are stated below for the various crystal systems. Triclinic: any orientation; monoclinic: $z$-axis is the symmetry axis or a normal to the reflection plane, where this exists, $x$- and $y$-axes are arbitrary; orthorhombic: of three perpendicular symmetry axes, at least two are equivalent; these are labelled $x$ and $y$; tetragonal, trigonal, hexagonal: $z$-axis is that of highest symmetry and the $y$-axis is a two-fold axis or a normal to a reflection plane, where this exists; cubic: of the three axes parallel to the cube edges, at least two are equivalent; these are labelled $x$ and $y$. Birss' tables may also be used for those molecules that have the same point-group symmetries as crystals.

(xii) An extended object with repeat units, like a crystal with its unit cells, possesses additional symmetries that arise from finite displacement of a unit. These symmetries, together with the point-group symmetry possessed by a unit, comprise the space–group symmetry of the extended object. Because a Cartesian tensor is defined in terms of the way its components transform under transformations of axes with the same origin, it is an object's point-group symmetry, and not its space group, that determines the number and nature of the components of its property tensors.

(xiii) Use of the symmetry of an object to determine the allowed components of a tensor is possible only if the tensor is a property tensor of that object. For instance, the Kronecker delta, although a tensor, is not the property of an object, so that symmetry may not be applied to it. Similarly for the Levi–Civita tensor. Furthermore, there are physical properties that either cannot be represented by a tensor, such as the dielectric breakdown strength of a crystal [10], or if they can be so represented, are not unique properties that reflect the symmetry of the object possessing it. An example of the latter is the magnetization of an iron crystal exhibiting hysteresis [11].

(xiv) Symmetry considerations do apply to property tensors which are defined by a relationship between one or more "cause" tensors and an "effect" tensor. Each multipole polarizability of Chapter 2, defined as the constant of proportionality between a multipole moment (the "effect") and a field or one of its derivatives (the "cause"), is a property tensor of this type.

(xv) The effect of the symmetry of an object on its property tensors is based on Neumann's Principle, according to which: each property tensor of an object must possess the space and time symmetry of that object, but it may also possess intrinsic symmetries unrelated to any point group (see Section 2.10). If, for example, an object is time symmetric, then its property tensors must be time even, apart from having to possess the point-group symmetry of the object. Thus the time-odd polarizability tensors in Table 3.2 vanish identically for a non-magnetic object. On the other hand, if the object is time antisymmetric, Neumann's Principle allows it to possess both time-odd and time-even property tensors. The former are obtained by imposing the particular magnetic point-group symmetry transformations of the object; the latter are found from the point-group transformations of the associated non-magnetic group.

Against the above background we now consider several examples to illustrate the use of point-group symmetry transformations and the application of Neumann's Principle for determining the components of a property tensor:

1. *Object with a centre of symmetry.* Both non-magnetic and magnetic objects may possess a centre of symmetry. On its own, inversion of axes through this centre is a time-symmetric operation of such an object. By Neumann's Principle its time-even tensors are unchanged under inversion. Because odd-rank polar and even-rank axial tensors change sign under inversion (Section 3.3), it follows that these property tensors are identically zero. This powerful and useful conclusion applies only to the time-even tensors of an object, whether it be non-magnetic or magnetic.

2. *Non-magnetic object with only a four-fold rotation axis.* A square table with four helical legs of the same hand at each corner possesses a four-fold rotation axis $z$, which is perpendicular to the table top and passes through its centre $O$. Figure 3.1 depicts this object with a right-hand set of axes, and shows by means of an arrow the sense of the $\pi/2$ rotation symmetry and also of each helix. Being non-magnetic this object possesses only time-even property tensors. Since reflection or inversion changes the hand of a helix, the only symmetry operations of this object are:

$$\text{the identity, rotations about the } z\text{-axis of } \pi/2, \pi, 3\pi/2 \text{ (or } -\pi/2). \quad (3.46)$$

These operations and the corresponding point-group class are shown below in the two commonly used conventions:

International:   $1, 2_z, \pm4_z$:   Class 4
Schönflies:   E, C$_2$, 2C$_4$:   Class C$_4$.

We now consider the effect of these symmetry operations on a second-rank polar property tensor $T_{ij}$. We denote the Cartesian axes before and after a transformation as $x, y, z$ and $x', y', z'$ respectively. From (3.14)

$$T'_{ij} = l_{ix}(l_{jx}T_{xx} + l_{jy}T_{xy} + l_{jz}T_{xz}) + l_{iy}(l_{jx}T_{yx} + l_{jy}T_{yy} + l_{jz}T_{yz})$$
$$+ l_{iz}(l_{jx}T_{zx} + l_{jy}T_{zy} + l_{jz}T_{zz}). \quad (3.47)$$

For all rotations about the $z$-axis, (3.5) yields

$$l_{xz} = l_{yz} = l_{zx} = l_{zy} = 0, \quad l_{zz} = 1, \quad (3.48)$$

while for a rotation through $\pi/2$

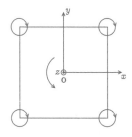

FIG. 3.1. A square table with four helical legs of the same hand.

$$l_{xx} = l_{yy} = 0, \quad l_{xy} = 1, \quad l_{yx} = -1. \tag{3.49}$$

It follows from (3.47)–(3.49) that the transformed tensor is

$$T'_{ij} = \begin{pmatrix} T_{yy} & -T_{yx} & T_{yz} \\ -T_{xy} & T_{xx} & -T_{xz} \\ T_{zy} & -T_{zx} & T_{zz} \end{pmatrix}. \tag{3.50}$$

Since $4_z$ is a symmetry operation, Neumann's Principle requires that

$$T'_{ij} = T_{ij} = \begin{pmatrix} T_{xx} & T_{xy} & T_{xz} \\ T_{yx} & T_{yy} & T_{yz} \\ T_{zx} & T_{zy} & T_{zz} \end{pmatrix}. \tag{3.51}$$

Comparison of (3.50) and (3.51) shows that, on the basis of the $4_z$ symmetry, the non-vanishing components of $T_{ij}$ relative to conventional axes are

$$T_{xx} = T_{yy}, \quad T_{zz}, \quad T_{xy} = -T_{yx}. \tag{3.52}$$

It is left to the reader to show that these findings are not changed by the other symmetry operations in (3.46).

We conclude that, before any intrinsic symmetry is allowed for, all time-even second-rank polar property tensors of a non-magnetic object with point-group symmetry 4 (or $C_4$) have the non-vanishing components in (3.52). Because this group is a subgroup of the magnetic point groups derived from it (see below), this conclusion also applies to the time-even second-rank polar tensors of those magnetic point groups. To illustrate the additional effect of intrinsic symmetry, we consider the electric polarizability, for which from (2.115), $\alpha_{ij} = \alpha_{ji}$. Then (3.52) shows that the non-vanishing components of $\alpha_{ij}$, relative to conventional axes, for the point group 4 are

$$\alpha_{xx} = \alpha_{yy}, \alpha_{zz}. \tag{3.53}$$

3. *Magnetic point group derived from a non-magnetic one.* We show by means of an example how a single non-magnetic point group may provide the basis for more than one magnetic point group. We begin with the non-magnetic object in Fig. 3.1 with its point-group symmetry 4, and place on each leg the same spin in the same sense. These spins are depicted by the broken arrows in Fig. 3.2. It is evident that this object is now magnetic but still has the same symmetry operations as Fig. 3.1, namely those in (3.46). Thus the object in Fig. 3.2 belongs to the magnetic point group 4. So does the object with all its spins reversed.

Next we imagine the object in Fig. 3.1 with opposite spins placed alternately on its four corners, as in Fig. 3.3 (a). Now its symmetry is no longer that of point group 4. For instance, a rotation of $\pi/2$ about the $z$-axis is no longer a symmetry operation, as Fig. 3.3 (b) shows. However, if this is followed by time reversal $T$ of the spins, one observes from Fig. 3.3 (a) and (c) that coincidence is obtained.

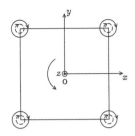

FIG. 3.2. As in Fig. 3.1, but with the same spin on each leg.

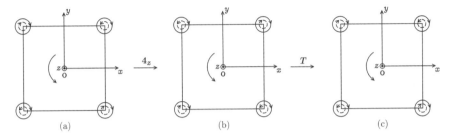

(a)                              (b)                              (c)

FIG. 3.3. (a) As in Fig. 3.2, but with alternating spins on the legs. (b) After a
rotation of $\pi/2$ about the $z$-axis. (c) After time reversal.

It is evident from Fig. 3.3 (a) that the rotation $2_z$ on its own is a symmetry
operation. In the International convention a bar is placed under a spatial operation to denote that combined with time reversal it is a symmetry operation.
These operations and the magnetic point group to which they belong for the
object in Fig. 3.3 (a) are denoted by

$$1,\, 2_z,\, \pm \underline{4}_z: \quad \text{Class } \underline{4}$$

In concluding this section we mention the useful tables given by Birss [10,12],
which list the allowed components of polar and axial property tensors, with rank
from 0 to 4, for the 32 non-magnetic and 90 magnetic crystalline classes. In using
these tables one must, of course, allow for the effect of any intrinsic symmetry
in a particular tensor.

## 3.7  Origin dependence of polarizability tensors

In Sections 1.2 and 1.7 it was shown that multipole moments of a charge distribution depend in general on the choice made for the origin of coordinates
to which the moments are referred. This remark applies also to the multipole
moment operators in (2.94)–(2.97), since these have the same forms as their classical counterparts. Because matrix elements of the moment operators enter the
quantum-mechanical expressions in (2.115)–(2.130) for polarizability tensors, the
latter may also be origin dependent. The purpose of this section is to determine
the origin dependence, if any, of these tensors. The reason for doing so is that

we will be interested in the translational behaviour of various physical properties of a medium, and these properties are expressed as linear combinations of polarizability tensors of the same multipole order (Chapters 4–9).

We determine first the origin dependence of the multipole moment operators in (2.94)–(2.96). (The magnetic octopole moment operator in (2.97) is not included as it does not contribute to the polarizability tensors in (2.115)–(2.130), which are of multipole orders up to electric octopole and magnetic quadrupole.) The change in the electric quadrupole moment operator, for example, is denoted

$$\Delta q_{ij} = \bar{q}_{ij} - q_{ij}, \tag{3.54}$$

where $q_{ij}$ is the operator relative to an origin $O$ in the charge distribution and $\bar{q}_{ij}$ is that relative to an origin $\bar{O}$ displaced by $\mathbf{d}$ from $O$. The position and momentum operators relative to $O$ and $\bar{O}$ are related by

$$\bar{\mathbf{r}} = \mathbf{r} - \mathbf{d}, \quad \bar{\mathbf{\Pi}} = \mathbf{\Pi}. \tag{3.55}$$

With (3.55) in (2.94)–(2.96) we obtain

$$\Delta p_i = -d_i \sum_{\alpha=1}^{N} q^{(\alpha)} \tag{3.56}$$

$$\Delta q_{ij} = -d_i p_j - d_j p_i + d_i d_j \sum_{\alpha=1}^{N} q^{(\alpha)} \tag{3.57}$$

$$\Delta q_{ijk} = -d_i q_{jk} - d_j q_{ik} - d_k q_{ij} + d_i d_j p_k + d_j d_k p_i + d_k d_i p_j - d_i d_j d_k \sum_{\alpha=1}^{N} q^{(\alpha)} \tag{3.58}$$

$$\Delta m_i = -\varepsilon_{ijk} d_j \sum_{\alpha=1}^{N} \frac{q^{(\alpha)}}{2m^{(\alpha)}} \Pi_k^{(\alpha)} \tag{3.59}$$

$$\Delta m_{ij} = -2d_j m_i + \frac{2}{3}\delta_{ij} d_k m_k$$
$$- \frac{1}{3}\varepsilon_{ikl} d_k \sum_{\alpha=1}^{N} \frac{q^{(\alpha)}}{m^{(\alpha)}} \left[ r_j^{(\alpha)} \Pi_l^{(\alpha)} + r_l^{(\alpha)} \Pi_j^{(\alpha)} - 2d_j \Pi_l^{(\alpha)} - i\hbar\delta_{jl} \right]. \tag{3.60}$$

In deriving (3.60) we used the commutation relation $r_i \Pi_j - \Pi_j r_i = i\hbar\delta_{ij}$ and

$$\varepsilon_{ikl} d_k (r_j \Pi_l - r_l \Pi_j) = \varepsilon_{ikl}\varepsilon_{mjl}\varepsilon_{mqr} d_k r_q \Pi_r = d_j l_i - \delta_{ij} d_k l_k. \tag{3.61}$$

For the present, we neglect any contribution from spin: the origin dependences in (3.59) and (3.60) were obtained from the magnetic dipole and quadrupole moment operators in (2.95) and (2.96), which exclude spin. Then we may use (2.65) and (2.80) to rewrite (3.59) and (3.60) as

$$\Delta m_i = -\frac{i}{2\hbar}\,\varepsilon_{ijk}\,d_j\left[H^{(0)}, p_k\right] \tag{3.62}$$

$$\Delta m_{ij} = -2d_j m_i + \frac{2}{3}\delta_{ij}d_k m_k + \frac{i}{3\hbar}\varepsilon_{ikl}d_k\left[H^{(0)}, 2d_j p_l - q_{jl}\right]. \tag{3.63}$$

We now use the origin dependences (3.56)–(3.58), (3.62), and (3.63) for multipole moment operators to determine the origin dependences of the polarizability tensors listed in Table 3.2 and given by (2.115)–(2.130). The method is illustrated for the tensor $L_{ijk}$ in Appendix F. The results are [13]

$$\Delta\alpha_{ij} = 0 \tag{3.64}$$

$$\Delta\alpha'_{ij} = 0 \tag{3.65}$$

$$\Delta a_{ijk} = -d_j\alpha_{ik} - d_k\alpha_{ij} \tag{3.66}$$

$$\Delta a'_{ijk} = -d_j\alpha'_{ik} - d_k\alpha'_{ij} \tag{3.67}$$

$$\Delta G_{ij} = -\frac{1}{2}\omega\varepsilon_{jkl}d_k\alpha'_{il} \tag{3.68}$$

$$\Delta G'_{ij} = \frac{1}{2}\omega\varepsilon_{jkl}d_k\alpha_{il} \tag{3.69}$$

$$\Delta b_{ijkl} = -d_j a_{ikl} - d_k a_{ijl} - d_l a_{ijk} + d_j d_k \alpha_{il} + d_j d_l \alpha_{ik} + d_k d_l \alpha_{ij} \tag{3.70}$$

$$\Delta b'_{ijkl} = -d_j a'_{ikl} - d_k a'_{ijl} - d_l a'_{ijk} + d_j d_k \alpha'_{il} + d_j d_l \alpha'_{ik} + d_k d_l \alpha'_{ij} \tag{3.71}$$

$$\Delta d_{ijkl} = -d_i a_{jkl} - d_j a_{ikl} - d_k a_{lij} - d_l a_{kij} + d_i d_k \alpha_{jl} + d_i d_l \alpha_{jk}$$
$$+ d_j d_k \alpha_{il} + d_j d_l \alpha_{ik} \tag{3.72}$$

$$\Delta d'_{ijkl} = -d_i a'_{jkl} - d_j a'_{ikl} + d_k a'_{lij} + d_l a'_{kij} + d_i d_k \alpha'_{jl} + d_i d_l \alpha'_{jk}$$
$$+ d_j d_k \alpha'_{il} + d_j d_l \alpha'_{ik} \tag{3.73}$$

$$\Delta H_{ijk} = -2d_k G_{ij} + \frac{2}{3}\delta_{jk}d_l G_{il} - \frac{1}{3}\omega\varepsilon_{jlm}d_l(a'_{ikm} - 2d_k\alpha'_{im}) \tag{3.74}$$

$$\Delta H'_{ijk} = -2d_k G'_{ij} + \frac{2}{3}\delta_{jk}d_l G'_{il} + \frac{1}{3}\omega\varepsilon_{jlm}d_l(a_{ikm} - 2d_k\alpha_{im}) \tag{3.75}$$

$$\Delta L_{ijk} = -d_i G_{jk} - d_j G_{ik} + \frac{1}{2}\omega\varepsilon_{klm}d_l(a'_{mij} + d_i\alpha'_{jm} + d_j\alpha'_{im}) \tag{3.76}$$

$$\Delta L'_{ijk} = -d_i G'_{jk} - d_j G'_{ik} + \frac{1}{2}\omega\varepsilon_{klm}d_l(a_{mij} - d_i\alpha_{jm} - d_j\alpha_{im}) \tag{3.77}$$

$$\Delta\chi_{ij} = \frac{1}{2}\omega(\varepsilon_{ikl}d_k G'_{lj} + \varepsilon_{jkl}d_k G'_{li}) + \frac{1}{4}\omega^2\varepsilon_{ikl}\varepsilon_{jmn}d_k d_m\alpha_{ln} \tag{3.78}$$

$$\Delta\chi'_{ij} = -\frac{1}{2}\omega(\varepsilon_{ikl}d_k G_{lj} - \varepsilon_{jkl}d_k G_{li}) + \frac{1}{4}\omega^2\varepsilon_{ikl}\varepsilon_{jmn}d_k d_m\alpha'_{ln}. \tag{3.79}$$

We note the following points concerning these results:

(i) Equations (3.64)–(3.79) also apply to the polarizability densities introduced in Section 2.11, and this property is used in the following chapters to study the translational behaviour of macroscopic multipole theory.

(ii) The origin dependences (3.66)–(3.79) are expressed in terms of the arbitrary shift of origin **d** and other polarizabilities. The latter depend on frequency and therefore so do (3.66)–(3.79).

(iii) As (3.64)–(3.79) show, the polarizabilities $\alpha_{ij}$ and $\alpha'_{ij}$ of electric dipole order are translationally invariant, whereas those of higher multipole order are not. Thus an origin-independent observable cannot be expressed in general in terms of a single polarizability of order above electric dipole. The expression must be an origin-independent linear combination of all or some polarizabilities of the same higher multipole order. The effect of the symmetry of the medium and of any intrinsic symmetry of its tensors may reduce the number of polarizabilities in the expression for the observable. For instance, the expression for the rotation angle $\phi$ of the plane of linearly polarized light in a non-magnetic optically active medium contains both $G'_{ij}$ and $a_{ijk}$ for all but cubic and isotropic symmetries (Section 5.3). For the latter two symmetries $\phi$ depends only on the trace $G'_{ii}$, which (3.69) and (2.73) show to be origin independent.

(iv) According to (3.78), the *dc* magnetic susceptibility $\chi_{ij}$ is independent of origin, in agreement with the finding of Van Vleck (see Section 2.4). The origin dependence of $\chi_{ij}$ at non-zero frequencies implies from (iii) that it is not an independent observable. It is known to contribute with other polarizabilities of electric octopole–magnetic quadrupole order to a light scattering effect [14]. The wave theory in Section 5.1 gives an example of how $\chi_{ij}$ combines with other polarizabilities to form an invariant tensor; see (5.10).

(v) We now consider the effect of including spin in the above results. Spin contributes to the magnetic dipole and quadrupole moment operators (but not, of course, to the electric moment operators). From (2.102) we have

$$m_i = \sum_{\alpha=1}^{N} \frac{q^{(\alpha)}}{2m^{(\alpha)}} \left( l_i^{(\alpha)} + g^{(\alpha)} s_i^{(\alpha)} \right) \tag{3.80}$$

$$m_{ij} = \sum_{\alpha=1}^{N} \frac{q^{(\alpha)}}{2m^{(\alpha)}} \left[ r_j^{(\alpha)} \left( \frac{2}{3} l_i^{(\alpha)} + g^{(\alpha)} s_i^{(\alpha)} \right) + \left( \frac{2}{3} l_i^{(\alpha)} + g^{(\alpha)} s_i^{(\alpha)} \right) r_j^{(\alpha)} \right]. \tag{3.81}$$

The origin dependence of $m_i$ obtained from (3.80) is still given by (3.59). However, the origin shift of $m_{ij}$ obtained from (3.81) differs from (3.60) in two respects: the spinless operator for **m** in (3.60) is to be replaced by that in (3.80), and to $\Delta m_{ij}$ in (3.60) is to be added the spin term

$$-\frac{1}{3} \delta_{ij} d_k \sum_{\alpha=1}^{N} \frac{q^{(\alpha)} g^{(\alpha)}}{m^{(\alpha)}} s_k^{(\alpha)}. \tag{3.82}$$

Hence, inclusion of spin affects just two of the results in (3.64)–(3.79), namely those for the polarizabilities $H_{ijk}$ and $H'_{ijk}$, which involve the matrix elements of $m_{jk}$ (see (2.125) and (2.126)), and which contribute at electric octopole–magnetic quadrupole order (Table 3.2).

(vi) Thus, to allow for spin, the only change required in (3.64)–(3.79) is to include the contribution of the matrix element of (3.82) in (3.74) and (3.75). This contribution differs from others in (3.66)–(3.79), which are all expressed in terms of polarizabilities, and we will not consider it further. For our purpose of determining the origin dependence of certain multipole results, the spinless expressions (3.74) and (3.75) are sufficient (Section 4.5).

(vii) Equations (3.64)–(3.79) are also valid for dissipative media because they are unaffected by the modification that must be made to the polarizabilities (2.115)–(2.130) to allow for dissipation (Section 2.8).

## 3.8   A pictorial determination of symmetry conditions

The symmetry conditions for the existence of an equilibrium effect in a macroscopic medium can be established from a pictorial representation of space and time transformations on the experiment in which the effect is exhibited [15]. It is assumed in this approach that if such an effect occurs, so does the corresponding transformed effect when the entire experiment is subjected to a transformation in space or time. The experiment under consideration is represented in a schematic diagram. Also depicted are the forms of the experiment under suitable space transformations and time reversal. Comparison of such figures enables one to decide whether the effect is vetoed or not by parity and time reversal. For allowed effects the method also enables one to determine symmetry conditions for the existence of an effect [15]. The method is applied on a macroscopic level, and does not require knowledge of the structure of a system, or its tensor relations, or the use of symmetry in tensor transformations.

In this section we apply the pictorial approach to two effects. This requires knowledge of the transformation behaviour under parity $P$ and time reversal $T$ of the physical quantities involved. This behaviour is given in Table 3.1.

1. *The Faraday effect in a fluid.* In this effect a magnetostatic field **B** applied parallel to the path of linearly polarized light in a fluid induces a rotation of the plane of polarization by an angle $\phi$ which is proportional to B. Figure 3.4 depicts the original experiment and its various transformed versions, together with the axes used. We remark that because of the random orientation of molecules, a rotated fluid is indistinguishable from the original, as is a time-reversed fluid, even when its molecules are time antisymmetric [16, 17]. If the fluid is optically active, its space-inverted form can be distinguished from the original and cannot be expected to behave in the same way.

From Fig. 3.4: (i) As there is no contradiction between (c) and (d), we conclude that time reversal does not veto the effect. (ii) Nor does parity, as a comparison of (d) and (e) reveals for an optically inactive fluid. (iii) If the fluid is optically active, then the media in (d) and (e) are not the same, and one cannot

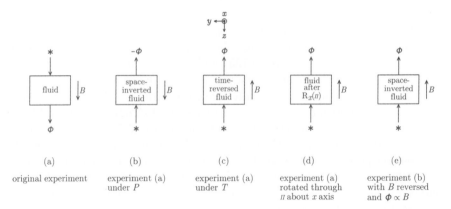

FIG. 3.4. Depiction of the Faraday effect experiment and its space and time transformations.

require $\phi$ to be either the same or different. Again there is no contradiction. From this analysis we conclude that since a rotation $\phi$ proportional to $B$ is not vetoed by $P$ or $T$, it may exist in any fluid, whether optically active or not.

It is of interest to contrast this conclusion with a quadratic effect where $\phi$ is proportional to $B^2$. Then in (b) one can reverse $B$ without changing the sign of $\phi$; the resulting diagram contradicts (d). Thus parity vetoes the quadratic effect.

The electric analogue of the Faraday effect has never been observed in a fluid. It is left to the reader to show, by an analysis similar to the above, that time reversal vetoes this effect. The same conclusion has been reached in a quantitative theory of the effect [17].

2. *The electric analogue of the Faraday effect in a crystal.* To avoid the complication of natural birefringence, we consider only uniaxial crystals with the light beam along the optic axis. We also ignore field-induced linear birefringence (Pockels effect); analysis of this effect using the pictorial approach is given in Ref. 15.

The relevant experiments are depicted in Fig. 3.5, in which the axes shown are crystallographic, with the $z$-axis along the optic axis. These experiments are sufficient for drawing the desired conclusions [15]. Consider first a non-magnetic crystal. From (b) and (e), and also from (b) and (c), we see that contradictory results are obtained if a crystal rotated by $\pi$ about its $x$-axis is indistinguishable from either the original or space-inverted crystal. If this is so, then $\phi = 0$. Thus non-magnetic uniaxial crystals which do not possess these symmetries may exhibit the electric analogue of the Faraday effect: the point-group symmetries which these crystals possess can be identified with the aid of tables [10]. They are $3$, $\bar{3}$, $4$, $\bar{4}$, $4/m$, $6$, $\bar{6}$, $6/m$.

For a magnetic crystal, comparison of (b) and (c), and also of (b) and (e), shows inconsistent results if its appearance after rotation by $\pi$ about its $x$-axis

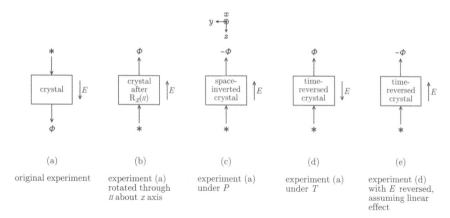

FIG. 3.5. Depiction of the electric analogue of the Faraday effect in a crystal, and its space and time transformations.

is the same as either its space-inverted or time-reversed form. The point-group symmetries of magnetic uniaxials which do not possess these symmetries (and therefore may exhibit the effect) can again be read off from tables [10] and they are 422, $4\underline{mm}$, $\bar{4}2\underline{m}$, $4/\underline{mmm}$, 32, $3\underline{m}$, $\bar{3}\underline{m}$, 622, $6\underline{mm}$, $\bar{6}\underline{m}2$, $6/\underline{mmm}$, 4, $\bar{4}$, $4/\underline{m}$, 3, $\bar{3}$, 6, $\bar{6}$, $6/\underline{m}$.

The above findings for magnetic and non-magnetic crystals are in agreement with results of a quantitative theory which expresses the effect in terms of polarizability tensors, and enables one to obtain the relevant point groups from tables of tensor components [17].

As we have seen above, in the Faraday effect a magnetic field **B** is able to induce in an optically inactive fluid a chiral effect that is linear in $B$, namely the rotation of the plane of linearly polarized light, whereas an electric field **E** is unable to do so. This difference in response is intuitively expected, in that for the Faraday effect a handed quantity, **B**, is imposed on the fluid, whereas in the electric analogue **E** lacks this property.

In this regard it is interesting to consider a scattering experiment in which right and left circularly polarized light beams of the same intensity are alternately incident on a fluid in a uniform electric field **E** perpendicular to the scattering plane. The scattered intensity component proportional to $E$ is different for the two polarizations [15, 18]. In view of the comment in the previous paragraph it may seem counter-intuitive that the application of an electric field to a fluid could induce, to first order in $E$, a chiral effect in a scattering experiment, namely the different scattered intensities for incident right and left circularly polarized light. The explanation is that for scattering, a three-dimensional set of directions is imposed on the fluid: the directions of the incident and scattered beams and, perpendicular to these, the electric field. Reversing the field changes the hand of this set of directions and thus the sign of the intensity difference for a linear response.

The pictorial approach has been applied to a range of light scattering effects in fluids [15]. Also, a verbal analysis, as a condensed form of the pictorial method, has been used to show that magneto-chiral birefringence in a fluid is not vetoed by parity or time reversal [19].

## 3.9   Discussion

In this chapter we first considered a Cartesian tensor as a member of a broad class of objects defined by the manner in which their components transform under rotation of the coordinate system. For tensors representing physical quantities it is useful to specify subcategories of this class according to their behaviour under space and time transformations:

(i) Thus we discussed also improper transformations and distinguished polar and axial tensors.

(ii) Similarly, consideration of time reversal introduces time-even and time-odd tensors. Examples of these types of tensor from mechanics, electromagnetism, and multipole theory are given in Tables 3.1 and 3.2.

(iii) We also considered translation of the coordinate system and defined origin-independent and origin-dependent tensors. For vectors, some examples are given in Section 3.2, while the translational properties of multipole moments and polarizability tensors are discussed in Section 3.7. The six types of tensor (polar or axial, time even or time odd, origin independent or origin dependent) introduced in this way play an important role in this book, particularly in the theory of transmission phenomena (Chapter 5) and in the transformation theory of Chapters 7 and 8.

A tensor describing a particular property of an object is either a so-called property tensor or it is not (Section 3.6). The former type possesses at least the space–time point-group symmetry of the object. Extensive use is made in this book of symmetry tables [10, 12] of property tensors to obtain expressions for observables that apply to a crystal or molecule with a particular point-group symmetry.

## References

[1] Arfken, G. B. and Weber, H. J. (1995). *Mathematical methods for physicists*, Sect. 3.3. Academic, San Diego.

[2] Jeffreys, H. (1963). *Cartesian tensors*, Ch. 7. Cambridge University Press, London.

[3] Sakurai, J. J. (1985). *Modern quantum mechanics*, Sect. 4.4. Benjamin, Menlo Park. This reference provides a particularly clear account of time reversal in quantum mechanics.

[4] Wigner, E. P. (1959). *Group theory and its application in the theory of atomic spectra*. Academic, New York.

[5] Galindo, A. and Pascual, P. (1990). *Quantum mechanics I*, Springer, Berlin.

[6] Sachs, R. G. (1987). *The physics of time reversal*, Ch. 3. University of Chicago Press, Chicago.

[7] Nye, J. F. (1985). *Physical properties of crystals*, p. 54. Clarendon, Oxford.

[8] Jackson, J. D. (1999). *Classical electrodynamics*, p. 271. Wiley, New York.

[9] Barron, L. D. and Buckingham, A. D. (2001). Time reversal and molecular properties. *Accounts of Chemical Research*, **34**, 781–789.

[10] Birss, R. R. (1966). *Symmetry and magnetism*. North-Holland, Amsterdam.

[11] Bhagavantam, S. (1966). *Crystal symmetry and physical properties*. Academic, London.

[12] Birss, R. R. (1963). Macroscopic symmetry in space-time. *Reports on Progress in Physics*, **26**, 307–360.

[13] Graham, E. B. and Raab, R. E. (2000). Multipole solution for the macroscopic electromagnetic boundary conditions at a vacuum-dielectric interface. *Proceedings of the Royal Society London* A, **456**, 1193–1215.

[14] de Figueiredo, I. M. B. and Raab, R. E. (1981). A molecular theory of new differential light-scattering effects in a fluid. *Proceedings of the Royal Society London* A, **375**, 425–441.

[15] de Figueiredo, I. M. B. and Raab, R. E. (1980). A pictorial approach to macroscopic space-time symmetry, with particular reference to light scattering. *Proceedings of the Royal Society London* A, **369**, 501–516.

[16] Van Vleck, J. H. (1932). *The theory of electric and magnetic susceptibilities*, p. 279. Clarendon, Oxford.

[17] Buckingham, A. D., Graham, C., and Raab, R. E. (1971). On the theory of linear electro-optical rotation. *Chemical Physics Letters*, **8**, 622-624.

[18] Buckingham, A. D. and Raab, R. E. (1975). Electric-field-induced differential scattering of right and left circularly polarized light. *Proceedings of the Royal Society London* A, **345**, 365–377.

[19] Barron, L. D. and Vrbancich, J. (1984). Magneto-chiral birefringence and dichroism. *Molecular Physics*, **51**, 715–730.

# 4

# LINEAR CONSTITUTIVE RELATIONS FROM MULTIPOLE THEORY

*These (Maxwell's) equations are, however, largely*
*useless until the relations between the quantities*
**D**, **B**, **E** *and* **H** *which appear in them*
*have been established.*
L. D. Landau and E. M. Lifshitz
(*Electrodynamics of continuous media*)

We use the results of multipole theory from previous chapters to obtain linear constitutive relations for the response fields **D** and **H** in homogeneous media in terms of the macroscopic electric and magnetic fields **E** and **B** (Section 4.1). We work to electric octopole–magnetic quadrupole order and express the material constants in the constitutive relations for harmonic plane wave fields in terms of the macroscopic polarizability tensors appropriate to each order. In Sections 4.2–4.4 we discuss conditions that apply to these material constants. These conditions relate to origin independence, symmetries, and the so-called "Post constraint". We then examine the extent to which the material constants calculated directly from multipole theory are consistent with these conditions for both non-dissipative and dissipative media (Section 4.5).

## 4.1 Constitutive relations

The procedure for obtaining constitutive relations for the response fields **D** and **H** from multipole theory is as follows.

(i) We assume the long-wavelength regime (see Section 1.12 and Ref. 1) and start with the multipole expressions (1.118) and (1.119) for $D_i$ and $H_i$ in terms of arbitrary macroscopic fields $E_i$ and $B_i$, and the macroscopic electric and magnetic multipole moment densities. The highest multipole order to which physical effects have been studied is electric octopole–magnetic quadrupole (Section 2.12 and Chapter 5). We therefore work to this order and take account of the moment densities $P_i$ (electric dipole), $Q_{ij}$ (electric quadrupole), $Q_{ijk}$ (electric octopole), $M_i$ (magnetic dipole), and $M_{ij}$ (magnetic quadrupole).

(ii) We consider homogeneous media with **E** and **B** represented by harmonic plane waves and use the electric and magnetic multipole moments calculated for a charge distribution in Section 2.7 (see (2.110)–(2.114)). The macroscopic moments are obtained from these as described in Section 2.11;

see the examples for $P_i$ and $Q_{ij}$ in (2.146) and (2.148). Similar expressions can be written down for $Q_{ijk}$, $M_i$, ... based on (2.112)–(2.114). These expressions are all linear functions of **E**, **B**, and their space and time derivatives.

(iii) For **E** and **B** we use the complex forms

$$\mathbf{F} = \mathbf{F}_0 e^{i(\mathbf{k}\cdot\mathbf{r}-\omega t)} \quad (\mathbf{F} = \mathbf{E} \text{ or } \mathbf{B}), \qquad (4.1)$$

where **k** is the wave vector and $\omega$ is the angular frequency of the wave. Then all space and time derivatives of a field can be expressed in terms of the field itself by making the replacements $\nabla_j \to ik_j$ and $\partial/\partial t \to -i\omega$. Consequently the multipole moment densities $P_i$, $M_i$, ... reduce to linear functions of **E** and **B**. Note that for dissipative media **k** is complex in (4.1).

(iv) The multipole moment densities are also linear functions of the polarizability densities (see, for example, (2.146) and (2.148)). In the multipole moment densities obtained from (2.111)–(2.114) we use the relations in (2.117)–(2.122) and (2.125)–(2.128) to eliminate the polarizability densities $\mathfrak{a}_{ijk}$, $\mathfrak{a}'_{ijk}$, $\mathcal{G}_{ij}$, $\mathcal{G}'_{ij}$, ... in favour of $a_{ijk}$, $a'_{ijk}$, $G_{ij}$, $G'_{ij}$, ... . Thus we obtain from (1.118) and (1.119) the linear constitutive relations

$$D_i = A_{ij}\, E_j + T_{ij}\, B_j \qquad (4.2)$$

and

$$H_i = U_{ij}\, E_j + X_{ij}\, B_j. \qquad (4.3)$$

The second-rank tensors in (4.2) and (4.3) are referred to as material constants: they are independent of **E** and **B**, and are given in terms of the polarizability densities by

$$A_{ij} = \varepsilon_0 \delta_{ij} + \alpha_{ij} - i\alpha'_{ij} + \frac{1}{2}\left[i(a_{ijk} - a_{jki}) + a'_{ijk} + a'_{jki}\right]k_k$$
$$+ \left[-\frac{1}{6}(b_{ijkl} + b_{jikl}) + \frac{i}{6}(b'_{ijkl} - b'_{jikl}) + \frac{1}{4}(d_{ikjl} - id'_{ikjl})\right]k_k k_l \qquad (4.4)$$

$$T_{ij} = G_{ij} - iG'_{ij} + \frac{1}{2}\left[i(H_{ijk} - L_{ikj}) + H'_{ijk} - L'_{ikj}\right]k_k \qquad (4.5)$$

$$U_{ij} = -G_{ji} - iG'_{ji} + \frac{1}{2}\left[i(H_{jik} - L_{jki}) - H'_{jik} + L'_{jki}\right]k_k \qquad (4.6)$$

$$X_{ij} = \mu_0^{-1}\delta_{ij} - \chi_{ij} + i\chi'_{ij}. \qquad (4.7)$$

In (4.4)–(4.7) we have included all contributions up to electric octopole–magnetic quadrupole order, and we have omitted the permanent multipole moment densities. For a homogeneous medium the latter make no contribution to the Maxwell equations (1.120) and (1.121).

It is appropriate to remark at this point that in general the dependence of the response fields on $\mathbf{E}(\mathbf{r}, t)$ and $\mathbf{B}(\mathbf{r}, t)$ will be spatially and temporally non-local, described by integral relations [2]: for a stationary, homogeneous, linear medium

$$D_i(\mathbf{r}, t) = \int d^3 r' \int dt' [A_{ij}(\mathbf{r}', t')E_j(\mathbf{r}-\mathbf{r}', t-t') + T_{ij}(\mathbf{r}', t')B_j(\mathbf{r}-\mathbf{r}', t-t')], \quad (4.8)$$

and similarly for $\mathbf{H}$. In this book we consider response fields given by the multipole expansion which, as discussed above, is a differential rather than an integral formulation, relying on various space and time derivatives of $\mathbf{E}$ and $\mathbf{B}$ to account for phenomena (see also Section 2.12). To our knowledge, the extent to which this differential formulation approximates integral relations such as (4.8) has, in general, not been established. We therefore adopt an empirical approach in this book, and compare the results of multipole theory with observed phenomena (Chapters 5 and 9).

Equations (4.2)–(4.7) are the constitutive relations given directly by multipole theory to electric octopole–magnetic quadrupole order. We now discuss conditions which should be satisfied by the material constants in (4.2) and (4.3) (Sections 4.2–4.4). In Section 4.5 we examine the extent to which the material constants (4.4)–(4.7) satisfy these conditions, and we consider the effect of dissipation on the multipole constitutive relations.

## 4.2 Origin independence

The material constants in (4.2) and (4.3) represent observables for the permittivity ($A_{ij}$), inverse permeability ($X_{ij}$), and the magnetoelectric effect ($T_{ij}$ and $U_{ij}$). It is therefore reasonable to require of any theoretical expressions for these observables that they should be independent of the choice of origin of coordinates. That is,

$$\Delta A_{ij} = \Delta T_{ij} = \Delta U_{ij} = \Delta X_{ij} = 0, \quad (4.9)$$

where $\Delta$ denotes the change due to a shift in the origin of coordinates. At this stage we simply state (4.9) as a requirement; detailed discussion within the context of multipole theory and its applications is given later (Chapters 7 and 8).

## 4.3 Symmetries

The following derivation of symmetries for material constants is based on application of the Lagrangian formulation of classical field theory in macroscopic electromagnetism. We first specify our notation [3]. A contravariant 4-vector having components $A^0$, $A^1$, $A^2$, and $A^3$ is written as

$$A^\mu = (A^0, \mathbf{A}). \quad (4.10)$$

The components of the corresponding covariant 4-vector are given by

$$A_\mu = g_{\mu\nu} A^\nu, \quad (4.11)$$

where the metric tensor $g_{\mu\nu}$ is diagonal with

$$g_{00} = 1, \qquad g_{11} = g_{22} = g_{33} = -1. \tag{4.12}$$

(Here, and in what follows, a repeated Greek subscript implies summation from 0 to 3.) Examples are the contravariant and covariant forms of the position 4-vector

$$x^{\mu} = (x^0, x^1, x^2, x^3) = (x^0, \mathbf{r}) \tag{4.13}$$

and

$$x_{\mu} = (x^0, -\mathbf{r}), \tag{4.14}$$

where $x^0 = ct$. For the 4-current density they are

$$J^{\mu} = (c\rho, \mathbf{J}) \tag{4.15}$$
$$J_{\mu} = (c\rho, -\mathbf{J}), \tag{4.16}$$

where $\rho$ and $\mathbf{J}$ are the charge and current densities, respectively.

A Lagrangian density $\mathcal{L}$ for the electromagnetic field has the functional dependence

$$\mathcal{L} = \mathcal{L}(A^{\mu}, \partial^{\nu} A^{\mu}), \tag{4.17}$$

where $A^{\mu}$ is the 4-vector potential

$$A^{\mu} = \left(\frac{1}{c}\Phi, \mathbf{A}\right) \tag{4.18}$$

and

$$\partial^{\nu} \equiv \frac{\partial}{\partial x_{\nu}} = \left(\frac{\partial}{\partial x^0}, -\nabla\right). \tag{4.19}$$

According to (4.17), the differential of the Lagrangian density can be written

$$d\mathcal{L} = \frac{\partial \mathcal{L}}{\partial A^{\mu}} dA^{\mu} + \frac{\partial \mathcal{L}}{\partial(\partial^{\nu} A^{\mu})} d(\partial^{\nu} A^{\mu}). \tag{4.20}$$

If (4.20) is used in the action principle

$$\delta \int \mathcal{L} \, d^4 x = 0, \tag{4.21}$$

one obtains (after an integration by parts) the Euler–Lagrange equations

$$\partial^{\nu} G_{\nu\mu} = -\frac{\partial \mathcal{L}}{\partial A^{\mu}}, \tag{4.22}$$

where

$$G_{\nu\mu} = -\frac{\partial \mathcal{L}}{\partial(\partial^{\nu} A^{\mu})}. \tag{4.23}$$

(Note: the 4-tensor $G_{\nu\mu}$ used in this section should not be confused with the Cartesian tensor $G_{ij}$ which denotes a multipole polarizability; see Section 2.7.)

In the following calculations leading to (4.28), no explicit form of the Lagrangian density is assumed. In fact, in macroscopic electromagnetism (unlike the situation in microscopic electromagnetism), the Lagrangian density is, in general, unknown until the constitutive relations $D_i = D_i(\mathbf{E},\mathbf{B})$ and $H_i = H_i(\mathbf{E},\mathbf{B})$ are specified (see below). Instead, the calculation proceeds in two steps. First, we assume that the dependence of $\mathcal{L}$ on $A^\mu$ is given by the usual interaction term [4] $\mathcal{L}_{int} = -J_\mu A^\mu$, so that

$$\frac{\partial \mathcal{L}}{\partial A^\mu} = -J_\mu. \tag{4.24}$$

Second, we note that if $G_{\nu\mu}$ is given in terms of the response fields by

$$G_{\nu\mu} = \begin{pmatrix} 0 & cD_1 & cD_2 & cD_3 \\ -cD_1 & 0 & -H_3 & H_2 \\ -cD_2 & H_3 & 0 & -H_1 \\ -cD_3 & -H_2 & H_1 & 0 \end{pmatrix}, \tag{4.25}$$

then (4.22), (4.24), (4.25), and (4.16) yield the inhomogeneous Maxwell equations (1.120) and (1.121) for $\mathbf{D}$ and $\mathbf{H}$. In making this identification we assume that $J_\mu$ in (4.24) is the free 4-current density $(c\rho_f, -\mathbf{J}_f)$.

Using (4.23) and (4.24) in (4.20), and the antisymmetry of $G_{\nu\mu}$ evident in (4.25), we have

$$d\mathcal{L} = -J_\mu dA^\mu - \frac{1}{2} G_{\nu\mu} dF^{\nu\mu}. \tag{4.26}$$

Here $F^{\nu\mu} = \partial^\nu A^\mu - \partial^\mu A^\nu$ is the usual electromagnetic field tensor with elements

$$F^{\nu\mu} = \begin{pmatrix} 0 & -E_1/c & -E_2/c & -E_3/c \\ E_1/c & 0 & -B_3 & B_2 \\ E_2/c & B_3 & 0 & -B_1 \\ E_3/c & -B_2 & B_1 & 0 \end{pmatrix}. \tag{4.27}$$

Substituting (4.16), (4.18), and (4.27) in (4.26) we obtain the desired result

$$d\mathcal{L} = \mathbf{J} \cdot d\mathbf{A} - \rho d\Phi + \mathbf{D} \cdot d\mathbf{E} - \mathbf{H} \cdot d\mathbf{B}. \tag{4.28}$$

(For the reader who wishes to perform the calculations leading to (4.28) using Minkowski space, with a fourth imaginary dimension $x_4 = ict$ rather than the metric (4.12), details can be found in Ref. 5.)

In the above derivation, (4.28) depends on (i) the functional dependence in (4.17), (ii) the form of the interaction term leading to (4.24), and (iii) the requirement that the action principle should yield the Maxwell equations for $\mathbf{D}$ and $\mathbf{H}$. We remark that an alternative form of (4.28) also appears in the literature [6]: this is obtained from the above Lagrangian density by the Legendre transformation $\mathcal{L}' = \mathcal{L} + \mathbf{H} \cdot \mathbf{B} - \mathbf{D} \cdot \mathbf{E}$, so that instead of (4.28) one has

$$d\mathcal{L}' = \mathbf{J} \cdot d\mathbf{A} - \rho d\Phi - \mathbf{E} \cdot d\mathbf{D} + \mathbf{B} \cdot d\mathbf{H}. \tag{4.29}$$

The latter differential is less convenient for our purposes than is (4.28).

To allow for the discussion of complex fields, we consider two sets of real electromagnetic fields $\mathbf{E}^{(k)}$, $\mathbf{B}^{(k)}$, $\mathbf{D}^{(k)}$, $\mathbf{H}^{(k)}$ ($k = 1$ or $2$). According to (4.28) the differentials of the Lagrangian densities are

$$d\mathcal{L}^{(k)} = \mathbf{D}^{(k)} \cdot d\mathbf{E}^{(k)} - \mathbf{H}^{(k)} \cdot d\mathbf{B}^{(k)}, \tag{4.30}$$

where, for brevity, the contributions of the first two terms in (4.28) have been omitted. Hence the differential $d\mathcal{L} = d\mathcal{L}^{(1)} + d\mathcal{L}^{(2)}$ can be expressed as

$$d\mathcal{L} = \frac{1}{2}(\mathbf{D}^* \cdot d\mathbf{E} + \mathbf{D} \cdot d\mathbf{E}^* - \mathbf{H}^* \cdot d\mathbf{B} - \mathbf{H} \cdot d\mathbf{B}^*). \tag{4.31}$$

Here $\mathbf{E}$, $\mathbf{B}$, $\mathbf{D}$, and $\mathbf{H}$ are the complex fields $\mathbf{E} = \mathbf{E}^{(1)} + i\mathbf{E}^{(2)}$, etc. In our application, $\mathbf{E}^{(1)}$ would be the real part of the complex plane wave (4.1), while $\mathbf{E}^{(2)}$ would be the imaginary part, etc. From (4.31)

$$D_i = 2\frac{\partial \mathcal{L}}{\partial E_i^*} \tag{4.32}$$

$$H_i = -2\frac{\partial \mathcal{L}}{\partial B_i^*}. \tag{4.33}$$

Finally, we use (4.32) and (4.33) to deduce symmetry relations for the material constants in linear constitutive relations: we assume that the $d\mathcal{L}^{(k)}$ of (4.30) are perfect differentials (see Section 4.5). Then so is $d\mathcal{L}$. Hence, from (4.31)

$$\frac{\partial^2 \mathcal{L}}{\partial E_j \partial E_i^*} = \frac{\partial^2 \mathcal{L}}{\partial E_i^* \partial E_j} \tag{4.34}$$

$$\frac{\partial^2 \mathcal{L}}{\partial E_j \partial B_i^*} = \frac{\partial^2 \mathcal{L}}{\partial B_i^* \partial E_j} \tag{4.35}$$

$$\frac{\partial^2 \mathcal{L}}{\partial B_j \partial B_i^*} = \frac{\partial^2 \mathcal{L}}{\partial B_i^* \partial B_j}. \tag{4.36}$$

From (4.32)–(4.36), and using the property that $\mathcal{L}$ is real, we obtain the Maxwell-type relations

$$\frac{\partial D_i}{\partial E_j} = \left(\frac{\partial D_j}{\partial E_i}\right)^* \tag{4.37}$$

$$\frac{\partial H_i}{\partial E_j} = -\left(\frac{\partial D_j}{\partial B_i}\right)^* \tag{4.38}$$

$$\frac{\partial H_i}{\partial B_j} = \left(\frac{\partial H_j}{\partial B_i}\right)^*. \tag{4.39}$$

Next, we apply (4.37)–(4.39) to the constitutive relations (4.2) and (4.3). Using (4.2) and (4.3) in (4.37)–(4.39) yields the symmetry relations for complex material constants

$$A_{ij} = A_{ji}^* \tag{4.40}$$

$$U_{ij} = -T_{ji}^* \tag{4.41}$$

$$X_{ij} = X_{ji}^*. \tag{4.42}$$

These symmetry relations have also been deduced for time-harmonic fields in non-dissipative media by using the complex Poynting theorem [7]. The discussion given above is based on the assumption that $\mathcal{L}$ is a unique function of the fields $\mathbf{E}$ and $\mathbf{B}$; in this regard it is similar to proofs of symmetry of the static permittivity and permeability tensors of crystals, based on the property that the free energy is a unique function of the fields [8]. The symmetries (4.40)–(4.42) apply only in non-dissipative media, as is evident from an explicit calculation of the dissipation for linear constitutive relations [7,9]: specifically, the terms in (4.2) and (4.3) involving the permittivity, the magnetoelectric coefficients, and the inverse permeability, make contributions to the total dissipation which are proportional to $A_{ij} - A_{ji}^*$, $U_{ij} + T_{ji}^*$, and $X_{ij} - X_{ji}^*$, respectively. In the context of multipole theory, this restriction to non-dissipative processes is discussed further in Section 4.5.

It is interesting to note that the symmetries (4.40)–(4.42) enable one to integrate (4.31) for linear constitutive relations: from (4.31), (4.2), (4.3), and (4.40)–(4.42) we have

$$\partial \mathcal{L} = \frac{1}{2}[A_{ij}\, d(E_i^* E_j) + T_{ij}\, d(E_i^* B_j) - U_{ij}\, d(B_i^* E_j) - X_{ij}\, d(B_i^* B_j)], \tag{4.43}$$

which can be integrated to yield

$$\mathcal{L} = \frac{1}{2}(A_{ij}\, E_i^* E_j + T_{ij}\, E_i^* B_j - U_{ij}\, B_i^* E_j - X_{ij}\, B_i^* B_j). \tag{4.44}$$

In terms of the response fields, (4.44) is simply

$$\mathcal{L} = \frac{1}{2}(\mathbf{E}^* \cdot \mathbf{D} - \mathbf{B}^* \cdot \mathbf{H}). \tag{4.45}$$

(By inspection, one checks that (4.44), (4.32), and (4.33) do yield the constitutive relations (4.2) and (4.3).) From (4.44) it is evident that, for $\mathbf{E}$ and $\mathbf{B}$ fields represented by complex harmonic plane waves, a sufficient condition for translational invariance of the Lagrangian density is the invariances (4.9) of the material constants. This is, however, not a necessary condition (Section 4.6).

## 4.4 The "Post constraint"

Post [10] considered linear constitutive relations in covariant form

$$G_{\nu\mu} = \frac{1}{2} \chi_{\nu\mu\lambda\rho} F^{\lambda\rho} \tag{4.46}$$

and argued that for uniform media the constitutive tensor $\chi_{\nu\mu\lambda\rho}$ should satisfy the constraint

$$\chi_{[\nu\mu\lambda\rho]} = 0. \qquad (4.47)$$

Here [ ] is the alternation sign: thus $\chi_{[\nu\mu\lambda\rho]}$ is a linear combination of all 24
permutations of $\chi_{\nu\mu\lambda\rho}$ with even (odd) permutations appearing with a $+(-)$
sign. Because of the antisymmetry of $F^{\lambda\rho}$ and $G_{\nu\mu}$ in (4.46), $\chi$ possesses the
obvious symmetries

$$\chi_{\nu\mu\lambda\rho} = -\chi_{\nu\mu\rho\lambda} = -\chi_{\mu\nu\lambda\rho}, \qquad (4.48)$$

and consequently (4.47) can be expressed as

$$\chi_{\nu\mu\lambda\rho} + \chi_{\nu\lambda\rho\mu} + \chi_{\nu\rho\mu\lambda} + \chi_{\mu\lambda\nu\rho} + \chi_{\mu\rho\lambda\nu} + \chi_{\lambda\rho\nu\mu} = 0. \qquad (4.49)$$

To justify (4.47), Post used the distinction between functional and structural
fields. The former specify the state of the medium: an example is the 4-vector
potential (i. e. the fields **E** and **B**), for which the corresponding Lagrangian
density yields the Maxwell equations for **D** and **H** (Section 4.3). Structural fields,
such as the constitutive tensor in (4.46), specify the properties of the medium.
Post argued that this field contributes a term

$$\mathcal{L}_s = \chi_{[\nu\mu\lambda\rho]} \partial^\nu A^\mu \partial^\lambda A^\rho \qquad (4.50)$$

to the Lagrangian density. The contribution of (4.50) to the Euler–Lagrange
equations (4.22) is

$$2\partial^\nu \chi_{[\nu\mu\lambda\rho]} \partial^\lambda A^\rho = 2\chi_{[\nu\mu\lambda\rho]} \partial^\nu \partial^\lambda A^\rho = 0, \qquad (4.51)$$

where we assumed a uniform medium (constant $\chi$) and used the symmetry of
$\partial^\nu \partial^\lambda$, and the antisymmetry of $\chi_{[\nu\mu\lambda\rho]}$, in $\nu$ and $\lambda$. Thus (4.50) has no effect on
the Maxwell equations for **D** and **H**. For this reason Post imposed (4.49).

In terms of the elements of the constitutive tensor

$$\begin{pmatrix} A_{ij} & T_{ij} \\ U_{ij} & X_{ij} \end{pmatrix} \qquad (4.52)$$

in (4.2) and (4.3), the constraint (4.49) does not involve the dielectric or purely
magnetic properties ($A_{ij}$ or $X_{ij}$), and takes the simple form

$$T_{ii} = U_{ii}. \qquad (4.53)$$

That is, equality of the traces of the magnetoelectric tensors. It is this form of
the Post constraint which is suitable for our purposes. We remark that Post also
outlined a proof of an additional symmetry of the constitutive tensor for complex
fields in non-dissipative media, namely [10]

$$\chi_{\nu\mu\lambda\rho} = (\chi_{\lambda\rho\nu\mu})^*. \qquad (4.54)$$

Taken together with (4.48), this is equivalent to the symmetries (4.40)–(4.42).

The constitutive tensor $\chi_{\nu\mu\lambda\rho}$ in (4.46) has $4^4 = 256$ components. The symmetries (4.48) reduce the number of independent components to 36, and thus the constitutive tensor can be represented by a $6 \times 6$ matrix as in (4.52). The symmetry relations (4.40)–(4.42) (that is, (4.54)) reduce the number of independent components to 21 for a non-dissipative medium, and the constraint (4.53) reduces it further to 20. For a dissipative medium the symmetries (4.40)–(4.42) are not valid, and the number of independent components is 35.

The derivation and applications of the Post constraint have attracted considerable attention in the literature. Alternative derivations have been presented, emphasizing that the constraint is a consequence of the mathematical structure of Maxwell's equations and not of the physical nature of the medium [11,12]. The derivation has also been extended to inhomogeneous media [13] and to spatially and temporally non-local media [14]. The constraint has played an important role in a debate on the existence of so-called Tellegen media (non-reciprocal bi-isotropic media) [15, 16].

## 4.5   Comparison with direct multipole results

In Section 4.1 we obtained multipole expressions for material constants in the constitutive relations (4.2) and (4.3), for the first three orders of the multipole expansion, in terms of various polarizability densities. We now examine the extent to which these results satisfy the conditions of origin independence, symmetries, and the Post constraint discussed in Sections 4.2–4.4. For this purpose it is convenient to consider separately the contributions from each multipole order (using the classification of Table 3.2), and to distinguish between non-magnetic and magnetic media. In making the latter distinction it is necessary to recall from Section 3.6 that for non-magnetic media the following polarizability tensors are zero

$$\alpha'_{ij}, \quad a'_{ijk}, \quad G_{ij}, \quad \chi'_{ij}, \quad H_{ijk}, \quad L_{ijk}, \quad b'_{ijkl}, \quad d'_{ijkl}. \tag{4.55}$$

We also consider separately non-dissipative and dissipative media. We start with non-dissipative media; for these the polarizability tensors are real (Sections 2.7 and 2.8), as is the wave vector $\mathbf{k}$.

### 4.5.1   *Electric dipole order*

For a non-magnetic medium the sum of the vacuum and electric dipole contributions to the multipole results (4.4)–(4.7) is

$$A^M_{ij} = \varepsilon_0 \delta_{ij} + \alpha_{ij}, \quad T^M_{ij} = 0, \quad U^M_{ij} = 0, \quad X^M_{ij} = \mu_0^{-1}\delta_{ij}. \tag{4.56}$$

Here, and in what follows, we introduce a superscript $M$ to indicate results obtained directly from multipole theory: this will enable us to distinguish the direct multipole results from others constructed by means of a transformation theory based on the non-uniqueness of $\mathbf{D}$ and $\mathbf{H}$ (Chapters 7 and 8). In this chapter we are concerned only with direct multipole results.

Because $\alpha_{ij}$ is origin independent (Section 3.7), symmetric (see (2.115)), and real in the absence of dissipation, it is clear that $A_{ij}^M$ satisfies the invariance in (4.9) and the symmetry (4.40). There is no contribution to the magnetoelectric coefficients from this order and thus the Post constraint (4.53) is not involved. The origin independence and symmetry of $X_{ij}^M$ in (4.56) are trivial. For a magnetic medium one has an additional contribution $A_{ij}^M = -i\alpha'_{ij}$ to (4.56) (see (4.4)). Here $\alpha'_{ij}$ is origin independent (Section 3.7), antisymmetric (see (2.116)), and real, so that (4.9) and (4.40) are still satisfied, while the Post constraint is again not involved.

### 4.5.2 Electric quadrupole–magnetic dipole order

For a non-magnetic medium the direct multipole contribution in (4.4)–(4.7) coming from this order is

$$A_{ij}^M = \frac{i}{2}(a_{ijk} - a_{jik})k_k, \quad T_{ij}^M = -iG'_{ij}, \quad U_{ij}^M = -iG'_{ji}, \quad X_{ij}^M = 0. \quad (4.57)$$

The polarizability tensors $a_{ijk}$ and $G'_{ij}$ are real for no dissipation and it therefore follows from (4.57) that the symmetries (4.40) and (4.41) are satisfied. Note that this conclusion does not depend on any intrinsic symmetries of the polarizability tensors in (4.57).

It is also clear that the Post constraint (4.53) is satisfied by the magneto-electric coefficients in (4.57). However, the first three invariances in (4.9) do not hold: in fact, using the expressions (3.66) and (3.69) for the origin dependence of $a_{ijk}$ and $G'_{ij}$ one has

$$\Delta A_{ij}^M = \frac{i}{2}(d_i\alpha_{jk} - d_j\alpha_{ik})k_k \quad (4.58)$$

$$\Delta T_{ij}^M = -\frac{i}{2}\omega\varepsilon_{jkl}\,d_k\alpha_{il} \quad (4.59)$$

$$\Delta U_{ij}^M = \Delta T_{ji}^M, \quad (4.60)$$

which, for an arbitrary shift $\mathbf{d}$ of origin, are not zero.

If the medium is magnetic we see from (4.4)–(4.7) that there is the additional contribution to (4.57)

$$A_{ij}^M = \frac{1}{2}(a'_{ijk} + a'_{jik})k_k, \quad T_{ij}^M = G_{ij}, \quad U_{ij}^M = -G_{ji}, \quad X_{ij}^M = 0. \quad (4.61)$$

The polarizability tensors in (4.61) are real and therefore (4.40) and (4.41) are satisfied. However, $T_{ii}^M = G_{ii}$ whereas $U_{ii}^M = -G_{ii}$, and therefore the Post constraint does not hold. Also, using (3.67) and (3.68) in (4.61) one again finds a non-zero dependence on origin shift

$$\Delta A_{ij}^M = -\frac{1}{2}(d_i \alpha'_{jk} + d_j \alpha'_{ik})k_k \tag{4.62}$$

$$\Delta T_{ij}^M = -\frac{1}{2}\omega \varepsilon_{jkl} d_k \alpha'_{il} \tag{4.63}$$

$$\Delta U_{ij}^M = -\Delta T_{ji}^M. \tag{4.64}$$

### 4.5.3 *Electric octopole–magnetic quadrupole order*

For a non-magnetic medium the direct multipole contribution in (4.4)–(4.7) coming from this order is

$$A_{ij}^M = \left[ -\frac{1}{6}\left( b_{ijkl} + b_{jikl} \right) + \frac{1}{4} d_{ikjl} \right] k_k k_l \tag{4.65}$$

$$T_{ij}^M = \frac{1}{2}\left( H'_{ijk} - L'_{ikj} \right) k_k \tag{4.66}$$

$$U_{ij}^M = -\frac{1}{2}\left( H'_{jik} - L'_{jki} \right) k_k \tag{4.67}$$

$$X_{ij}^M = -\chi_{ij}. \tag{4.68}$$

The polarizability tensors in (4.65)–(4.68) are real for no dissipation, and $d_{ijkl}$ and $\chi_{ij}$ possess the intrinsic symmetries (2.143) and (2.129): hence the symmetries (4.40)–(4.42) are satisfied. However, from (4.66) and (4.67)

$$T_{ii}^M = \frac{1}{2}(H'_{iik} - L'_{iki})k_k \quad \text{and} \quad U_{ii}^M = -\frac{1}{2}(H'_{iik} - L'_{iki})k_k,$$

and hence the Post constraint does not hold. Also, the invariances (4.9) are not valid: using (3.70), (3.72), (3.75), (3.77), and (3.78) in (4.65)–(4.68) one finds

$$\Delta A_{ij}^M = \left[ \frac{1}{6}\left( d_i a_{jkl} + d_j a_{ikl} \right) - \frac{1}{4}\left( d_i a_{kjl} + d_j a_{kil} - d_i d_j \alpha_{kl} \right) \right.$$
$$\left. + \frac{1}{12}\left( d_l a_{ijk} + d_l a_{jik} - d_i d_k \alpha_{jl} - d_j d_k \alpha_{il} - d_k d_l \alpha_{ij} \right) \right] k_k k_l \tag{4.69}$$

$$\Delta T_{ij}^M = \frac{1}{2}\left[ d_i G'_{kj} - d_k G'_{ij} + \frac{2}{3}\delta_{jk} d_l G'_{il} + \omega \varepsilon_{jlm} d_l \left( \frac{1}{3} a_{ikm} \right. \right.$$
$$\left. \left. - \frac{1}{2} a_{mik} + \frac{1}{2} d_i \alpha_{km} - \frac{1}{6} d_k \alpha_{im} \right) \right] k_k \tag{4.70}$$

$$\Delta U_{ij}^M = -\Delta T_{ji}^M \tag{4.71}$$

$$\Delta X_{ij}^M = -\frac{1}{2}\omega\left( \varepsilon_{ikl} d_k G'_{lj} + \varepsilon_{jkl} d_k G'_{li} \right) - \frac{1}{4}\omega^2 \varepsilon_{ikl} \varepsilon_{jmn} d_k d_m \alpha_{ln}, \tag{4.72}$$

for an arbitrary shift **d** of origin.

If the medium is magnetic the additional contributions to (4.65)–(4.68) are, from (4.4)–(4.7),

$$A_{ij}^M = i \left[ \frac{1}{6} \left( b'_{ijkl} - b'_{jikl} \right) - \frac{1}{4} d'_{ikjl} \right] k_k k_l \tag{4.73}$$

$$T_{ij}^M = \frac{i}{2} \left( H_{ijk} - L_{ikj} \right) k_k \tag{4.74}$$

$$U_{ij}^M = \frac{i}{2} \left( H_{jik} - L_{jki} \right) k_k \tag{4.75}$$

$$X_{ij}^M = i \chi'_{ij}. \tag{4.76}$$

These satisfy the symmetries (4.40)–(4.42) and the Post constraint (4.53), but are not origin independent. The origin dependences obtained by using (3.71), (3.73), (3.74), (3.76), and (3.79) in (4.73)–(4.76) are

$$\Delta A_{ij}^M = i \left[ \frac{1}{6} \left( d_i a'_{jkl} - d_j a'_{ikl} \right) + \frac{1}{4} \left( d_i a'_{kjl} - d_j a'_{kil} \right) \right.$$
$$\left. - \frac{1}{12} \left( d_l a'_{ijk} - d_l a'_{jik} + d_i d_k \alpha'_{jl} - d_j d_k \alpha'_{il} - d_k d_l \alpha'_{lj} \right) \right] k_k k_l \tag{4.77}$$

$$\Delta T_{ij}^M = \frac{i}{2} \left[ d_i G_{kj} - d_k G_{ij} + \frac{2}{3} \delta_{jk} d_l G_{il} - \omega \varepsilon_{jlm} d_l \left( \frac{1}{3} a'_{ikm} \right. \right.$$
$$\left. \left. + \frac{1}{2} a'_{mik} + \frac{1}{2} d_i \alpha'_{km} - \frac{1}{6} d_k \alpha'_{im} \right) \right] k_k \tag{4.78}$$

$$\Delta U_{ij}^M = \Delta T_{ji}^M \tag{4.79}$$

$$\Delta X_{ij}^M = i \left[ -\frac{1}{2} \omega (\varepsilon_{ikl} d_k G_{lj} - \varepsilon_{jkl} d_k G_{li}) + \frac{1}{4} \omega^2 \varepsilon_{ikl} \varepsilon_{jmn} d_k d_m \alpha'_{ln} \right]. \tag{4.80}$$

## 4.6   Discussion

We remark that the above expressions for the origin dependence of material constants can be used to show that the Lagrangian density (4.44) is origin independent (for **E** and **B** fields represented by complex harmonic plane waves). Thus the invariances (4.9) are not necessary conditions for this property of $\mathcal{L}$.

In the previous section we have discussed the multipole constitutive relations (4.2)–(4.7) for non-dissipative media. The discussion can be extended to dissipative media by taking into account two additional features. The first is that all polarizability tensors in (4.4)–(4.7) become complex in the manner described in Section 2.8. This does not affect the origin dependence of these tensors (Section 3.7). The second feature is that the wave vector **k** in (4.1) and (4.4)–(4.7) also becomes complex, with the imaginary part of **k** describing attenuation of the wave ( [17] and Chapter 5). We remark that the wave (4.1) with complex **k** is also referred to as a plane wave; it is homogeneous if the real and imaginary parts of **k** are parallel, and inhomogeneous otherwise [17].

The results in Section 4.5 regarding origin dependence of multipole material constants are not affected by the above changes to the polarizability tensors and wave vector; thus they apply also in dissipative media. Furthermore, the conclusions regarding the validity of the Post constraint are also unaltered. However, the material constants (4.4)–(4.7) with complex polarizabilities and wave vector clearly do not satisfy the symmetries (4.40)–(4.42). For example, in (4.56) because the real (dispersive) and imaginary (absorptive) parts of $\alpha_{ij}$ necessarily have the same intrinsic symmetry (Section 2.8), we see that $A_{ij}^M$ does not satisfy (4.40). This conclusion is not surprising in view of the general restriction of the symmetries (4.40)–(4.42) to non-dissipative media [9]. Despite this restriction, we will find that these symmetries play a useful role in the further development of multipole theory (Chapter 8). It is interesting to note the point at which our derivation in Section 4.3 of the symmetries (4.40)–(4.42) fails for a dissipative medium. It is that the Maxwell relations (4.37)–(4.39) are not valid for the multipole fields given by (4.2)–(4.7), if there is dissipation; consequently, the differential (4.31) of the Lagrangian density for these fields is not perfect.

As emphasized in Section 1.15, our discussion of macroscopic electromagnetism is based on primitive, rather than traceless, macroscopic multipole moments. Thus the results in this chapter are for material constants deduced using primitive moments (Section 4.1). In Section 1.15 we pointed out that the response fields $\mathbf{D}$ and $\mathbf{H}$ can also be expressed in terms of traceless moments, although this is done at the cost of incorporating traces of the macroscopic moments into the source densities. Nevertheless, one can ask: if this formulation in terms of traceless moments is used in the derivation of constitutive relations in Section 4.1, are the resulting material constants physically acceptable? The answer is no: one can show that the material constants deduced in this manner are also origin dependent.

It is instructive to elaborate on the unphysical nature of the direct multipole constitutive relations (4.2)–(4.7). From (4.1)–(4.3), and using Faraday's law, we see that $\mathbf{D}$ and $\mathbf{H}$ are harmonic plane wave fields with complex amplitudes $\mathbf{D}_0$ and $\mathbf{H}_0$ given by

$$D_{0i} = A_{ij}E_{0j} + \omega^{-1}T_{ij}\,\varepsilon_{jmn}k_m E_{0n} \qquad (4.81)$$

$$H_{0i} = U_{ij}E_{0j} + \omega^{-1}X_{ij}\,\varepsilon_{jmn}k_m E_{0n}. \qquad (4.82)$$

The results of the previous section show that these amplitudes are unphysical because they depend on the choice of coordinate origin. For example, for a non-magnetic medium and to electric quadrupole–magnetic dipole order, we see from (4.82), (4.57), (4.59), and (4.60) that

$$\Delta H_{0i} = \Delta U_{ij}E_{0j} = -\frac{i}{2}\,\omega\varepsilon_{ikl}d_k\alpha_{jl}E_{0j}, \qquad (4.83)$$

which is non-zero in general. Clearly, these conclusions apply also in the electric dipole–magnetic dipole approximation mentioned in Section 1.14, for which (4.57) and (4.61) are $A_{ij}^M = 0$, $T_{ij}^M = G_{ij} - iG'_{ij}$, $U_{ij}^M = -G_{ji} - iG'_{ji}$, $X_{ij}^M = 0$.

The above conclusions are summarized in Table 4.1 which shows that for the direct multipole contributions beyond electric dipole order, the expressions for material constants are all origin dependent and sometimes violate the Post constraint. These failures, particularly the lack of translational invariance, lead one to question whether the current formulation of multipole theory, when taken beyond electric dipole order, can yield physically acceptable results. This question is studied in more detail in the next two chapters where we consider transmission, scattering, and reflection phenomena.

**Table 4.1** *Origin independence, symmetries, and Post constraint of the direct multipole results for material constants. The first three multipole orders are considered, and the symmetries are for non-dissipative media.*

|  | Electric dipole | |
|---|:---:|:---:|
|  | Non-magnetic | Magnetic |
| Origin independence (4.9) | Yes | Yes |
| Symmetries (4.40)–(4.42) | Yes | Yes |
| Post constraint (4.53) | — | — |

|  | Electric quadrupole–magnetic dipole | | Electric octopole–magnetic quadrupole | |
|---|:---:|:---:|:---:|:---:|
|  | Non-magnetic | Magnetic | Non-magnetic | Magnetic |
| Origin independence (4.9) | No | No | No | No |
| Symmetries (4.40)–(4.42) | Yes | Yes | Yes | Yes |
| Post constraint (4.53) | Yes | No | No | Yes |

Before proceeding, it is helpful to contrast the origin dependences in multipole theory of a charge distribution with those for a macroscopic medium. For a charge distribution, the choice of coordinate origin within the distribution (the location of $O$ in Figs. 1.1 and 1.2) is arbitrary, and the origin dependences of physical quantities are evident from the start. Thus, except for the leading non-zero moments, all electric and magnetic multipole moments are origin dependent (Sections 1.2 and 1.7). For example, the electric quadrupole moment of a dipolar molecule is an origin-dependent observable, whose measured value has meaning only with respect to a stated choice of origin. How this can be done is elucidated in the pioneering work of Buckingham and others (Section 5.11). Furthermore, a finite multipole expansion for a charge distribution, such as a truncated version of the potential (1.3), is an origin-dependent approximation to the manifestly invariant quantities (1.1) or (1.8). However, we do not reject such an expansion as being unphysical because the origin dependence is negligible due to the convergence condition: dimensions of the charge distribution must be small compared to the distance to the field point (Fig. 1.1).

For macroscopic media the situation is different. Here one uses macroscopic fields and multipole moment densities which are obtained by a certain averaging process (Sections 1.12 and 1.13). The condition for the validity of the theory is that these averaged quantities should vary on a scale which is large compared to molecular dimensions. The choice of coordinate origin in macroscopic theory is not constrained by this convergence condition. Instead, the choice of origin is completely arbitrary and consequently the unphysical origin dependences, demonstrated above for origin-independent observables such as material constants and field amplitudes, are not negligible in general.

## References

[1] Robinson, F. N. H. (1973). *Macroscopic electromagnetism.* Pergamon, Oxford.

[2] Jackson, J. D. (1999). *Classical electrodynamics*, Sect. 1.4. Wiley, New York.

[3] For an introduction to the Lagrangian approach to classical electromagnetism, see Ref. 2, Sect. 12.7.

[4] Landau, L. D. and Lifshitz, E. M. (1975). *The classical theory of fields*, Ch. 4. Pergamon, Oxford.

[5] Raab, R. E. and de Lange, O. L. (2001). Symmetry constraints for electromagnetic constitutive relations. *Journal of Optics* A: *Pure and Applied Optics*, **3**, 446–451.

[6] Guggenheim, E. A. (1936). On magnetic and electrostatic energy. *Proceedings of the Royal Society London* A, **155**, 49–70.

[7] Kong, J. A. (1986). *Electromagnetic wave theory*, Sects. 2.2 and 7.10. Wiley, New York.

[8] Nye, J. F. (1985). *Physical properties of crystals*, Ch. 3. Clarendon, Oxford.

[9] Landau, L. D. and Lifshitz, E. M. (1960). *Electrodynamics of continuous media*, Sects. 76 and 82. Pergamon, Oxford.

[10] Post, E. J. (1962). *Formal structure of electromagnetics*, Chs. 5 and 6. North-Holland, Amsterdam. (Reprinted 1997. Dover, New York.)

[11] Lakhtakia, A. and Weiglhofer, W. S. (1994). Constraint on linear, homogeneous, constitutive relations. *Physical Review* E, **50**, 5017–5019.

[12] Weiglhofer, W. S. (1994). On a medium constraint arising directly from Maxwell's equations. *Journal of Physics* A: *Mathematical and General*, **27**, L871–L874.

[13] Lakhtakia, A. and Weiglhofer, W. S. (1995). On a constraint on the electromagnetic constitutive relations of nonhomogeneous linear media. *IMA Journal of Applied Mathematics*, **54**, 301–306.

[14] Lakhtakia, A. and Weiglhofer, W. S. (1996). Lorentz covariance, Occam's razor, and the constraint on linear constitutive relations. *Physics Letters*, **A213**, 107–111.

[15] Weiglhofer, W. S. and Lakhtakia, A. (1998). The Post constraint revisited. *Arch Electron Übertr*, **52**, 276–279, and references therein.

[16] Raab, R. E. and Sihvola, A. H. (1997). On the existence of linear non-reciprocal bi-isotropic (NRBI) media. *Journal of Physics* A: *Mathematical and General*, **30**, 1335–1344.

[17] See Ref. 9, Sect. 63.

# TRANSMISSION AND SCATTERING EFFECTS: DIRECT MULTIPOLE RESULTS

> *The bearings of this observation*
> *lays in the application of it.*
> Charles Dickens
> (*Dombey and son*)

We apply multipole theory to certain transmission and scattering phenomena. In Section 5.1 we use the constitutive relations for **D** and **H** obtained in Chapter 4 to derive a wave equation for the propagation of electromagnetic waves in a homogeneous dielectric. An important feature of this wave equation is its translational invariance, and therefore all results obtained from it are origin independent. In Sections 5.2–5.8 we use the wave equation to describe several transmission effects of various multipole orders: intrinsic Faraday rotation and circular dichroism in a uniaxial magnetic crystal (electric dipole order); natural optical activity, time-odd linear birefringence, and gyrotropic birefringence (electric quadrupole–magnetic dipole order); Lorentz birefringence and intrinsic Faraday rotation in magnetic cubics (electric octopole–magnetic quadrupole order). An induced effect in a gas (Kerr effect) is analyzed in Section 5.9 by means of wave theory , and in Section 5.10 in terms of scattering theory. We conclude the chapter with a discussion of an effect for which the medium is inhomogeneous, namely electric-field-gradient-induced birefringence in a gas. This is treated using scattering theory in Section 5.11 and wave theory in Section 5.12.

## 5.1 The wave equation

The theory of an effect which occurs when a harmonic plane electromagnetic wave propagates in a dielectric may be derived using Maxwell's equations. To do this we assume the long-wavelength regime, which allows a multipole approach to be employed.

The wave (or propagation) equation is obtained from the Maxwell equation (1.121), which for a dielectric is

$$\nabla \times \mathbf{H} = \dot{\mathbf{D}}. \tag{5.1}$$

In this chapter we use for **D** and **H** the direct multipole forms of Section 4.1, which describe a linear response of homogeneous non-magnetic and magnetic media to the fields of the electromagnetic wave. We work to the order of electric

octopole and magnetic quadrupole, for which the response fields are given by
(4.2)–(4.7). It is convenient to take the complex fields of the wave in the form

$$\mathbf{F} = \mathbf{F}_0 e^{-i\omega(t - n\mathbf{r}\cdot\boldsymbol{\sigma}/c)}, \quad (\mathbf{F} = \mathbf{E} \text{ or } \mathbf{B}). \tag{5.2}$$

For a dissipative medium, $n$ is complex and its real part represents the refractive
index for the polarization state described by the amplitude $\mathbf{E}_0$ when propagation
is along the unit-wave-normal $\boldsymbol{\sigma}$ . Except for linear polarization, $\mathbf{E}_0$ is complex
(see below). From (5.2) and the Maxwell equation

$$\nabla \times \mathbf{E} = -\dot{\mathbf{B}} \tag{5.3}$$

one can express the contributions to $\mathbf{D}$ and $\mathbf{H}$ which involve $\mathbf{B}$ in (4.2) and (4.3)
in terms of $\mathbf{E}$.

Use of these expressions for $\mathbf{D}$ and $\mathbf{H}$ in (5.1) yields the wave equation for
homogeneous media in tensor form. We express this wave equation as

$$\{n^2\sigma_i\sigma_j - (n^2 - 1)\delta_{ij} + \varepsilon_0^{-1}[\tilde{\alpha}_{ij} + nc^{-1}\tilde{\beta}_{ij} + n^2c^{-2}\tilde{\gamma}_{ij}]\}E_{0j} = 0, \tag{5.4}$$

where $\mathbf{E}_0$ is the amplitude of the electric field. The tensors denoted by a tilde in
(5.4) are material properties of the dielectric, each of a different multipole order
and having the complex form

$$\tilde{t}_{ij} = t^s_{ij} - it^a_{ij}. \tag{5.5}$$

From their forms below and (2.115), (2.116), (2.123), (2.124), (2.129), and (2.130)
it can be seen that the real part is symmetric (superscript $s$) and the imaginary
part antisymmetric (superscript $a$) on interchange of the subscripts $i$ and $j$.

$$\alpha^s_{ij} = \alpha_{ij} \tag{5.6}$$

$$\alpha^a_{ij} = \alpha'_{ij} \tag{5.7}$$

$$\beta^s_{ij} = \sigma_k\left[-\varepsilon_{ikl}G_{jl} - \varepsilon_{jkl}G_{il} + \frac{1}{2}\omega(a'_{ijk} + a'_{jik})\right] \tag{5.8}$$

$$\beta^a_{ij} = \sigma_k\left[\varepsilon_{ikl}G'_{jl} - \varepsilon_{jkl}G'_{il} - \frac{1}{2}\omega(a_{ijk} - a_{jik})\right] \tag{5.9}$$

$$\gamma^s_{ij} = \sigma_k\sigma_l\left[-\frac{1}{6}\omega^2(b_{ijkl} + b_{jikl}) + \frac{1}{4}\omega^2 d_{ikjl} - \frac{1}{2}\omega(\varepsilon_{ikm}H'_{jml} + \varepsilon_{jkm}H'_{iml})\right.$$
$$\left. + \frac{1}{2}\omega(\varepsilon_{ikm}L'_{jlm} + \varepsilon_{jkm}L'_{ilm}) + \varepsilon_{ikm}\varepsilon_{jln}\chi_{mn}\right] \tag{5.10}$$

$$\gamma^a_{ij} = \sigma_k\sigma_l\left[-\frac{1}{6}\omega^2(b'_{ijkl} - b'_{jikl}) + \frac{1}{4}\omega^2 d'_{ikjl} - \frac{1}{2}\omega(\varepsilon_{ikm}H_{jml} - \varepsilon_{jkm}H_{iml})\right.$$
$$\left. + \frac{1}{2}\omega(\varepsilon_{ikm}L_{jlm} - \varepsilon_{jkm}L_{ilm}) + \varepsilon_{ikm}\varepsilon_{jln}\chi'_{mn}\right]. \tag{5.11}$$

Note that the symmetries of $\alpha^s_{ij}$ and $\alpha^a_{ij}$ follow from the intrinsic symmetries of
$\alpha_{ij}$ and $\alpha'_{ij}$. The symmetries of $\beta^s_{ij}$ and $\beta^a_{ij}$ are a consequence of the structures

of these tensors, and not of any intrinsic symmetries. The symmetries of $\gamma_{ij}^s$ and $\gamma_{ij}^a$ are based on both their structure and the intrinsic symmetries of $d_{ijkl}$, $d'_{ijkl}$, $\chi_{mn}$, and $\chi'_{mn}$.

Inspection of Table 3.2 shows the multipole order and behaviour under time reversal of (5.6)–(5.11):

  *electric dipole*: $\alpha_{ij}^s$ is time even; $\alpha_{ij}^a$ is time odd;
  *electric quadrupole–magnetic dipole*: $\beta_{ij}^s$ is time odd; $\beta_{ij}^a$ is time even;
  *electric octopole–magnetic quadrupole*: $\gamma_{ij}^s$ is time even; $\gamma_{ij}^a$ is time odd.

In addition, one can show from the origin dependences of the polarizability tensors in (3.64)–(3.79) that each of the expressions in (5.6)–(5.11) is independent of the choice of coordinate origin. Thus the wave equation (5.4) does not depend on the choice of this origin, and nor does the Maxwell equation (5.1) for response fields given by the multipole expressions (4.2)–(4.7) [1]. We emphasize that this translational invariance of the tensors in the wave equation (5.4) is in contrast with the origin dependence of the material constants (beyond electric dipole order) in the constitutive relations (4.2) and (4.3); see Section 4.5. It is somewhat surprising that the use of origin-dependent constitutive relations in (5.1) leads to an origin-independent wave equation, and we discuss this further in Section 9.1.

In terms of the lowest multipole order that is required to explain an effect (cf. the hierarchy of contributions (1.122)), each of the expressions (5.6)–(5.11) has been shown to describe one or more effects. Some of these multipole theories for both non-magnetic and magnetic media are detailed in this chapter.

The solution of the wave equation (5.4) yields, for each wave that propagates in the medium, its refractive index and polarization state for the components of its electric field amplitude. To obtain these solutions we choose laboratory Cartesian axes which coincide with crystallographic axes (Section 3.6). The choice of axes for a field-free fluid is immaterial. We label tensor components with $x$, $y$, $z$ rather than $1, 2, 3$ to facilitate reference to Birss' tables [2,3].

By setting $i = x$, $y$, $z$ in turn in (5.4) one obtains three homogeneous linear equations in the components of the field amplitude $\mathbf{E}_0$. The condition that these components are not all zero is that the determinant of their coefficients should vanish. This is the secular equation

$$\begin{vmatrix} a_{xx} & a_{xy} & a_{xz} \\ a_{yx} & a_{yy} & a_{yz} \\ a_{zx} & a_{zy} & a_{zz} \end{vmatrix} = 0, \tag{5.12}$$

where from (5.4)

$$a_{ij} = \left[ n^2 \sigma_i \sigma_j - (n^2 - 1)\delta_{ij} \right] + \left[ \varepsilon_0^{-1} \tilde{\alpha}_{ij} \right] + \left[ \varepsilon_0^{-1} nc^{-1} \tilde{\beta}_{ij} \right] + \left[ \varepsilon_0^{-1} n^2 c^{-2} \tilde{\gamma}_{ij} \right]. \tag{5.13}$$

It is of interest to note that of the four terms in square brackets in (5.13), the first describes wave propagation in a vacuum and the others allow successively for the contributions of multipoles of increasing order, starting with electric dipole.

It is instructive to rearrange the wave equation (5.4) by placing the term $\delta_{ij}$ on the right-hand side [4]. Then

$$\begin{pmatrix} b_{xx} & b_{xy} & b_{xz} \\ b_{yx} & b_{yy} & b_{yz} \\ b_{zx} & b_{zy} & b_{zz} \end{pmatrix} \begin{pmatrix} E_{0x} \\ E_{0y} \\ E_{0z} \end{pmatrix} = \begin{pmatrix} E_{0x} \\ E_{0y} \\ E_{0z} \end{pmatrix} \tag{5.14}$$

with

$$b_{ij} = \delta_{ij} - a_{ij}. \tag{5.15}$$

The eigenvectors in (5.14) specify the polarization states that the crystal supports for a given direction of phase propagation $\boldsymbol{\sigma}$. Thus Maxwell's equation (5.1) combined with the multipole forms of $\mathbf{D}$ and $\mathbf{H}$ for a harmonic plane wave leads to an important conclusion: waves propagate in a dielectric medium as eigenpolarizations.

In practice, it is usually more convenient to find first the refractive indices $n$ from the secular equation (5.12). Then by substituting each value of $n$ in turn into the three equations (5.4) on which (5.12) is based, one solves these for the amplitude components $E_{0j}$. This approach is used in the theories of transmission effects in Sections 5.2–5.9.

## 5.2   Intrinsic Faraday rotation in a ferromagnetic crystal

The only polarizability densities of electric dipole order, namely $\alpha_{ij}$ and $\alpha'_{ij}$, are both origin independent (Section 3.7). Thus different physical effects may exist due to each. Those described by $\alpha'_{ij}$, a time-odd tensor, occur only in magnetic media. Effects that are explained in terms of $\alpha_{ij}$ include refraction and the familiar linear birefringence in uniaxial and biaxial crystals. The dependence of this birefringence on propagation direction may be determined using the Fresnel ellipsoid model [5]. An algebraic alternative to this geometric approach is offered by the eigenvector theory of Section 5.1 [6].

In this section we discuss an electric dipole effect due to $\alpha'_{ij}$. Consider a harmonic plane electromagnetic wave propagating in a uniaxial magnetic crystal. Except for members of the triclinic and monoclinic classes, crystallographic axes serve as principal axes for all symmetric second-rank polar tensors that are time even. This is evident from the tables of Birss [2,3]. As $\alpha_{ij}$ is a tensor of this type (see (2.115) and Table 3.2), it follows that

$$\alpha_{ij} = 0 \quad \text{for } i \neq j. \tag{5.16}$$

Birss' tables also show that

$$\alpha_{xx} = \alpha_{yy} \tag{5.17}$$

for uniaxials. From (2.116) $\alpha'_{ij}$ is antisymmetric, so that

$$\alpha'_{xx} = \alpha'_{yy} = \alpha'_{zz} = 0, \qquad \alpha'_{xy} = -\alpha'_{yx}, \qquad \text{etc.} \tag{5.18}$$

Also, for all uniaxials [2,3]

$$\alpha'_{xz} = \alpha'_{yz} = 0. \tag{5.19}$$

Then from (5.5)–(5.7), (5.12), (5.13), and (5.16)–(5.19) we obtain

$$\begin{vmatrix} n^2(\sigma_x^2-1)+1+\varepsilon_0^{-1}\alpha_{xx} & n^2\sigma_x\sigma_y-i\varepsilon_0^{-1}\alpha'_{xy} & n^2\sigma_x\sigma_z \\ n^2\sigma_y\sigma_x+i\varepsilon_0^{-1}\alpha'_{xy} & n^2(\sigma_y^2-1)+1+\varepsilon_0^{-1}\alpha_{xx} & n^2\sigma_y\sigma_z \\ n^2\sigma_z\sigma_x & n^2\sigma_z\sigma_y & n^2(\sigma_z^2-1)+1+\varepsilon_0^{-1}\alpha_{zz} \end{vmatrix} = 0. \tag{5.20}$$

For simplicity we consider propagation along the optic axis, which in a uniaxial crystal is the main symmetry axis, $z$. Thus

$$\boldsymbol{\sigma} = (0,\, 0,\, 1) \tag{5.21}$$

and (5.20) becomes

$$\begin{vmatrix} -n^2 + 1 + \varepsilon_0^{-1}\alpha_{xx} & -i\varepsilon_0^{-1}\alpha'_{xy} & 0 \\ i\varepsilon_0^{-1}\alpha'_{xy} & -n^2 + 1 + \varepsilon_0^{-1}\alpha_{xx} & 0 \\ 0 & 0 & 1+\varepsilon_0^{-1}\alpha_{zz} \end{vmatrix} = 0. \tag{5.22}$$

The third row of the determinant in (5.22) is obtained from the equation

$$(1 + \varepsilon_0^{-1}\alpha_{zz})E_{0z} = 0. \tag{5.23}$$

Since $1 + \varepsilon_0^{-1}\alpha_{zz} \neq 0$ for all wavelengths for a given crystal, then

$$E_{0z} = 0. \tag{5.24}$$

Thus when propagation is along the optic axis of a magnetic uniaxial, the wavefronts of the eigenpolarizations are perpendicular to the propagation direction. From (5.22) the refractive indices of these eigenpolarizations are given by

$$-n_1^2 + 1 + \varepsilon_0^{-1}\alpha_{xx} = \varepsilon_0^{-1}\alpha'_{xy} \tag{5.25}$$

$$-n_2^2 + 1 + \varepsilon_0^{-1}\alpha_{xx} = -\varepsilon_0^{-1}\alpha'_{xy}. \tag{5.26}$$

Substituting these in turn into the two equations on which the first two rows of the determinant in (5.22) are based yields the corresponding normalized eigenpolarizations

$$n_1: \quad \mathbf{E}_0^{(1)} = (1,\, -i,\, 0)/\sqrt{2} \tag{5.27}$$

$$n_2: \quad \mathbf{E}_0^{(1)} = (1,\, i,\, 0)/\sqrt{2}. \tag{5.28}$$

Because (5.27) and (5.28) describe right and left circularly polarized waves respectively, we write

$$n_1 = n_R, \qquad n_2 = n_L. \tag{5.29}$$

Thus there is circular birefringence, which from (5.25) and (5.26) is given by [7]

$$n_L - n_R = \frac{2\alpha'_{xy}}{\varepsilon_0(n_R + n_L)} \tag{5.30}$$

$$\approx \frac{\alpha'_{xy}}{\varepsilon_0 n}. \tag{5.31}$$

We note that the right-hand side of (5.30) is an origin-independent quantity, as required by the translational invariance of the wave equation (5.4). The quantity that is directly observed is the rotation $\phi$ of the plane of a linearly polarized wave traversing a length $l$ of the crystal. With the convention that $\phi$ is positive for a clockwise rotation as seen by an observer receiving the light, we have [8]

$$\phi = \pi l(n_L - n_R)/\lambda. \tag{5.32}$$

The magnetic crystals that possess $\alpha'_{xy}$ can be identified from tables [2, 3]. Their symmetry classes are [7]

$$\left.\begin{array}{c} 4,\ \bar{4},\ 4/m,\ 4\underline{22},\ 4\underline{mm},\ \bar{4}2\underline{m},\ 4/m\underline{mm}, \\ 3,\ \bar{3},\ 3\underline{2},\ 3\underline{m},\ \bar{3}\underline{m},\ 6,\ \bar{6},\ 6/m,\ 6\underline{22}, \\ 6\underline{mm},\ \bar{6}\underline{m}2,\ 6/m\underline{mm}. \end{array}\right\} \tag{5.33}$$

These tables show that the classes (5.33) also have a net magnetization, and hence an internal field, along their main symmetry axis. They are thus ferromagnetic, and their ability to rotate the plane of a linearly polarized wave is due to an intrinsic Faraday effect. Some of the classes (5.33) exhibit optical activity along their main symmetry axis (Section 5.3). They are

$$4,\ 4\underline{22},\ 3,\ 3\underline{2},\ 6,\ 6\underline{22}. \tag{5.34}$$

For these the rotation $\phi$ is due to two effects: one from the time-odd tensor $\alpha'_{xy}$ and the other from the time-even tensors of Section 5.3. As the part in $\alpha'_{xy}$ changes sign on reversal of the light path, the two contributions may be separated experimentally.

We mention that a microscopic theory by Argyres [9] yielded after averaging an antisymmetric second-rank tensor which served to explain the rotation in a ferromagnetic crystal. Measurements of the effect have been reported in a number of crystals containing the $Eu^{++}$ ion [10].

As discussed in Section 2.8, a property tensor like $\alpha'_{ij}$ in (5.31) becomes complex when the material absorbs radiation. Thus from (5.25) and (5.26), $n$ is similarly affected. It is now the real part of (5.30) that is the frequency-dependent circular birefringence, while the imaginary part, denoted

$$k_L - k_R = \frac{2}{\varepsilon_0} \mathcal{I}m\left\{\frac{\alpha'_{xy}}{n_R + n_L}\right\}, \tag{5.35}$$

is an example of the phenomenon of circular dichroism (see Section 5.5). This property is discussed further in the next section on optical activity. Many magnetic crystals are strongly coloured due to their absorption, and therefore circular dichroism in them is not uncommon in the visible spectrum.

## 5.3  Natural optical activity

Consider a harmonic plane electromagnetic wave propagating in a non-magnetic medium. We seek an effect different from those explained in terms of the electric dipole tensor $\alpha_{ij}$, and therefore work to the order of electric quadrupole and magnetic dipole. Thus we neglect the higher-multipole tensor $\tilde{\gamma}_{ij}$ in (5.13). Also, for a non-magnetic medium the time-odd tensors $\alpha'_{ij}$ of (5.7) and $\beta^s_{ij}$ of (5.8) are zero. Then from (5.12), (5.13), and (5.5) we have

$$\begin{vmatrix} n^2(\sigma_x^2-1)+1+\varepsilon_0^{-1}\alpha_{xx} & n^2\sigma_x\sigma_y-in\varepsilon_0^{-1}c^{-1}\beta^a_{xy} & n^2\sigma_x\sigma_z-in\varepsilon_0^{-1}c^{-1}\beta^a_{xz} \\ n^2\sigma_y\sigma_x+in\varepsilon_0^{-1}c^{-1}\beta^a_{xy} & n^2(\sigma_y^2-1)+1+\varepsilon_0^{-1}\alpha_{yy} & n^2\sigma_y\sigma_z-in\varepsilon_0^{-1}c^{-1}\beta^a_{yz} \\ n^2\sigma_z\sigma_x+in\varepsilon_0^{-1}c^{-1}\beta^a_{xz} & n^2\sigma_z\sigma_y+in\varepsilon_0^{-1}c^{-1}\beta^a_{yz} & n^2(\sigma_z^2-1)+1+\varepsilon_0^{-1}\alpha_{zz} \end{vmatrix} = 0, \tag{5.36}$$

in which use was made of the antisymmetry of $\beta^a_{ij}$.

It is known that to electric dipole order linear birefringence is not manifest along the optic ($z$) axis in uniaxial crystals and in any direction for cubics [11]. To determine whether there is any effect of electric quadrupole–magnetic dipole order that can be inferred from (5.36), we therefore consider propagation along the $z$ axis in a uniaxial or cubic crystal. From Birss' tables [2,3]

$$\alpha_{xx} = \alpha_{yy}, \qquad \beta^a_{xz} = \beta^a_{yz} = 0. \tag{5.37}$$

Then (5.36) becomes

$$\begin{vmatrix} -n^2+1+\varepsilon_0^{-1}\alpha_{xx} & -in\varepsilon_0^{-1}c^{-1}\beta^a_{xy} & 0 \\ in\varepsilon_0^{-1}c^{-1}\beta^a_{xy} & -n^2+1+\varepsilon_0^{-1}\alpha_{xx} & 0 \\ 0 & 0 & 1+\varepsilon_0^{-1}\alpha_{zz} \end{vmatrix} = 0. \tag{5.38}$$

This has the form of (5.22) and thus describes optical activity along the $z$-axis in those non-magnetic uniaxial and cubic crystals for which $\beta^a_{xy}$ exists.

The refractive indices for right and left circularly polarized waves can be found in similar manner to those in Section 5.2 and are given by

$$n_R^2 + n_R\varepsilon_0^{-1}c^{-1}\beta^a_{xy} - 1 - \varepsilon_0^{-1}\alpha_{xx} = 0 \tag{5.39}$$

$$n_L^2 - n_L\varepsilon_0^{-1}c^{-1}\beta^a_{xy} - 1 - \varepsilon_0^{-1}\alpha_{xx} = 0. \tag{5.40}$$

From these one obtains the circular birefringence

$$n_L - n_R = (\varepsilon_0 c)^{-1}\beta^a_{xy}. \tag{5.41}$$

Thus it is at the order of electric quadrupole–magnetic dipole that natural optical activity is explained. The origin independence of $\beta^a_{ij}$ (Section 5.1) ensures that (5.41) yields an origin-independent observable.

From Birss' tables [2,3] one is able to identify the non-magnetic uniaxial and cubic crystal classes for which $\beta^a_{xy}$ exists and to determine the corresponding form of (5.9). For propagation along the $z$-axis the results are uniaxials:

$$4,422,3,32,6,622 \atop \beta_{xy}^a = -2G'_{xx} - \omega a_{xyz} \Bigg\}$$ (5.42)

cubics:

$$23,432 \atop \beta_{xy}^a = -2G'_{xx}. \Bigg\}$$ (5.43)

We note the following points in regard to optical activity.

(i) The order of magnitude of the circular birefringence in non-magnetic crystals is typically $10^{-4}$ [8]. For instance, in quartz it is $7.1 \times 10^{-5}$ at 589 nm for propagation along the optic axis ($z$) [8], while from (5.32) the corresponding rotation angle $\phi$ is about $220°$ for a path length of 1 cm. By contrast, the order of magnitude of the linear birefringence $n_O - n_E$, where $n_O$ and $n_E$ are respectively the ordinary and extraordinary refractive indices, is $10^{-1} - 10^{-3}$ [8]. In quartz it is $9.1 \times 10^{-3}$ and in calcite $1.7 \times 10^{-1}$. Comparison of these magnitudes of circular and linear birefringence shows that the electric quadrupole–magnetic dipole effect is about $10^{-2} - 10^{-3}$ of that of electric dipole order, consistent with the hierarchy in (1.122).

(ii) Since each magnetic point group contains as a subgroup the associated non-magnetic point group (Section 3.6), a magnetic crystal exhibits the same optical activity as that of a non-magnetic crystal having the symmetry of the related subgroup.

(iii) The time-even tensors $a_{ijk}$ and $G'_{ij}$, being respectively third-rank polar and second-rank axial (Table 3.2), vanish for centrosymmetric crystals (Section 3.6), which therefore cannot be optically active. There are also symmetry classes for which $a_{ijk}$ and $G'_{ij}$ exist but which do not exhibit optical activity (because $\beta_{ij}^a = 0$), for example, $4mm$ and $3m$ [2,3]. Thus the existence of both $a_{ijk}$ and $G'_{ij}$ is a necessary but not sufficient condition for optical activity in a uniaxial crystal.

(iv) For a cubic crystal [2,3]

$$\alpha_{xx} = \alpha_{yy} = \alpha_{zz}.$$ (5.44)

Also, for an arbitrary propagation direction (5.9) yields

$$\beta_{ij}^a = -2\varepsilon_{ijk}\sigma_k G'_{xx}.$$ (5.45)

It can be shown from (5.44) and (5.45) that (5.36) reduces to

$$n^2 \pm 2n\varepsilon_0^{-1}c^{-1}G'_{xx} - 1 - \varepsilon_0^{-1}\alpha_{xx} = 0.$$ (5.46)

Since (5.46) applies for an arbitrary propagation direction, it follows that optical activity in a cubic crystal is isotropic, at least to electric quadrupole–magnetic dipole order. The circular birefringence found from (5.46) is

$$n_L - n_R = -2(\varepsilon_0 c)^{-1}G'_{xx},$$ (5.47)

the sign of which is established on the basis of (5.41) and (5.43).

(v) To illustrate a number of points we consider propagation along the $x$-axis of a uniaxial crystal of the class 422. The only non-vanishing components of $G'_{ij}$ and $a_{ijk}$ are [2, 3]

$$\left.\begin{array}{c} G'_{xx} = G'_{yy}, \quad G'_{zz} \\ a_{xyz} = a_{xzy} = -a_{yxz} = -a_{yzx}. \end{array}\right\} \tag{5.48}$$

From (5.9) and (5.48)

$$\beta^a_{xy} = \beta^a_{xz} = 0, \qquad \beta^a_{yz} = -G'_{xx} - G'_{zz} + \frac{1}{2}\omega a_{xyz}, \tag{5.49}$$

so that (5.36) becomes

$$\begin{vmatrix} 1+\varepsilon_0^{-1}\alpha_{xx} & 0 & 0 \\ 0 & -n^2+1+\varepsilon_0^{-1}\alpha_{xx} & -in\varepsilon_0^{-1}c^{-1}\beta^a_{yz} \\ 0 & in\varepsilon_0^{-1}c^{-1}\beta^a_{yz} & -n^2+1+\varepsilon_0^{-1}\alpha_{zz} \end{vmatrix} = 0, \tag{5.50}$$

where we have also used (5.17).

The solution of (5.50) yields expressions for two refractive indices $n_1$ and $n_2$, which contain contributions of electric dipole and electric quadrupole–magnetic dipole orders. The birefringence can be shown to have magnitude

$$|n_1 - n_2| = (n_O - n_E)\left[1 + \frac{1}{2}\left(\frac{\varepsilon_0^{-1}c^{-1}\beta^a_{yz}}{n_O - n_E}\right)^2\right], \tag{5.51}$$

where [6]

$$n_O^2 = 1 + \varepsilon_0^{-1}\alpha_{xx} \tag{5.52}$$
$$n_E^2 = 1 + \varepsilon_0^{-1}\alpha_{zz}. \tag{5.53}$$

From (5.41) and the discussion in (i), it is evident that the circular birefringence term involving $\beta^a_{yz}$ in (5.51) is very much smaller than the linear birefringence when propagation is perpendicular to the optic axis. Nevertheless, various techniques have been developed to measure the optical activity in such cases [12, 13]. The first of these measurements confirmed the existence of optical activity in a crystal with symmetry $\bar{4}2m$ [12], which does not exhibit optical activity along its optic axis [8].

(vi) Equation (5.36) may be applied to an optically active gas by replacing the macroscopic property tensors $\alpha_{ij}$ and $\beta^a_{ij}$ by the corresponding averaged molecular properties. To do this we make the following assumptions: the gas is ideal so that molecular interactions may be neglected; the electromagnetic wave is of low intensity so that its fields do not bias a molecule's orientation; and orientation is continuous. Thus the orientational averaging of a molecular tensor $t$ is isotropic, denoted by $\langle t \rangle$.

Since the gas itself is isotropic, all propagation directions are equivalent. Accordingly, we may use the result for propagation along the $z$ axis in (5.41), which now becomes

$$n_L - n_R = (\varepsilon_0 c)^{-1} N \langle \bar{\beta}^a_{xy} \rangle, \tag{5.54}$$

where $N$ is the number density of molecules and $\bar{\beta}^a_{xy}$ is the corresponding molecular property and not the macroscopic property density (see Section 2.11). From (5.9) we have

$$\beta^a_{xy} = -G'_{xx} - G'_{yy} - \frac{1}{2}\omega(a_{xyz} - a_{yxz}). \tag{5.55}$$

For the evaluation of isotropic averages, the reader is referred to Ref. 14. We note that for the components of $G'_{ij}$ and $a_{ijk}$ in (5.55) the results of this averaging are proportional to the isotropic tensors $\delta_{ij}$ and $\varepsilon_{ijk}$, respectively. Since $a_{ijk} = a_{ikj}$ (see (2.141)) and $\varepsilon_{ijk}$ is antisymmetric in any two subscripts (see (1.32)), the contribution of $a_{ijk}$ to $n_L - n_R$ vanishes.

(vii) To show the necessity of including the electric quadrupole contribution for an anisotropic medium, Buckingham and Dunn derived an expression for the optical rotation $\phi$ by a fluid of aligned molecules [15]. In this theory the averaging is no longer isotropic and the terms in $a_{ijk}$ in (5.55) also contribute to $\phi$. The resulting expression for $\phi$ was shown to be origin independent [15]. We note that a fluid of aligned molecules may be regarded as a uniaxial medium and consequently the expression for circular birefringence is obtained by using (5.42) in (5.54).

(viii) Experimental confirmation of the necessity to include the electric quadrupole contribution in the theoretical expression for circular birefringence of an anisotropic chiral medium comes from microwave measurements by Theron and Cloete [16, 17]. These workers assembled chiral metal objects having dimensions of about 3 mm in a regular array supported by a dielectric foam to create a 2-m-long "crystal" with point-group symmetry 422. Their measured values at various wavelengths of the rotation of linearly polarized microwaves emerging from the medium agreed to within 13% with the theoretical values based on a consistent multipole computation for their chiral objects. When the quadrupole contribution was omitted, the theoretical values for the rotation decreased by approximately one-half. This microwave study provides significant support for the multipole description of electromagnetic effects at long wavelengths. The equivalent calculation of rotation in a crystal for visible wavelengths would present a considerable challenge.

(ix) We emphasize that the above results for circular birefringence are origin independent because they depend on the invariant tensor $\beta^a_{ij}$ (Section 5.1). This can readily be checked in specific cases: for example, the result (5.47) for cubics is origin independent because for crystallographic axes in these crystals $\alpha_{ij}$ is isotropic and hence from (3.69) $\Delta G'_{xx} = 0$. Similarly, for uniaxials each term in (5.42) is origin independent. The general question of origin independence of linear combinations of polarizability tensors is discussed further in Chapter 8.

(x) Dissipation in an optically active medium results in circular dichroism, which is expressed by the imaginary part of (5.41). According to (5.2) and (5.41), the left and right circular eigenpolarizations are absorbed differently. On emerging with different amplitudes, the two waves combine to produce an elliptically polarized wave with major axis rotated from the plane of the incident wave. This rotation as a function of wavelength is the basis of an analytical technique known as optical rotatory dispersion which is used for structural determinations by organic chemists [18].

## 5.4   Time-odd linear birefringence in magnetic cubics

Having identified in optical activity a time-even transmission effect of electric quadrupole–magnetic dipole order, we now investigate whether, at the same multipole order, any time-odd effects exist in transmission. For simplicity, we restrict ourselves to magnetic cubic crystals. Birss' tables [2,3] show that there are eleven magnetic cubic classes, of which three are optically active and thus not considered further. The time-even tensor $\beta_{ij}^a$ in (5.9) vanishes for the remaining eight classes, while the time-odd tensor $\beta_{ij}^s$ in (5.8) exists only for the three classes

$$\underline{m}3, \ \bar{4}3m, \ \underline{m}3m \tag{5.56}$$

and has the form

$$\beta_{ij}^s = \left|\omega\varepsilon_{ijk}\sigma_k a'_{xyz}\right|. \tag{5.57}$$

Of the electric dipole polarizabilities $\alpha_{ij}$ and $\alpha'_{ij}$, the latter is zero for all magnetic cubics and $\alpha_{ij}$ has only the components

$$\alpha_{xx} = \alpha_{yy} = \alpha_{zz}. \tag{5.58}$$

From (5.12), (5.13), (5.52), (5.57), and (5.58), and for propagation along the $z$-axis, one obtains

$$\begin{vmatrix} -n^2 + n_O^2 & n(\varepsilon_0 c)^{-1}\omega a'_{xyz} & 0 \\ n(\varepsilon_0 c)^{-1}\omega a'_{xyz} & -n^2 + n_O^2 & 0 \\ 0 & 0 & n_O^2 \end{vmatrix} = 0. \tag{5.59}$$

Thus we have

$$n_+^2 - n_+(\varepsilon_0 c)^{-1}\omega a'_{xyz} - n_O^2 = 0, \qquad \mathbf{E}_0^{(+)} = (1,1,0)/\sqrt{2} \tag{5.60}$$

$$n_-^2 + n_-(\varepsilon_0 c)^{-1}\omega a'_{xyz} - n_O^2 = 0, \qquad \mathbf{E}_0^{(-)} = (1,-1,0)/\sqrt{2}. \tag{5.61}$$

The real eigenpolarizations in (5.60) and (5.61) describe linearly polarized waves. Equivalent results are obtained for propagation along the $x$- and $y$-axes. From (5.60) and (5.61)

$$n_+ - n_- = (\varepsilon_0 c)^{-1} \omega a'_{xyz}, \tag{5.62}$$

which is a linear birefringence relative to fast–slow axes along the bisectors of the $x$ and $y$ crystallographic axes. Note that because $a'_{ij} = 0$, it follows from (3.67) that (5.62) is origin independent.

The result (5.62) is of particular interest because it shows that, in terms of multipole theory, the magnetic cubics (5.56) will exhibit the linear birefringence (5.62) only if the electric quadrupole contribution is included in $\mathbf{D}$ in (1.118). As far as is known, this birefringence has not been observed. Its magnitude should be comparable to that of circular birefringence in a crystal, since they are both of electric quadrupole–magnetic dipole order and their expressions in (5.41)–(5.43), and (5.62) have the same dimensions.

## 5.5 Optical properties in the Jones calculus

Before continuing with our discussion of transmission phenomena, we give a brief outline of the optical properties of a non-depolarizing, homogeneous, anisotropic medium that were deduced by R. C. Jones in the formulation of the optical calculus that bears his name [19,20]. Jones represented a perfectly polarized harmonic plane wave incident in a given direction on a medium by a $2 \times 1$ column vector whose elements are, in complex notation, the orthogonal components of the electric field of the wave. When the wave emerges in the same direction from the medium, its polarization state has in general been changed so that it is now described by a different Jones vector. The formal means of transforming the incident $2 \times 1$ column vector into the emergent one is by means of a $2 \times 2$ matrix with complex elements (the Jones matrix).

Through the real and imaginary parts of its four elements, a Jones matrix represents in general the effect of eight physical properties of the medium [19]. These properties are: refractive index and absorption coefficient; linear birefringence and dichroism relative to orthogonal fast and slow axes; circular birefringence and dichroism; a second independent linear birefringence and dichroism relative to fast and slow axes that bisect the first pair. Jones assumed that each of these properties could be allocated to a different thin slice of the medium (platelet) to which macroscopic electromagnetism is applicable. An explicit form for the Jones matrix was obtained by considering the integrated effect of these platelets. Detailed accounts of the Jones calculus are available in the literature [19,21].

Where both of the above linear birefringences coexist, one is termed Jones birefringence, usually that relative to the bisectors of the natural axes (the crystallographic axes or those along and perpendicular to an applied field) [19,22]. At the time, Jones gave no indication of systems in which both linear birefringences might coexist. Such systems have since been identified in multipole theories of wave propagation [4,22], one system being a fluid to which parallel electric and magnetic fields are applied [22]. Jones birefringence induced in this way has been measured in a liquid subjected to large electric and magnetic fields, confirming for the first time Jones's prediction of some 50 years previously of this new birefringence [23]. Both linear birefringences can also be induced in non-magnetic crystals having the symmetries $\bar{4}$ and $\bar{6}$, by an electric field acting perpendicular to the light path along the optic axis [24, 25].

Yet other theories of Jones birefringence induced by parallel electric and magnetic fields have appeared in the literature. One of these presents *ab initio* computer calculations to evaluate the effect in a range of gases [26], while its existence in the quantum vacuum has been described in a theory based on the Heisenberg–Euler Lagrangian [27].

## 5.6   Gyrotropic birefringence

By applying space inversion and time reversal to a form of linear constitutive relations which included a spatial dispersion term (see Section 2.12), Brown, Shtrikman, and Treves predicted a new optical phenomenon in certain magnetic crystals, which they termed gyrotropic birefringence [28]. This effect is a time-odd linear birefringence relative to orthogonal fast and slow axes which are rotated from the crystallographic axes, the sense of rotation being reversed in a domain of opposite spins. Support for this prediction was provided by a theory of wave propagation in magnetic crystals by Birss and Shrubsall [29]. Subsequently, Hornreich and Shtrikman produced a quantitative theory of gyrotropic birefringence, in which electric quadrupole contributions were allowed for [30]. It was subsequently shown that this allowance was incomplete in that, while the quadrupole contribution to the induced electric dipole moment density $P_i$ was included through the term $a'_{ijk}$ in (2.146), the quadrupole polarization contribution $Q_{ij}$ to $D_i$ in (1.118) was not [31, 32]. The latter theory, which was presented consistently to electric quadrupole–magnetic dipole order, also showed that gyrotropic birefringence is a combination of the normal linear birefringence and Jones birefringence (Section 5.5), which in certain magnetic crystals is explained within the above multipole order.

We consider wave propagation along the optic axis of a uniaxial magnetic crystal with symmetry $4/m$ [31]. (The reason for this choice will become apparent below.) For this class [2, 3]

$$\alpha_{xx} = \alpha_{yy}, \qquad \alpha'_{ij} = 0, \qquad \beta^a_{ij} = 0, \tag{5.63}$$

while for $\boldsymbol{\sigma} = (0,\,0,\,1)$ the only non-vanishing components of $\beta^s_{ij}$ are [2, 3]

$$\beta^s_{xx} = -\beta^s_{yy} = 2\left(G_{xy} + \frac{1}{2}\omega a'_{xxz}\right) \tag{5.64}$$

$$\beta^s_{xy} = \beta^s_{yx} = 2\left(-G_{xx} + \frac{1}{2}\omega a'_{xyz}\right). \tag{5.65}$$

From (5.5), (5.8), (5.13), (5.52), (5.53), and (5.63)–(5.65) the secular equation (5.12) becomes

$$\begin{vmatrix} -n^2 + n_O^2 + n\varepsilon_0^{-1}c^{-1}\beta^s_{xx} & n\varepsilon_0^{-1}c^{-1}\beta^s_{xy} & 0 \\ n\varepsilon_0^{-1}c^{-1}\beta^s_{xy} & -n^2 + n_O^2 - n\varepsilon_0^{-1}c^{-1}\beta^s_{xx} & 0 \\ 0 & 0 & n_E^2 \end{vmatrix} = 0 \tag{5.66}$$

to the order of electric quadrupole and magnetic dipole. For simplicity we neglect dissipation, and hence $\beta_{ij}^s$ is real. The solutions of (5.66) to first order in $\beta_{ij}^s$, together with the corresponding normalized eigenvectors, are

$$n_1 = n_O + \frac{1}{2}\beta, \qquad \mathbf{E}_0^{(1)} = \left(1, \frac{\beta - \beta_1}{\beta_2}, 0\right) \frac{\beta_2}{\sqrt{2\beta(\beta - \beta_1)}} \qquad (5.67)$$

$$n_2 = n_O - \frac{1}{2}\beta, \qquad \mathbf{E}_0^{(2)} = \left(1, -\frac{\beta + \beta_1}{\beta_2}, 0\right) \frac{\beta_2}{\sqrt{2\beta(\beta + \beta_1)}}, \qquad (5.68)$$

where

$$\beta_1 = \varepsilon_0^{-1}c^{-1}\beta_{xx}^s, \qquad \beta_2 = \varepsilon_0^{-1}c^{-1}\beta_{xy}^s, \qquad \beta = (\beta_1^2 + \beta_2^2)^{1/2}. \qquad (5.69)$$

Being real, the eigenvectors in (5.67) and (5.68) are linearly polarized. They are also orthogonal. Thus from (5.64), (5.65), (5.67)–(5.69) a linear birefringence

$$n_1 - n_2 = \beta = 2\varepsilon_0^{-1}c^{-1}\left[\left(G_{xx} - \frac{1}{2}\omega a'_{xyz}\right)^2 + \left(G_{xy} + \frac{1}{2}\omega a'_{xxz}\right)^2\right]^{1/2} \qquad (5.70)$$

exists relative to orthogonal fast and slow axes along the directions of $\mathbf{E}_0^{(2)}$ and $\mathbf{E}_0^{(1)}$ respectively. Because these axes do not coincide with the crystallographic $x$- and $y$-axes, this birefringence is gyrotropic. The angle between the crystallographic $x$-axis and $\mathbf{E}_0^{(1)}$ has magnitude

$$\left|\cos^{-1}\frac{\beta_2}{\sqrt{2\beta(\beta - \beta_1)}}\right|. \qquad (5.71)$$

From (5.64) and (5.65), $\beta_1$ and $\beta_2$ are both of electric quadrupole–magnetic dipole order, so that for a crystal with $\underline{4/m}$ symmetry the rotation of the fast and slow axes of the gyrotropic birefringence from the crystallographic axes may be large and easily measurable.

Equation (5.66) is now solved in an alternative way that yields Jones' platelet properties (Section 5.5) appropriate to propagation along the optic axis in a $\underline{4/m}$ crystal. We proceed as follows:

(i) Consider a platelet which possesses only the property $\alpha_{xx}$. Then from (5.52) and (5.66)

$$\begin{vmatrix} -n^2 + 1 + \varepsilon_0^{-1}\alpha_{xx} & 0 & 0 \\ 0 & -n^2 + 1 + \varepsilon_0^{-1}\alpha_{xx} & 0 \\ 0 & 0 & n_E^2 \end{vmatrix} = 0. \qquad (5.72)$$

The solutions of this are given by

$$n_1^2 = 1 + \varepsilon_0^{-1}\alpha_{xx} = n_O^2, \qquad \mathbf{E}_0^{(1)} = (1, 0, 0) \qquad (5.73)$$

$$n_2^2 = 1 + \varepsilon_0^{-1}\alpha_{xx} = n_O^2, \qquad \mathbf{E}_0^{(2)} = (0, 1, 0). \qquad (5.74)$$

Thus this platelet possesses only one refractive index $n_O$ and behaves isotropically.

(ii) Let a second platelet possess only the property $\beta_{xx}^s$ given by (5.64). Then (5.66) becomes

$$\begin{vmatrix} -n^2+1+n\varepsilon_0^{-1}c^{-1}\beta_{xx}^s & 0 & 0 \\ 0 & -n^2+1-n\varepsilon_0^{-1}c^{-1}\beta_{xx}^s & 0 \\ 0 & 0 & n_E^2 \end{vmatrix} = 0, \qquad (5.75)$$

the solutions of which, with their corresponding eigenpolarizations, are

$$n_x = 1 + \frac{1}{2}\varepsilon_0^{-1}c^{-1}\beta_{xx}^s, \qquad \mathbf{E}_0^{(x)} = (1,\,0,\,0) \qquad (5.76)$$

$$n_y = 1 - \frac{1}{2}\varepsilon_0^{-1}c^{-1}\beta_{xx}^s, \qquad \mathbf{E}_0^{(y)} = (0,\,1,\,0). \qquad (5.77)$$

Thus there is the linear birefringence

$$n_x - n_y = \varepsilon_0^{-1}c^{-1}\beta_{xx}^s \qquad (5.78)$$

with fast and slow axes along the $x$ and $y$ crystallographic axes.

(iii) Now let a third platelet possess only the property $\beta_{xy}^s$ in (5.65), so that from (5.66)

$$\begin{vmatrix} -n^2+1 & n\varepsilon_0^{-1}c^{-1}\beta_{xy}^s & 0 \\ n\varepsilon_0^{-1}c^{-1}\beta_{xy}^s & -n^2+1 & 0 \\ 0 & 0 & n_E^2 \end{vmatrix} = 0. \qquad (5.79)$$

The solutions and their eigenpolarizations are

$$n_+ = 1 + \frac{1}{2}\varepsilon_0^{-1}c^{-1}\beta_{xy}^s, \qquad \mathbf{E}_0^{(+)} = (1,\,1,\,0)/\sqrt{2} \qquad (5.80)$$

$$n_- = 1 - \frac{1}{2}\varepsilon_0^{-1}c^{-1}\beta_{xy}^s, \qquad \mathbf{E}_0^{(-)} = (1,\,-1,\,0)/\sqrt{2}. \qquad (5.81)$$

From these one obtains the linear birefringence

$$n_+ - n_- = \varepsilon_0^{-1}c^{-1}\beta_{xy}^s \qquad (5.82)$$

relative to the bisectors of the $x$ and $y$ crystallographic axes. This would be deemed a Jones birefringence according to the earlier explanation of this term (Section 5.5).

From the alternative ways of solving the wave equation in (5.66) it is evident that gyrotropic birefringence is the result of two linear birefringences, each relative to a pair of orthogonal fast and slow axes that bisect those of the other pair. Because gyrotropy is sometimes used as a synonym for optical activity [33], the term gyrotropic birefringence may be misleading. It has no connection with optical activity (e.g. a 4/m crystal is optically inactive) and it is a linear, not a circular, birefringence.

A magnetic crystal which has been studied, both experimentally and theo-retically, for its gyrotropic birefringence is $Cr_2O_3$ in its room-temperature anti-ferromagnetic phase [32, 34]. The solution of the wave equation for propagation along the $y$-axis in this crystal can be shown to yield three platelet birefringences: the familiar time-even effect in a uniaxial crystal, of electric dipole order, and two time-odd birefringences of electric quadrupole–magnetic dipole order, one of them a Jones birefringence [32]. Thus gyrotropic birefringence exists for $y$-propagation in $Cr_2O_3$ but the presence of the much larger electric dipole effect renders it difficult to measure. Nevertheless, Pisarev and co-workers were able to measure a rotation of the axes of about $1.2 \times 10^{-3}$ radians at 1156 nm, but could not detect a difference in the birefringence for the two antiferromagnetic domains related by time reversal [34]. Of all the optically inactive uniaxial anti-ferromagnetics, only $\underline{4/m}$ crystals exhibit gyrotropic birefringence along an optic axis, where the dominant familiar birefringence is absent [31]. It appears that the effect has not yet been measured in such a crystal.

Seen as the resultant of two platelet linear birefringences relative to different axes, gyrotropic birefringence may also exist in those non-magnetic crystals in which, for propagation perpendicular to an optic axis, both the familiar linear birefringence and a Jones birefringence occur. As the latter has been shown to be of electric octopole–magnetic quadrupole order [4, 24], the rotation angle between the gyrotropic and crystallographic axes is probably too small to measure in a non-magnetic crystal.

## 5.7   Linear birefringence in non-magnetic cubic crystals (Lorentz birefringence)

Relative to crystallographic axes the second-rank tensors of all cubic crystals, whether magnetic or non-magnetic, are either zero or isotropic [2, 3]. For in-stance, the only non-vanishing components of the electric polarizability for a cubic crystal are

$$\alpha_{xx} = \alpha_{yy} = \alpha_{zz}. \tag{5.83}$$

It follows from (5.36) and (5.83) that, to the order of electric dipole (i.e. $\beta_{ij}^a = 0$), the refractive index of a cubic crystal is independent of direction of propagation and polarization, and therefore does not exhibit linear birefringence. However, even though its second-rank tensors are isotropic, a cubic crystal is itself not an isotropic system. Its higher-rank tensors are not in general isotropic (Section 3.3) and thus may account for a dependence of refractive index on the direction of propagation of the wave and on its polarization. To investigate this possibility, but without the complications of circular birefringence and time-odd effects, we limit consideration to optically inactive non-magnetic cubics, namely the classes [2, 3]

$$m3, \quad \bar{4}3m, \quad m3m \tag{5.84}$$

(see Section 5.3). Thus, of the material contributions to the wave equation, we include only the terms in $\alpha_{ij}$ in (5.6) and $\gamma_{ij}^s$ in (5.10). As the latter is of electric octopole–magnetic quadrupole order, it is associated in part with the spatial dispersion term $\nabla_k \nabla_j E_i$ (Section 2.12). In this regard it is of interest to note that in the earliest theory of linear birefringence in cubic crystals, H. A. Lorentz allowed for the non-uniformity of the wave field because of its finite wavelength [35]. In a subsequent theory Condon and Seitz explained Lorentz birefringence in terms of the contribution of $\nabla_k \nabla_j E_i$ to the polarization density $\mathbf{P}$ [36]. A consistent multipole theory which includes all relevant contributions has since appeared [37], and this is outlined below.

Table 5.1 lists the components of $\gamma_{ij}^s$ found from (5.10) and Birss' tables [2,3], which are required for solution of the wave equation. For propagation along the $z$-axis, $\boldsymbol{\sigma} = (0, 0, 1)$ and then from (5.12), (5.13), (5.5), (5.10), and Table 5.1 one obtains

$$\begin{vmatrix} -n^2+n_O^2+\varepsilon_0^{-1}c^{-2}n^2\gamma_{xx}^s & 0 & 0 \\ 0 & -n^2+n_O^2+\varepsilon_0^{-1}c^{-2}n^2\gamma_{yy}^s & 0 \\ 0 & 0 & n_O^2+\varepsilon_0^{-1}c^{-2}n^2\gamma_{zz}^s \end{vmatrix} = 0. \quad (5.85)$$

The solutions of (5.85) are

$$n_1^2 = n_O^2(1 - \varepsilon_0^{-1}c^{-2}\gamma_{xx}^s)^{-1}, \quad \mathbf{E}_0^{(1)} = (1, 0, 0) \quad (5.86)$$

$$n_2^2 = n_O^2(1 - \varepsilon_0^{-1}c^{-2}\gamma_{yy}^s)^{-1}, \quad \mathbf{E}_0^{(2)} = (0, 1, 0) \quad (5.87)$$

$$n_3^2 = -n_O^2(\varepsilon_0^{-1}c^{-2}\gamma_{zz}^s)^{-1}, \quad \mathbf{E}_0^{(3)} = (0, 0, 1). \quad (5.88)$$

Equivalent solutions are obtained for propagation along the $x$- and $y$-axes.

The solution (5.88) reveals an unusual feature, namely the occurrence of a longitudinal wave with its electric field parallel to the propagation direction. Such a wave is an example of a so-called additional wave, and is a consequence of spatial dispersion [38,39]. Furthermore, if $\gamma_{zz}^s$ is negative then associated with this additional wave is an extraordinarily large refractive index $n_3$. The reason is that the terms in $\gamma_{ii}^s$ in (5.86)–(5.88), being of electric octopole–magnetic quadrupole order, are very much smaller than unity, probably of order $10^{-6}$ (see below). Thus $n_3$ in (5.88) is about $10^3$ times larger than a typical refractive index.

It follows from (5.86) and (5.87) that to first order in $\gamma_{ii}^s$

$$n_1 - n_2 = \frac{1}{2}\varepsilon_0^{-1}c^{-2}n_O(\gamma_{xx}^s - \gamma_{yy}^s). \quad (5.89)$$

From this and Table 5.1 it is evident that in $\bar{4}3m$ and $m3m$ crystals there is (to this order) no linear birefringence for $z$ propagation; nor is there any along the other cube edges from the equivalent results for $x$ and $y$ propagation. This conclusion is in agreement with previous theoretical and experimental findings [36, 37, 40]. By contrast, linear birefringence exists along a cube edge in a crystal with $m3$ symmetry. Its magnitude, found from (5.89) and Table 5.1, is

**Table 5.1** *Components of the electric octopole–magnetic quadrupole term $\gamma_{ij}^s$ in the wave equation for the optically inactive non-magnetic cubic classes shown.*

| Class | $\gamma_{xx}^s$ | $\gamma_{yy}^s$ | $\gamma_{zz}^s$ | $\gamma_{ij}^s, i \neq j$ |
|---|---|---|---|---|
| $m3$ | $\sigma_x^2\gamma_1 + \sigma_y^2\gamma_2 + \sigma_z^2\gamma_3$ | $\sigma_x^2\gamma_3 + \sigma_y^2\gamma_1 + \sigma_z^2\gamma_2$ | $\sigma_x^2\gamma_2 + \sigma_y^2\gamma_3 + \sigma_z^2\gamma_1$ | $\sigma_i\sigma_j\gamma_4$ |
| $\bar{4}3m, m3m$ | $\sigma_x^2\gamma_1 + \sigma_y^2\gamma_5 + \sigma_z^2\gamma_5$ | $\sigma_x^2\gamma_5 + \sigma_y^2\gamma_1 + \sigma_z^2\gamma_5$ | $\sigma_x^2\gamma_5 + \sigma_y^2\gamma_5 + \sigma_z^2\gamma_1$ | $\sigma_i\sigma_j\gamma_6$ |

$$\gamma_1 = -\frac{1}{3}\omega^2 b_{xxxx} + \frac{1}{4}\omega^2 d_{xxxx}$$

$$\gamma_2 = -\frac{1}{3}\omega^2 b_{xxyy} + \frac{1}{4}\omega^2 d_{xyxy} - \omega(H'_{xzy} - L'_{xyz}) + \chi_{xx}$$

$$\gamma_3 = -\frac{1}{3}\omega^2 b_{yyxx} + \frac{1}{4}\omega^2 d_{xyxy} + \omega(H'_{xyz} - L'_{xyz}) + \chi_{xx}$$

$$\gamma_4 = -\frac{1}{3}\omega^2(b_{xxyy} + b_{yyxx}) + \frac{1}{4}\omega^2(d_{xxyy} + d_{xyxy}) - \frac{1}{2}\omega(H'_{xyz} - H'_{xzy}) - \chi_{xx}$$

$$\gamma_5 = -\frac{1}{3}\omega^2 b_{xxyy} + \frac{1}{4}\omega^2 d_{xyxy} + \omega H'_{xyz} + \chi_{xx}$$

$$\gamma_6 = -\frac{2}{3}\omega^2 b_{xxyy} + \frac{1}{4}\omega^2(d_{xxyy} + d_{xyxy}) - \omega H'_{xyz} - \chi_{xx}$$

$$\Delta n = \frac{1}{2}\varepsilon_0^{-1}c^{-2}n_O\left[\frac{1}{3}\omega^2(b_{xxyy} - b_{yyxx}) + \omega(H'_{xyz} + H'_{xzy} - 2L'_{xyz})\right]. \quad (5.90)$$

Origin independence of this expression follows from that of $\gamma_{ij}^s$, Section 5.1.

The solution of the wave equation for propagation along a body diagonal, say $\sigma = (1, 1, 1)/\sqrt{3}$, can be shown to yield identical roots for orthogonal waves for all three symmetries in (5.84). Thus these cubics exhibit no linear birefringence along a body diagonal, whereas they do along a face diagonal. For $\bar{4}3m$ and $m3m$ crystals the solutions of the wave equation for propagation along a face diagonal are readily found to be

$$n_1^2 = n_O^2[1 - \varepsilon_0^{-1}c^{-2}\gamma_5]^{-1} \quad (5.91)$$

with a linear eigenpolarization perpendicular to the face containing the propagation direction, and

$$n_2^2 = n_O^2\left[1 + \frac{1}{2}\varepsilon_0^{-1}c^{-2}(\gamma_1 + \gamma_5 - \gamma_6)\right] \quad (5.92)$$

with its eigenpolarization parallel to the other diagonal in the same face. There is no additional wave in this case. From (5.91) and (5.92), and to first order in the polarizabilities, the birefringence is

$$n_1 - n_2 = \frac{1}{4}\varepsilon_0^{-1}c^{-2}n_O\omega^2 \left[\frac{1}{3}(b_{xxxx} - 3b_{xxyy}) - \frac{1}{4}(d_{xxxx} - 2d_{xyxy} - d_{xxyy})\right].$$

$$(5.93)$$

In their measurements on a single crystal of silicon (symmetry $m3m$) Pastrnak and Vedam [40] found that there was no linear birefringence along a cube edge or a body diagonal, whereas along a face diagonal they obtained the value $5.04 \times 10^{-6}$ at a wavelength of 1150 nm. The above theoretical conclusions for a cubic crystal with symmetry $m3m$ accord with their observations . The measured value of the linear birefringence for silicon was used to calculate the order of magnitude for the refractive index of the additional wave given by (5.88). With respect to orders of magnitude, it is instructive to compare these for the measured linear and circular birefringences of different multipole order:

electric dipole: $10^{-1}$–$10^{-3}$ [8]
electric quadrupole–magnetic dipole: $10^{-4}$ [8]
electric octopole–magnetic quadrupole: $10^{-6}$ [40].

These values lend support to the hierarchy (1.122) of multipole contributions to an electromagnetic effect, such as birefringence.

In addition to the theories of Lorentz birefringence mentioned above [35–37], two further accounts exist in the literature, in which the effect is explained in terms of electric quadrupole transitions [41, 42]. However, the birefringence expressions in (5.90) and (5.93), together with the quantum-mechanical expressions in (2.121), (2.123), (2.126), and (2.128) for the relevant polarizabilities, show that electric octopole, magnetic dipole, and magnetic quadrupole transitions are also involved. Without a consistent multipole treatment the birefringence expressions would be origin dependent.

## 5.8 Intrinsic Faraday rotation in magnetic cubics

Also of electric octopole–magnetic quadrupole order in the wave equation (5.4) is the time-odd tensor $\gamma_{ij}^a$ in (5.11). We now investigate whether this term is responsible for a physical effect when a wave propagates in a magnetic crystal [43].

Depending on the crystal symmetry and the propagation direction, all six material properties in (5.6)–(5.11) could contribute in principle to the wave equation, thereby greatly complicating its solution. To isolate, as far as possible, the effect of $\gamma_{ij}^a$, we consider magnetic cubics. We note immediately from Birss' tables [2,3] that the time-odd tensors of which $\gamma_{ij}^a$ consists, namely evenrank polar and odd-rank axial, vanish identically for 3 of the 11 magnetic cubics. Of the remaining eight classes the independent components of $\gamma_{ij}^a$ have been determined from (5.11) and tables, and these are listed in Table 5.2. The tensor $d'_{ijkl}$ possesses the intrinsic symmetry (2.124), while $L_{ijk} = L_{jik}$ (see (2.127)): consequently $\gamma_{ij}^a$ vanishes for three of the classes, as indicated in Table 5.2.

It also follows from Birss' tables [2,3] that both $\beta_{ij}^s$ in (5.8) and $\beta_{ij}^a$ in (5.9) vanish for the classes $m3$, $\bar{4}3m$, and $m3m$ in Table 5.2. Thus it is in these classes that we seek an effect due to $\gamma_{ij}^a$. As Table 5.1 shows, the time-even tensor $\gamma_{ij}^s$ (also of electric octopole–magnetic quadrupole order) exists for the non-magnetic

**Table 5.2** *Components of the electric octopole–magnetic quadrupole tensor $\gamma_{ij}^a$ in the wave equation for magnetic cubic classes.*

| | | | Class | | | | | |
|---|---|---|---|---|---|---|---|---|
| | 23 | $m3$ | 432 | $\bar{4}32$ | $\bar{4}3m$ | $\bar{4}3m$ | $m3m$ | $m3m$ |
| $\gamma_{ij}^a, i \neq j$ | $\sigma_i\sigma_j\gamma_7$ | $\sigma_i\sigma_j\gamma_7$ | 0 | $\sigma_i\sigma_j\gamma_8$ | 0 | $\sigma_i\sigma_j\gamma_8$ | 0 | $\sigma_i\sigma_j\gamma_8$ |

$$\gamma_7 = -\frac{1}{3}\omega^2(b'_{xxyy} - b'_{yyxx}) + \frac{1}{4}\omega^2 d'_{xxyy} - \frac{1}{2}\omega(H_{xyz} + H_{xzy} - 2L_{xyz})$$

$$\gamma_8 = -\frac{2}{3}\omega^2 b'_{xxyy} + \frac{1}{4}\omega^2 d'_{xxyy} - \omega(H_{xyz} - L_{xyz})$$

subgroups of these classes, namely $m3$, $\bar{4}3m$, and $m3m$. Accordingly, both $\gamma_{ij}^a$ and $\gamma_{ij}^s$ enter the wave equation. However, as shown in Section 5.7, $\gamma_{ij}^s$ produces no birefringence along a body diagonal in these three subgroups or along a cube edge in $\bar{4}3m$ and $m3m$. In respect of propagation along a cube edge, we note from Table 5.2 that $\gamma_{ij}^a$ is zero in such a case. Thus it is along a body diagonal in one of the cubic classes

$$m3, \quad \bar{4}3\underline{m}, \quad m3\underline{m} \tag{5.94}$$

that a possible effect due only to $\gamma_{ij}^a$ is to be sought.

From (5.12), (5.13), and Tables 5.1 and 5.2 we find that the wave equation for $\sigma = (1, 1, 1)/\sqrt{3}$ is

$$\begin{vmatrix} n^2(-\frac{2}{3} + u) + n_O^2 & n^2(\frac{1}{3} + v^*) & n^2(\frac{1}{3} + v) \\ n^2(\frac{1}{3} + v) & n^2(-\frac{2}{3} + u) + n_O^2 & n^2(\frac{1}{3} + v^*) \\ n^2(\frac{1}{3} + v^*) & n^2(\frac{1}{3} + v) & n^2(-\frac{2}{3} + u) + n_O^2 \end{vmatrix} = 0. \tag{5.95}$$

In this

$$u = \frac{1}{3}\varepsilon_0^{-1}c^{-2}\gamma, \qquad v = \frac{1}{3}\varepsilon_0^{-1}c^{-2}(\gamma' + i\gamma''), \tag{5.96}$$

where for

$$m3: \quad \gamma = \gamma_1 + \gamma_2 + \gamma_3, \quad \gamma' = \gamma_5, \quad \gamma'' = \gamma_7 \tag{5.97}$$

$$\bar{4}3\underline{m}, \ m3\underline{m}: \quad \gamma = \gamma_1 + 2\gamma_4, \quad \gamma' = \gamma_4, \quad \gamma'' = \gamma_8. \tag{5.98}$$

The expressions for $\gamma_i$ appear in Tables 5.1 and 5.2.

Equation (5.95) has three solutions, one of which, with its eigenvector, is

$$n^2 = -\frac{n_O^2}{u + v + v^*}, \qquad \mathbf{E}_0 = (1, 1, 1)/\sqrt{3}. \tag{5.99}$$

Thus there is an additional wave (Section 5.7). The other two solutions, to first order in the polarizability densities, are

$$n_1^2 = n_O^2\left[1 + u - \frac{1}{2}(v + v^*) + i\frac{1}{2}\sqrt{3}(v - v^*)\right], \qquad \mathbf{E}_0^{(1)} = (1 + i\sqrt{3},\ 1 - i\sqrt{3},\ -2)/\sqrt{12}$$

$$\text{(5.100)}$$

$$n_2^2 = n_O^2\left[1 + u - \frac{1}{2}(v + v^*) - i\frac{1}{2}\sqrt{3}(v - v^*)\right], \qquad \mathbf{E}_0^{(2)} = (1 - i\sqrt{3},\ 1 + i\sqrt{3},\ -2)/\sqrt{12}.$$

$$\text{(5.101)}$$

To interpret the polarization types of the eigenvectors in (5.100) and (5.101), we transform them by means of (3.4) from crystallographic axes, where $\boldsymbol{\sigma} = (1, 1, 1)/\sqrt{3}$, to a new right-hand system in which $\boldsymbol{\sigma} = (0, 0, 1)$. Then one finds the right and left circularly polarized forms $\mathbf{E}_0^{(1)} = (1, -i, 0)/\sqrt{2}$ and $\mathbf{E}_0^{(2)} = (1, i, 0)/\sqrt{2}$, respectively. Thus $n_1 = n_R$ and $n_2 = n_L$, so that from (5.100), (5.101), and (5.96) there is a circular birefringence given by

$$n_L - n_R = i\frac{\sqrt{3}}{2}\,n_O(v - v^*) = -\frac{1}{\sqrt{3}}\,\varepsilon_0^{-1}c^{-2}n_O\gamma''. \qquad \text{(5.102)}$$

There are a number of points to be noted concerning this birefringence:

(i) It is time odd, as follows from (5.97), (5.98), and Tables 5.2 and 3.2. Thus the associated rotation of the plane of a linearly polarized wave is a Faraday-type, which reverses its sense under path reversal.

(ii) As no magnetic cubic crystal possesses a net magnetization [2, 3], it is surprising that an intrinsic rotation is expected in the classes in (5.94). The explanation of this warrants further consideration.

(iii) The circular birefringence in (5.102) should have the same order of magnitude $(10^{-6})$ as the linear birefringence described in Section 5.7, since their multipole order and dimensions are the same.

(iv) Despite its small magnitude, the circular birefringence in (5.102) offers the prospect of a measurable rotation, which can be enhanced by working at a small wavelength, see (5.32). For $\lambda = 600\,\text{nm}$, $l = 1\,\text{cm}$, and $|n_L - n_R| = 10^{-6}$, one obtains a rotation of about $3°$. Thus an experimental test of multipole theory to the relatively high order of electric octopole–magnetic quadrupole would be to measure the rotation along the body diagonal in a magnetic cubic with one of the symmetries in (5.94), rather than to measure a Lorentz birefringence in a non-magnetic cubic.

## 5.9    The Kerr effect in an ideal gas

The transmission effects described in Sections 5.2–5.8 occur naturally. Various types of induced effects are known to be exhibited by crystals and fluids, and here we consider one of these, namely the Kerr effect in an ideal gas. This is a linear birefringence induced by a uniform electric field that acts perpendicular to the light path through the gas. For weak applied fields, the effect is quadratic in the field. (Linear birefringence may also be induced in a crystal, depending on its symmetry, by an electric field applied along or transverse to the light path, and has a field dependence which may be linear or quadratic.)

An electric field in a fluid defines a preferred direction parallel to the field, rendering the fluid uniaxial. As linear birefringence in a uniaxial crystal is due to anisotropy in its electric polarizability (Section 5.2), which is an electric dipole tensor (Table 3.2), we consider a theory to this multipole order. Thus we write

$$\mathbf{D} = \varepsilon_0(\boldsymbol{\mathcal{E}} + \mathbf{E}) + \mathbf{P} \tag{5.103}$$

$$\mathbf{H} = \mu_0^{-1}\boldsymbol{\mathcal{B}}. \tag{5.104}$$

Here we adopt the notation that $\boldsymbol{\mathcal{E}}$ and $\boldsymbol{\mathcal{B}}$ are the electric and magnetic fields of the light wave, and $\mathbf{E}$ is the static (or low-frequency) applied field. In (5.103)

$$\mathbf{P} = N\bar{\mathbf{p}}, \tag{5.105}$$

where $N$ is the number density of molecules and $\bar{\mathbf{p}}$ is the orientational average of the electric dipole moment of a molecule. To evaluate this average we make the following assumptions: the light wave fields are sufficiently weak to have no orienting effect on a molecule; the orientational variable $\tau$ is continuous; a Boltzmann distribution applies; and molecules are independent. Then the orientational average of a quantity $X$ in the configuration described by $\tau$ in the presence of the field $\mathbf{E}$ is

$$\overline{X(\tau,\mathbf{E})} = \frac{\int X(\tau,\mathbf{E})\exp\{-W(\tau,\mathbf{E})/kT\}\,d\tau}{\int \exp\{-W(\tau,\mathbf{E})/kT\}\,d\tau}, \tag{5.106}$$

where $W(\tau,\mathbf{E})$ is the energy of a molecule.

In the theory that follows, and on the basis of the experimental dependence on $\mathbf{E}$ of the Kerr birefringence in a fluid, we work to quadratic order in the applied field $\mathbf{E}$. (Alternatively, it can be readily shown by the pictorial symmetry method of Section 3.8 that only an even dependence of linear birefringence on field is allowed in a fluid.) From (5.106) and the expansion

$$X(\tau,\mathbf{E}) = X(\tau,0) + (\partial X/\partial E_i)_0 E_i + \frac{1}{2}(\partial^2 X/\partial E_i\partial E_j)_0 E_i E_j + \cdots, \tag{5.107}$$

and a similar expansion for $W(\tau,\mathbf{E})$, it can be shown that

$$\begin{aligned}
\overline{X(\tau,\mathbf{E})} = \langle X\rangle &+ E_i\Big[\langle\partial X/\partial E_i\rangle - (kT)^{-1}\{\langle X\partial W/\partial E_i\rangle - \langle X\rangle\langle\partial W/\partial E_i\rangle\}\Big]\\
&+ \frac{1}{2}E_i E_j\Big[\langle\partial^2 X/\partial E_i\partial E_j\rangle - (kT)^{-1}\{2\langle(\partial X/\partial E_i)(\partial W/\partial E_j)\rangle\\
&+ \langle X\partial^2 W/\partial E_i\partial E_j\rangle - 2\langle\partial X/\partial E_i\rangle\langle\partial W/\partial E_j\rangle - \langle X\rangle\langle\partial^2 W/\partial E_i\partial E_j\rangle\}\\
&+ (kT)^{-2}\{\langle X(\partial W/\partial E_i)(\partial W/\partial E_j)\rangle - 2\langle X\partial W/\partial E_i\rangle\langle\partial W/\partial E_j\rangle\\
&- \langle X\rangle\langle(\partial W/\partial E_i)(\partial W/\partial E_j)\rangle + 2\langle X\rangle\langle\partial W/\partial E_i\rangle\langle\partial W/\partial E_j\rangle\}\Big] + \cdots.
\end{aligned} \tag{5.108}$$

In this $\langle \cdots \rangle$ denotes an orientational average in the absence of the applied field: that is, (5.106) with $\mathbf{E} = 0$.

The total electric dipole moment in the presence of a weak light-wave field $\mathcal{E}$ and relatively strong applied field $\mathbf{E}$ is

$$p_i = p_i^{(0)} + \alpha_{ij}\mathcal{E}_j + \alpha_{ij}^{(0)}E_j + \frac{1}{2}\beta_{ijk}\mathcal{E}_j E_k + \frac{1}{2}\beta_{ijk}^{(0)}E_j E_k + \frac{1}{6}\gamma_{ijkl}\mathcal{E}_j E_k E_l + \cdots . \quad (5.109)$$

Here the superscript (0) on a polarizability indicates the static value. The molecular tensors $\beta_{ijk}$, $\beta_{ijk}^{(0)}$, $\gamma_{ijkl}$, ... are hyperpolarizabilities, first introduced by Buckingham and Pople [44] in their theory of the Kerr effect. From (1.35), for a neutral molecule in a uniform field,

$$W = -\int_0^{\mathbf{E}} p_i \, dE_i. \quad (5.110)$$

Using (5.109) in this, but omitting terms in $\mathcal{E}$ since according to our assumption $\mathcal{E}$ does not orient a molecule, we obtain

$$W = W^{(0)} - p_i^{(0)} E_i - \frac{1}{2}\alpha_{ij}^{(0)} E_i E_j - \frac{1}{6}\beta_{ijk}^{(0)} E_i E_j E_k - \cdots . \quad (5.111)$$

To evaluate $\bar{p}_i$ in (5.105) we require for use in (5.108) the expressions (5.112)–(5.119) below, which are obtained from (5.109) and (5.111). As these expressions enter the various field-free (or isotropic) averages in (5.108), terms in $\mathbf{E}$ may be omitted in them. Thus

$$\partial p_i / \partial E_j = \alpha_{ij}^{(0)} + \frac{1}{2}\beta_{ikj}\mathcal{E}_k \quad (5.112)$$

$$\partial^2 p_i / \partial E_j \partial E_k = \beta_{ijk}^{(0)} + \frac{1}{3}\gamma_{iljk}\mathcal{E}_l \quad (5.113)$$

$$\partial W / \partial E_j = -p_j^{(0)} \quad (5.114)$$

$$\partial^2 W / \partial E_j \partial E_k = -\alpha_{jk}^{(0)} \quad (5.115)$$

$$p_i \partial W / \partial E_j = -(p_i^{(0)} + \alpha_{ik}\mathcal{E}_k)p_j^{(0)} \quad (5.116)$$

$$(\partial p_i / \partial E_j)(\partial W / \partial E_k) = -\left(\alpha_{ij}^{(0)} + \frac{1}{2}\beta_{ilj}\mathcal{E}_l\right)p_k^{(0)} \quad (5.117)$$

$$p_i \partial^2 W / \partial E_j \partial E_k = -(p_i^{(0)} + \alpha_{il}\mathcal{E}_l)\alpha_{jk}^{(0)} \quad (5.118)$$

$$p_i(\partial W / \partial E_j)(\partial W / \partial E_k) = (p_i^{(0)} + \alpha_{il}\mathcal{E}_l)p_j^{(0)}p_k^{(0)}. \quad (5.119)$$

In obtaining these use was made of the intrinsic symmetries

$$\alpha_{ij}^{(0)} = \alpha_{ji}^{(0)}, \qquad \beta_{ijk}^{(0)} = \beta_{ikj}^{(0)}, \qquad \gamma_{ijkl} = \gamma_{ijlk}. \quad (5.120)$$

That for $\alpha_{ij}^{(0)}$ is given in (2.22), while the last two follow from (5.109).

From (5.112)–(5.119) we note that (5.108) contains the isotropic averages $\langle p_i^{(0)} \rangle$, $\langle \beta_{ikj} \rangle$, $\langle \alpha_{ik} p_j^{(0)} \rangle$, $\langle p_i^{(0)} p_j^{(0)} p_k^{(0)} \rangle$. Each of these is zero: for $p_i^{(0)}$ this is obvious, while the others vanish because they are symmetric in at least two of their subscripts (see (2.115), (2.145), and the discussion below (5.55)). Thus we obtain

$$\bar{p}_i = \langle \alpha_{ij} \rangle \mathcal{E}_j + E_k \left[ \langle \alpha_{ik}^{(0)} \rangle + (kT)^{-1} \langle p_i^{(0)} p_k^{(0)} \rangle \right]$$

$$+ \frac{1}{2} \mathcal{E}_j E_k E_l \left[ \frac{1}{3} \langle \gamma_{ijkl} \rangle + (kT)^{-1} \left\{ \langle \alpha_{ij} \alpha_{kl}^{(0)} \rangle - \langle \alpha_{ij} \rangle \langle \alpha_{kl}^{(0)} \rangle + \langle p_l^{(0)} \beta_{ijk} \rangle \right\} \right.$$

$$\left. + (kT)^{-2} \left\{ \langle p_k^{(0)} p_l^{(0)} \alpha_{ij} \rangle - \langle p_k^{(0)} p_l^{(0)} \rangle \langle \alpha_{ij} \rangle \right\} \right]. \tag{5.121}$$

As the wave equation is obtained from $\nabla \times \mathbf{H} = \dot{\mathbf{D}}$, only time-dependent quantities need to be retained in $\mathbf{D}$ in (5.103), which from (5.105) and (5.121) is then written as

$$D_i = (\varepsilon_0 \delta_{ij} + N d_{ij}) \mathcal{E}_j, \tag{5.122}$$

where

$$d_{ij} = \langle \alpha_{ij} \rangle + \frac{1}{2} E_k E_l \left[ \frac{1}{3} \langle \gamma_{ijkl} \rangle + (kT)^{-1} \left\{ \langle \alpha_{ij} \alpha_{kl}^{(0)} \rangle - \langle \alpha_{ij} \rangle \langle \alpha_{kl}^{(0)} \rangle + \langle p_l^{(0)} \beta_{ijk} \rangle \right\} \right.$$

$$\left. + (kT)^{-2} \left\{ \langle p_k^{(0)} p_l^{(0)} \alpha_{ij} \rangle - \langle p_k^{(0)} p_l^{(0)} \rangle \langle \alpha_{ij} \rangle \right\} \right]. \tag{5.123}$$

From (5.1)–(5.3), (5.104), and (5.122) the secular equation for propagation along the laboratory $z$-axis can be shown to be

$$\begin{vmatrix} -n^2 + 1 + \varepsilon_0^{-1} N d_{xx} & \varepsilon_0^{-1} N d_{xy} & \varepsilon_0^{-1} N d_{xz} \\ \varepsilon_0^{-1} N d_{yx} & -n^2 + 1 + \varepsilon_0^{-1} N d_{yy} & \varepsilon_0^{-1} N d_{yz} \\ \varepsilon_0^{-1} N d_{zx} & \varepsilon_0^{-1} N d_{zy} & 1 + \varepsilon_0^{-1} N d_{zz} \end{vmatrix} = 0. \tag{5.124}$$

The procedure for evaluating the isotropic averages in $d_{ij}$ of (5.123) is detailed by Barron [14], to whose account the interested reader is referred. For the transverse field $\mathbf{E} = (E, 0, 0)$ the required results are

$$d_{ij} = 0 \quad (i \neq j) \tag{5.125}$$

$$d_{xx} = \frac{1}{3} \alpha_{ii} + \frac{1}{90} E^2 \left[ \gamma_{iijj} + 2\gamma_{ijij} + (kT)^{-1} \left\{ 6\alpha_{ij} \alpha_{ij}^{(0)} - 2\alpha_{ii} \alpha_{jj}^{(0)} \right. \right.$$

$$\left. \left. + 6 p_i^{(0)} \beta_{ijj} + 3 p_i^{(0)} \beta_{jji} \right\} + (kT)^{-2} \left\{ 6 p_i^{(0)} p_j^{(0)} \alpha_{ij} - 2 p_i^{(0)} p_i^{(0)} \alpha_{jj} \right\} \right] \tag{5.126}$$

$$d_{yy} = \frac{1}{3} \alpha_{ii} + \frac{1}{90} E^2 \left[ 2\gamma_{iijj} - \gamma_{ijij} + (kT)^{-1} \left\{ \alpha_{ii} \alpha_{jj}^{(0)} - 3\alpha_{ij} \alpha_{ij}^{(0)} \right. \right.$$

$$\left. \left. + 6 p_i^{(0)} \beta_{jji} - 3 p_i^{(0)} \beta_{ijj} \right\} + (kT)^{-2} \left\{ p_i^{(0)} p_i^{(0)} \alpha_{jj} - 3 p_i^{(0)} p_j^{(0)} \alpha_{ij} \right\} \right], \tag{5.127}$$

where, as a consequence of the averaging, the Cartesian components of the dipole moment and polarizability tensors are now with respect to molecular axes.

From (5.124) and (5.125), we have the solutions

$$n_x^2 = 1 + \varepsilon_0^{-1} N d_{xx}, \qquad \mathcal{E}_x^{(0)} = (1, 0, 0) \tag{5.128}$$

$$n_y^2 = 1 + \varepsilon_0^{-1} N d_{yy}, \qquad \mathcal{E}_y^{(0)} = (0, 1, 0). \tag{5.129}$$

Thus there is a linear birefringence given, to first order in the polarizabilities, by

$$
\begin{aligned}
n_x - n_y &= (2\varepsilon_0)^{-1} N (d_{xx} - d_{yy}) \\
&= (180\varepsilon_0)^{-1} N E^2 \left[ 3\gamma_{ijij} - \gamma_{iijj} + 3(kT)^{-1} \left\{ 3\alpha_{ij}\alpha_{ij}^{(0)} - \alpha_{ii}\alpha_{jj}^{(0)} \right. \right. \\
&\quad \left. \left. + p_i^{(0)}(3\beta_{ijj} - \beta_{jji}) \right\} + 3(kT)^{-2} \left\{ 3p_i^{(0)}p_j^{(0)}\alpha_{ij} - p_i^{(0)}p_i^{(0)}\alpha_{jj} \right\} \right].
\end{aligned}
\tag{5.130}
$$

This is the Kerr birefringence, quadratic in the applied field $\mathbf{E}$, that is induced in an ideal gas.

In their paper on the Kerr effect, Buckingham and Pople [44] assumed full permutation symmetry of subscripts for $\beta_{ijk}$ and $\gamma_{ijkl}$ in (5.109). This assumption is not made in a later paper by Buckingham [45], in which the expression for the Kerr effect and ours in (5.130) are in agreement.

## 5.10    Forward scattering theory of the Kerr effect

The approach adopted so far in this chapter to describe the propagation of a harmonic plane electromagnetic wave in a homogeneous medium uses Maxwell's macroscopic equations. Here we present an alternative, microscopic calculation based on forward scattering by molecules in the path of the incident wave.

Consider a thin lamina of homogeneous dielectric in the form of a square of side $2L$, centered on the origin and lying in the $xy$ plane. The lamina has thickness $\Delta z$. A harmonic plane wave propagating along the $+z$-axis in vacuum is incident on the lamina. Due to this incident wave, the charges in each molecule of the lamina are set oscillating with the angular frequency $\omega$ of the wave. A molecule at the point $S(x, y, 0)$ radiates to a distant point $P(0, 0, z)$ on the axis of the lamina. According to (1.85), the electric field at $P$ of the wave from $S$ is, to electric dipole order,

$$\mathcal{E}_i^{(r)}(\mathbf{R}, t) = \frac{1}{4\pi\varepsilon_0 c^2 R^3} (R_i R_j - R^2 \delta_{ij}) \ddot{p}_j(t'). \tag{5.131}$$

Here $\mathbf{R}$ is the displacement from $S$ to $P$, given by

$$\mathbf{R} = (-x, -y, z), \tag{5.132}$$

and $t' = t - R/c$ is the retarded time at $S$. Since the incident wave on the molecule is harmonic, of the form in (5.2),

$$\ddot{\mathbf{p}}(t') = -\omega^2 \mathbf{p}(t') = -\omega^2 \mathbf{p}_o e^{-i\omega(t - R/c)}. \tag{5.133}$$

From (5.131)–(5.133) the field at $(0, 0, z)$ due to a molecule at $(x, y, 0)$ has components

$$\mathcal{E}_x^{(r)}(\mathbf{R},t) =$$
$$\frac{\omega^2}{4\pi\varepsilon_0 c^2 R^3}\left[x(-xp_{ox} - yp_{oy} + zp_{oz}) + (x^2 + y^2 + z^2)p_{ox}\right]e^{-i\omega(t-R/c)}$$

$$(5.134)$$

$$\mathcal{E}_y^{(r)}(\mathbf{R},t) =$$
$$\frac{\omega^2}{4\pi\varepsilon_0 c^2 R^3}\left[y(-xp_{ox} - yp_{oy} + zp_{oz}) + (x^2 + y^2 + z^2)p_{oy}\right]e^{-i\omega(t-R/c)}.$$

$$(5.135)$$

Now

$$R = (x^2 + y^2 + z^2)^{1/2} = z\left[1 + \frac{1}{2}(x^2 + y^2)z^{-2} + \cdots\right].\qquad(5.136)$$

From (5.134)–(5.136) it can be seen that

$$\mathcal{E}_x^{(r)} = \frac{\omega^2}{4\pi\varepsilon_0 c^2 z}\left[p_{ox} + xz^{-1}p_{oz} - xyz^{-2}p_{oy} + y^2z^{-2}p_{ox}\right.$$
$$\left. - \frac{3}{2}(x^2 + y^2)z^{-2}p_{ox} + \cdots\right]e^{-i\omega(t-z/c)}e^{i\omega(x^2+y^2)/2zc}\quad(5.137)$$

$$\mathcal{E}_y^{(r)} = \frac{\omega^2}{4\pi\varepsilon_0 c^2 z}\left[p_{oy} + yz^{-1}p_{oz} - xyz^{-2}p_{ox} + x^2z^{-2}p_{oy}.\right.$$
$$\left. - \frac{3}{2}(x^2 + y^2)z^{-2}p_{oy} + \cdots\right]e^{-i\omega(t-z/c)}e^{i\omega(x^2+y^2)/2zc}.\quad(5.138)$$

The square brackets in (5.137) and (5.138) contain expansions in powers of $z^{-1}$, which have been written explicitly to terms in $z^{-2}$.

We now take the medium to be an ideal gas to which a uniform electric field $\mathbf{E} = (E, 0, 0)$ is applied. To determine the total field at $P$ that is radiated by all the molecules in the lamina, we require the orientational average of $\mathbf{p}_o$ for a single molecule in the presence of the field. Denoted $\bar{\mathbf{p}}_o$, this can be related, for the same assumptions as those made in Section 5.9, to the field-free orientational average. This result is available in (5.121), of which we consider the amplitude, and from which we omit the static term as irrelevant to a radiated field. Thus

$$\bar{p}_{oi} = d_{ij}\mathcal{E}_{oj},\qquad(5.139)$$

where $\mathcal{E}_o$ is the amplitude of the incident wave and the $d_{ij}$ are expressed in terms of field-free averages by (5.123), and in terms of molecular polarizabilities by (5.125)–(5.127).

From (5.137) and (5.138) the field radiated to $P$ by all the gas molecules in the lamina at $z = 0$ has components

$$\Delta \mathcal{E}_x^{(r)} =$$

$$\frac{N\omega^2}{4\pi\varepsilon_0 c^2 z} \bar{P}_{ox} e^{-i\omega(t-z/c)} \Delta z \iint_{-L}^{L} \left[ 1 - \frac{3x^2 + y^2}{2z^2} + \cdots \right] e^{i\omega(x^2+y^2)/2zc} \, dx \, dy \quad (5.140)$$

$$\Delta \mathcal{E}_y^{(r)} =$$

$$\frac{N\omega^2}{4\pi\varepsilon_0 c^2 z} \bar{P}_{oy} e^{-i\omega(t-z/c)} \Delta z \iint_{-L}^{L} \left[ 1 - \frac{x^2 + 3y^2}{2z^2} + \cdots \right] e^{i\omega(x^2+y^2)/2zc} \, dx \, dy, \quad (5.141)$$

where $N$ is the number density of molecules. In (5.140) and (5.141) we have omitted terms which are odd in $x$ and/or $y$ because these integrate to zero.

The components (5.140) and (5.141) are, in general, complicated functions of $z$, $L$, and the wavelength $\lambda = 2\pi c/\omega$. We are interested in the limiting values for $z$ and $L$ much greater than $\lambda$. We also suppose $z \approx L$. Then in (5.140) and (5.141) we use the asymptotic values

$$\int_{-a}^{a} e^{iu^2} \, du \rightarrow (1+i)\sqrt{\frac{\pi}{2}} \quad (5.142)$$

and

$$\int_{-a}^{a} u^2 e^{iu^2} \, du \rightarrow -iae^{ia^2} \quad (5.143)$$

for

$$a = \sqrt{\pi L^2/\lambda z} \approx \sqrt{\pi z/\lambda} \gg 1. \quad (5.144)$$

The limit (5.142) is a standard result in the theory of Fresnel integrals, while (5.143) is obtained from an integration-by-parts and (5.142). From (5.140)–(5.143), (5.139), and (5.125) we find

$$\Delta \mathcal{E}_x^{(r)} = i \frac{N\omega}{2\varepsilon_0 c} d_{xx} \mathcal{E}_{ox} e^{-i\omega(t-z/c)} \Delta z (1 + O(\sqrt{\lambda/z})) \quad (5.145)$$

$$\Delta \mathcal{E}_y^{(r)} = i \frac{N\omega}{2\varepsilon_0 c} d_{yy} \mathcal{E}_{oy} e^{-i\omega(t-z/c)} \Delta z (1 + O(\sqrt{\lambda/z})). \quad (5.146)$$

Thus we have shown that if $L$ and $z$ are comparable and $z \gg \lambda$, we can make a leading-term approximation to (5.140) and (5.141) by retaining only the first term in square brackets in these equations.

Now consider the field of the wave which emerges from the lamina. This is the combined field of the incident and scattered waves, and for $z \gg \lambda$ it has $x$ component

$$\left[1 + i\left(\frac{N\omega}{2\varepsilon_0 c}d_{xx}\right)\Delta z\right]\mathcal{E}_{ox}\exp[-i\omega(t - z/c)]$$

$$\approx \mathcal{E}_{ox}\exp\left[-i\omega\left(t - \frac{N}{2\varepsilon_0 c}d_{xx}\Delta z - \frac{z}{c}\right)\right], \quad (5.147)$$

and similarly for the $y$ component. This combined wave has the same amplitude as the incident wave, but its phase is modified in such a way that the wave is retarded: for the wavefront associated with the $x$ component of the field, it takes an additional time $(N/2\varepsilon_0 c)d_{xx}\Delta z$ to propagate from the lamina at $z = 0$ to the field point at $z$. This retardation proportional to $\Delta z$ is attributed to a slowing down of the wave in the lamina. Instead of traversing the lamina in a time $\Delta t = \Delta z/c$, it takes

$$\Delta t = \left(1 + \frac{N}{2\varepsilon_0}d_{xx}\right)\frac{\Delta z}{c} \quad (5.148)$$

to do so. In terms of the refractive index $n_x$ of the gas for propagation of this wavefront,

$$\Delta t = \frac{\Delta z}{c/n_x}. \quad (5.149)$$

It follows from (5.148) and (5.149) that

$$n_x = 1 + \frac{N}{2\varepsilon_0}d_{xx}. \quad (5.150)$$

Similarly, by considering the $y$ component of the combined field, we have

$$n_y = 1 + \frac{N}{2\varepsilon_0}d_{yy}. \quad (5.151)$$

Equations (5.150) and (5.151) show that a linear birefringence exists given by

$$n_x - n_y = \frac{N}{2\varepsilon_0}(d_{xx} - d_{yy}), \quad (5.152)$$

which from (5.126) and (5.127) agrees with (5.130). Note that in the absence of the static field **E**, we have from (5.126) and (5.127) that $d_{xx} = d_{yy} = \alpha_{ii}/3 = \alpha$, where $\alpha$ is the mean polarizability. Then $n_x = n_y = n = 1 + N\alpha/2\varepsilon_0$. This is a well-known result which follows from the Lorenz–Lorentz equation [46].

## 5.11 Birefringence induced in a gas by an electric field gradient: forward scattering theory

As shown in (1.31), a non-uniform electrostatic field exerts a torque on an electric quadrupole, producing partial alignment of quadrupolar molecules in a fluid and hence anisotropy. This gives rise to a linear birefringence known as electric-field-gradient-induced birefringence. For a gas this effect, when interpreted by means of a suitable theory, provides a direct method for determining the electric quadrupole moment of a molecule [47, 48]

The first such theory, due to Buckingham [47], applied only to non-dipolar molecules. If the molecule possesses both an electric dipole and a quadrupole, as does CO, for example, the quadrupole moment depends on the coordinate origin used to specify it (Section 1.2). This raises the question: to which origin in a dipolar molecule is its measured quadrupole moment referred? To address this, Buckingham and Longuet-Higgins [49] produced a new theory of the effect based on forward scattering of radiation by dipolar gas molecules in a non-uniform applied field. This theory was applied to electric quadrupole–magnetic dipole order and in it the traceless quadrupole moment of (1.14) was used. The molecular expression obtained was shown to be independent of origin as required.

A subsequent theory by Imrie and Raab [50] of this induced birefringence in a dilute gas used the wave theory, based on Maxwell's equations, that is described in Section 5.1. Working with the primitive (or traced) quadrupole moment (1.11), these authors derived a molecular expression for the effect which differs from that of Buckingham and Longuet-Higgins. To reconcile the two theories, the present authors have revisited both and shown that a careful treatment of wave theory yields the result obtained from scattering theory [51, 52]. In this section we present the forward scattering theory of electric-field-gradient-induced birefringence in a gas [51]. The wave theory [52] is described in the following section.

### 5.11.1  *Forward scattering by a molecule*

We begin with the electric field $\mathcal{E}^{(r)}$ radiated in free space by a molecule driven by an incident light wave. We suppose that the distance to the field point is much greater than the dimensions of the molecule, and work to electric quadrupole–magnetic dipole order. Then from (1.85)

$$\mathcal{E}_i^{(r)}(\mathbf{R}, t) = \frac{1}{4\pi\varepsilon_0 c^2 R} \left[ \frac{R_i R_j - R^2 \delta_{ij}}{R^2} \left\{ \ddot{p}_j(t') + \frac{R_k}{2cR} \dddot{q}_{jk}(t') \right\} + \varepsilon_{ijk} \frac{R_j}{cR} \ddot{m}_k(t') \right].$$

(5.153)

Here $p_i$, $q_{ij}$, and $m_i$ are given by (1.10), (1.11), and (1.50), and $\mathbf{R}$ is the displacement of the field point from the source point, at which $t' = t - R/c$ is the retarded time.

With the molecule at $(x, y, 0)$ and the field point $P$ at $(0, 0, z)$, then

$$\mathbf{R} = (-x, -y, z).$$

(5.154)

Consider an incident harmonic plane wave propagating in the $+z$ direction. It is important to take account of the finite width of the beam (see the evaluation of (5.202) and (5.203)) , and thus we express the incident electric field as

$$\mathcal{E} = u\mathcal{E}_0 e^{-i\omega(t - z/c)},$$

(5.155)

where $u = u(x, y)$ is a smoothly varying beam profile function and $\mathcal{E}_0$ is a constant. For a beam centered on the $z$-axis and with square cross-section of side $2L$, $u$ is an even function of $x$ and $y$, and

$$u(0,0) = 1 \tag{5.156}$$

$$u(x,y) = 0 \quad \text{for } |x| \text{ or } |y| = L. \tag{5.157}$$

The response in (5.153) of the electric dipole moment to the field (5.155) is

$$\ddot{p}_i(t') = -\omega^2 p_{oi} e^{-i\omega(t - R/c)}, \tag{5.158}$$

and similarly for $\ddot{q}_{ij}$ and $\ddot{m}_i$. (The "amplitudes" $p_{oi}$, $q_{oij}$, and $m_{oi}$ are functions of $x$ and $y$ through their dependence on $u$ and an applied electric field $\mathbf{E}$: see (5.167)–(5.169).) Thus, from (5.153) and (5.154) we have

$$\mathcal{E}_x^{(r)}(\mathbf{R}, t) = \frac{\omega^2}{4\pi\varepsilon_0 c^2 R} \left[ \frac{1}{R^2} \{x(-xp_{ox} - yp_{oy} + zp_{oz}) + (x^2 + y^2 + z^2)p_{ox}\} \right.$$
$$- \frac{i\omega}{2cR^3} \{x(x^2 q_{oxx} + y^2 q_{oyy} + z^2 q_{ozz} + 2xy q_{oxy} - 2yz q_{oyz} - 2zx q_{ozx})$$
$$\left. + (x^2 + y^2 + z^2)(-x q_{oxx} - y q_{oxy} + z q_{oxz})\} + \frac{1}{cR}(y m_{oz} + z m_{oy}) \right] e^{-i\omega(t - R/c)}, \tag{5.159}$$

and an analogous expression for $\mathcal{E}_y^{(r)}$, which we write down later in its final form (5.174). For $|x|, |y| \ll z$ use of (5.136) yields, to lowest order in $x/z$ and $y/z$,

$$\mathcal{E}_x^{(r)} = \frac{\omega^2}{4\pi\varepsilon_0 c^2 R} \left[ p_{ox} + \frac{x}{z} p_{oz} - \frac{i\omega}{2c} \left( q_{oxz} + \frac{x}{z} q_{ozz} - \frac{x}{z} q_{oxx} - \frac{y}{z} q_{oxy} \right) \right.$$
$$\left. + \frac{1}{c} \left( m_{oy} + \frac{y}{z} m_{oz} \right) \right] e^{-i\omega(t - z/c)} e^{i\omega(x^2 + y^2)/2zc}. \tag{5.160}$$

## 5.11.2  *Induced moments*

For a diamagnetic molecule the moments in the presence of the incident fields $\mathcal{E}$ and $\mathcal{B}$ are (see Section 2.7)

$$p_i = p_i^{(0)} + \alpha_{ij}\mathcal{E}_j + \frac{1}{2}a_{ijk}\nabla_k\mathcal{E}_j + \omega^{-1}G_{ij}'\dot{\mathcal{B}}_j \tag{5.161}$$

$$q_{ij} = q_{ij}^{(0)} + a_{kij}\mathcal{E}_k \tag{5.162}$$

$$m_i = -\omega^{-1}G_{ji}'\dot{\mathcal{E}}_j, \tag{5.163}$$

where $p_i^{(0)}$ and $q_{ij}^{(0)}$ are permanent moments. A diamagnetic molecule cannot possess time-antisymmetric properties (Section 3.6), and therefore the polarizability tensors in (5.161)–(5.163) are the only ones which occur to electric quadrupole–magnetic dipole order.

If a non-uniform electrostatic field $\mathbf{E}$ is applied to the molecule, the effect on the above polarizability tensors to first order in the field terms is given by

$$\alpha_{ij}(\mathbf{E}, \boldsymbol{\nabla}\mathbf{E}) = \alpha_{ij} + \beta_{ijk}E_k + \frac{1}{2}b_{ijkl}\nabla_l E_k \tag{5.164}$$

$$a_{ijk}(\mathbf{E}, \boldsymbol{\nabla}\mathbf{E}) = a_{ijk} + \mathfrak{b}_{iljk}E_l \tag{5.165}$$

$$G'_{ij}(\mathbf{E}, \boldsymbol{\nabla}\mathbf{E}) = G'_{ij} + J'_{ijk}E_k. \tag{5.166}$$

We remark that the manner in which (5.165) has been written is consistent with the usage in Refs. 49 and 51, but not with Ref. 50. From (5.161)–(5.166) the moments in the presence of the incident and electrostatic fields are, to electric quadrupole–magnetic dipole order,

$$p_i = p_i^{(0)} + \alpha_{ij}\mathcal{E}_j + \beta_{ijk}\mathcal{E}_j E_k + \frac{1}{2}b_{ijkl}\mathcal{E}_j \nabla_l E_k + \frac{1}{2}a_{ijk}\nabla_k\mathcal{E}_j$$

$$+ \frac{1}{2}\mathfrak{b}_{iljk}(\nabla_k\mathcal{E}_j)E_l + \omega^{-1}G'_{ij}\dot{\mathcal{B}}_j + \omega^{-1}\dot{\mathcal{B}}_j E_k \tag{5.167}$$

$$q_{ij} = q_{ij}^{(0)} + a_{kij}\mathcal{E}_k + \mathfrak{b}_{klij}\mathcal{E}_k E_l \tag{5.168}$$

$$m_i = -\omega^{-1}G'_{ji}\dot{\mathcal{E}}_j - \omega^{-1}J_{jik}\dot{\mathcal{E}}_j E_k. \tag{5.169}$$

In (5.167)–(5.169) we have omitted terms involving only the electrostatic field since these cannot contribute to the scattered field (5.153). Quantum expressions and symmetries for $\alpha_{ij}$, $a_{ijk}$, and $G'_{ij}$ have been obtained in Section 2.7; for the additional tensors $\beta_{ijk}$, $b_{ijkl}$, $\mathfrak{b}_{ijkl}$, and $J'_{ijk}$ they are given in Ref. 50.

Next it is necessary to average the moments (5.167)–(5.169) over all orientations of the molecule. The calculation, which is similar to that for the Kerr effect (Section 5.9), yields [50, 51]

$$\bar{p}_i = \langle\alpha_{ij}\rangle\mathcal{E}_j + \omega^{-1}\langle G'_{ij}\rangle\dot{\mathcal{B}}_j + E_l\left[\frac{1}{2}\langle\mathfrak{b}_{iljk}\rangle\nabla_k\mathcal{E}_j + \omega^{-1}\langle J'_{ijl}\rangle\dot{\mathcal{B}}_j\right.$$

$$+ \frac{1}{kT}\left(\frac{1}{2}\langle p_l^{(0)}a_{ijk}\rangle\nabla_k\mathcal{E}_j + \omega^{-1}\langle p_l^{(0)}G'_{ij}\rangle\dot{\mathcal{B}}_j\right)\right]$$

$$+ \frac{1}{2}\nabla_m E_l\left[\langle\mathfrak{b}_{ijlm}\rangle + \frac{1}{kT}\left(\langle\alpha_{ij}q_{lm}^{(0)}\rangle - \langle\alpha_{ij}\rangle\langle q_{lm}^{(0)}\rangle\right)\right]\mathcal{E}_j \tag{5.170}$$

$$\bar{q}_{ij} = E_l\left[\langle\mathfrak{b}_{klij}\rangle + \frac{1}{kT}\langle p_l^{(0)}a_{kij}\rangle\right]\mathcal{E}_k \tag{5.171}$$

$$\bar{m}_i = -\omega^{-1}\left[\langle G'_{ji}\rangle + E_k\left(\langle J'_{jik}\rangle + \frac{1}{kT}\langle p_k^{(0)}G'_{ji}\rangle\right)\right]\dot{\mathcal{E}}_j. \tag{5.172}$$

Here $\langle\cdots\rangle$ denotes an orientational average relative to laboratory axes and for zero field and field gradient.

### 5.11.3   Forward scattering by a lamina

We consider a thin extended lamina of ideal gas lying in the $xy$ plane and of thickness $\Delta z$. The radiated field $\Delta\mathcal{E}_x^{(r)}$ at $P(0,0,z)$ due to molecules in the lamina, that are excited by the field (5.155) of the beam, is obtained by summing the average field $\overline{\mathcal{E}}_x^{(r)}$ for all these molecules. Now $\overline{\mathcal{E}}_x^{(r)}$ is given by (5.160) with $p_i$, $q_{ij}$, and $m_i$ replaced by the average moments (5.170)–(5.172). Thus

$$\Delta \mathcal{E}_x^{(r)} = \frac{N\omega^2}{4\pi\varepsilon_0 c^2} \frac{\Delta z}{z} e^{-i\omega(t-z/c)} \iint_{-L}^{L} \left[ \overline{p}_{ox} + \frac{x}{z}\overline{p}_{oz} - \frac{i\omega}{2c}\left( \overline{q}_{oxz} + \frac{x}{z}\overline{q}_{ozz} \right) \right.$$
$$\left. - \frac{x}{z}\overline{q}_{oxx} - \frac{y}{z}\overline{q}_{oxy} \right) + \frac{1}{c}\left( \overline{m}_{oy} + \frac{y}{z}\overline{m}_{oz} \right) \right] e^{i\omega(x^2+y^2)/2zc}\, dx\, dy, \quad (5.173)$$

where $N$ is the number density of molecules, assumed constant over the cross-section of the beam. The expression for $\Delta\mathcal{E}_y^{(r)}$, obtained in a similar way, is

$$\Delta \mathcal{E}_y^{(r)} = \frac{N\omega^2}{4\pi\varepsilon_0 c^2} \frac{\Delta z}{z} e^{-i\omega(t-z/c)} \iint_{-L}^{L} \left[ \overline{p}_{oy} + \frac{y}{z}\overline{p}_{oz} - \frac{i\omega}{2c}\left( \overline{q}_{oyz} + \frac{y}{z}\overline{q}_{ozz} \right) \right.$$
$$\left. - \frac{y}{z}\overline{q}_{oyy} - \frac{x}{z}\overline{q}_{oxy} \right) - \frac{1}{c}\left( \overline{m}_{ox} + \frac{x}{z}\overline{m}_{oz} \right) \right] e^{i\omega(x^2+y^2)/2zc}\, dx\, dy. \quad (5.174)$$

To proceed further, and evaluate (5.173) and (5.174) with multipole moments given by (5.170)–(5.172), we need to specify the electrostatic field **E**.

### 5.11.4   *The electrostatic field*

Buckingham's quadrupole experiment [48, 53] employs a gas cell in the form of a long metal cylinder, along the length of which run two thin parallel wires equidistant from the axis of the cylinder. With the wires at the same potential relative to the cylinder, the electrostatic field **E** on the axis is zero but not the field gradient. A laser beam parallel to the axis, centered on it, and of width smaller than the wire separation, is used to measure the induced birefringence.

With $z$-axis along the axis of the cylinder, the $x, y$ coordinates of the two wires are taken to be $(a, 0)$ and $(-a, 0)$. For infinite wires of negligible thickness compared to their separation $2a$, the non-zero components of the field and its gradient are

$$E_x = K\left[ \frac{x+a}{(x+a)^2 + y^2} + \frac{x-a}{(x-a)^2 + y^2} \right] \quad (5.175)$$

$$E_y = K\left[ \frac{y}{(x+a)^2 + y^2} + \frac{y}{(x-a)^2 + y^2} \right] \quad (5.176)$$

$$\nabla_x E_y = -\nabla_y E_y = K\left[ \frac{y^2 - (x+a)^2}{\{(x+a)^2 + y^2\}^2} + \frac{y^2 - (x-a)^2}{\{(x-a)^2 + y^2\}^2} \right] \quad (5.177)$$

$$\nabla_x E_y = \nabla_y E_x = -2K\left[ \frac{(x+a)y}{\{(x+a)^2 + y^2\}^2} + \frac{(x-a)y}{\{(x-a)^2 + y^2\}^2} \right]. \quad (5.178)$$

Here $K = (2\pi\varepsilon_0)^{-1}\Lambda$, where $\Lambda$ is the line density of charge on the wires. The parity of these components is evident by inspection: thus $E_x$ is an odd function of $x$ and an even function of $y$, and so on.

### 5.11.5  *Radiated field for linearly polarized light*

Consider first an incident beam (5.155) linearly polarized parallel to the $x$-axis

$$\boldsymbol{\mathcal{E}} = \hat{\mathbf{x}} u \mathcal{E}_0 e^{-i\omega(t-z/c)}. \tag{5.179}$$

Now the gradients $\nabla_x u$ and $\nabla_y u$ are of order $u/L$ or less, whereas the exponential in (5.179) varies on a scale $\lambda \ll L$. Thus we neglect gradients of $u$ and take account only of

$$\mathcal{E}_x, \qquad B_y = \mathcal{E}_x/c, \qquad \nabla_z \mathcal{E}_x = i\omega \mathcal{E}_x/c. \tag{5.180}$$

From (5.158), (5.170)–(5.174), and (5.180) we have for an $x$-polarized beam

$$\Delta \mathcal{E}_x^{(r)} = \frac{N\omega^2}{4\pi\varepsilon_0 c^2} \frac{\Delta z}{z} \mathcal{E}_0 e^{-i\omega(t-z/c)} (a_1 I_1 + a_2 I_2 + a_3 I_3 + a_4 I_4) \tag{5.181}$$

$$\Delta \mathcal{E}_y^{(r)} = 0, \tag{5.182}$$

where

$$a_1 = \langle \alpha_{xx} \rangle \tag{5.183}$$

$$a_2 = \frac{1}{2} \left( \langle b_{xxxx} \rangle - \langle b_{xxyy} \rangle + \frac{1}{kT} \langle \alpha_{xx} (q_{xx}^{(0)} - q_{yy}^{(0)}) \rangle \right) \tag{5.184}$$

$$a_3 = \frac{1}{2} \left[ \langle \mathfrak{b}_{xxxx} \rangle - \langle \mathfrak{b}_{xxzz} \rangle + \langle \mathfrak{b}_{zxzx} \rangle + 2\omega^{-1} \langle J'_{xyz} \rangle \right.$$
$$\left. + \frac{1}{kT} \left( \langle p_x^{(0)} a_{xxx} \rangle - \langle p_x^{(0)} a_{xzz} \rangle + \langle p_x^{(0)} a_{zzx} \rangle + 2\omega^{-1} \langle p_x^{(0)} G'_{yz} \rangle \right) \right] \tag{5.185}$$

$$a_4 = \frac{1}{2} \left[ \langle \mathfrak{b}_{xyxy} \rangle - 2\omega^{-1} \langle J'_{xyz} \rangle + \frac{1}{kT} \left( \langle p_y^{(0)} a_{xxy} \rangle - 2\omega^{-1} \langle p_x^{(0)} G'_{yz} \rangle \right) \right] \tag{5.186}$$

$$I_1 = \iint_{-L}^{L} u e^{i\omega(x^2+y^2)/2zc} dx\, dy \tag{5.187}$$

$$I_2 = \iint_{-L}^{L} u \nabla_x E_x e^{i\omega(x^2+y^2)/2zc} dx\, dy \tag{5.188}$$

$$I_3 = \frac{i\omega}{zc} \iint_{-L}^{L} u x E_x e^{i\omega(x^2+y^2)/2zc} dx\, dy \tag{5.189}$$

$$I_4 = \frac{i\omega}{zc} \iint_{-L}^{L} u y E_y e^{i\omega(x^2+y^2)/2zc} dx\, dy. \tag{5.190}$$

In writing down (5.181)–(5.190) we have not used the explicit expressions (5.175)–(5.178) for the fields and gradients, but only their parity to omit integrals of quantities which are odd functions of $x$ and/or $y$. We have also omitted $\langle G'_{xx} \rangle$ and $\langle G'_{yy} \rangle$, which exist only for chiral molecules [14, 15]. Thus our theory does not apply to an optically active gas.

Next we consider an incident beam linearly polarized parallel to the $y$-axis and with the same intensity as the $x$-polarized beam (5.179). Thus

$$\boldsymbol{\mathcal{E}} = \hat{\boldsymbol{y}} u \mathcal{E}_0 e^{-i\omega(t-z/c)}. \tag{5.191}$$

Proceeding as before we obtain

$$\Delta \mathcal{E}_x^{(r)} = 0 \tag{5.192}$$

$$\Delta \mathcal{E}_y^{(r)} = \frac{N\omega^2}{4\pi\varepsilon_0 c^2} \frac{\Delta z}{z} \mathcal{E}_0 e^{-i\omega(t-z/c)} (a_1 I_1 - a_2 I_2 + a_3 I_3 + a_4 I_4). \tag{5.193}$$

The $a_i$ in (5.183)–(5.186) are in terms of averages relative to laboratory axes. They can be expressed in terms of molecule-fixed axes using the results and procedure in Ref. 14. Then

$$a_1 = \frac{1}{3} \alpha_{ii} \tag{5.194}$$

$$a_2 = \frac{1}{30} \left[ 3b_{ijij} - b_{iijj} + \frac{1}{kT} \left( 3\alpha_{ij} q_{ij}^{(0)} - \alpha_{ii} q_{jj}^{(0)} \right) \right] \tag{5.195}$$

$$a_3 = \frac{1}{60} \left[ 9\mathbf{b}_{ijij} - 3\mathbf{b}_{iijj} + 10\,\omega^{-1} \varepsilon_{ijk} J'_{ijk} \right.$$
$$\left. + \frac{1}{kT} p_i^{(0)} \left( 9a_{jji} - 3a_{ijj} + 10\,\omega^{-1} \varepsilon_{ijk} G'_{jk} \right) \right] \tag{5.196}$$

$$a_4 = \frac{1}{60} \left[ 3\mathbf{b}_{ijij} - \mathbf{b}_{iijj} - 10\,\omega^{-1} \varepsilon_{ijk} J'_{ijk} \right.$$
$$\left. + \frac{1}{kT} p_i^{(0)} \left( 3a_{jji} - a_{ijj} - 10\,\omega^{-1} \varepsilon_{ijk} G'_{jk} \right) \right]. \tag{5.197}$$

### 5.11.6   Field-gradient-induced birefringence

The double integrals $I_i$ in (5.187)–(5.190) cannot be evaluated in closed form. They are, in general, complicated complex functions of four lengths: the wavelength $\lambda = 2\pi c/\omega$, the distance to the field point $z$, the beam width $2L$, and the wire separation $2a$. There is, however, a suitable limit in which they can all be expressed in terms of a single integral related to the Fresnel integrals, namely

$$\int_{-\beta}^{\beta} e^{iv^2}\, dv \approx (1+i)\sqrt{\pi/2}, \tag{5.198}$$

where

$$\beta = \sqrt{\omega L^2/2zc} = \sqrt{\pi L^2/\lambda z} \gg 1. \tag{5.199}$$

In what follows we assume (5.199): thus $z$ is small compared to the ratio $L^2/\lambda$.

Consider first $I_1$ in (5.187). Because of (5.199) the exponential in the integral oscillates strongly except for regions close to the origin. The beam profile $u(x,y)$

is smoothly varying, and therefore it can be replaced in the integrand by $u(0,0) = 1$ (see (5.156)). Then from (5.198)

$$I_1 \approx i2\pi zc/\omega. \qquad (5.200)$$

Similarly, in $I_2$ we can replace $u\nabla_x E_x$ by its value at the origin $\eta = -2\mathcal{K}/a^2$ (see (5.177)). Then

$$I_2 \approx \eta I_1. \qquad (5.201)$$

To determine $I_3$ we start with (5.188) and integrate by parts with respect to $x$. In this process, a term evaluated on the perimeter of the beam (at $x = \pm L$) is zero because of (5.157). Also, $\nabla_x u$ is small compared to $\nabla_x$ of the exponential in (5.188) because of (5.199). (For example, if $u \sim \exp[-(x^2 + y^2)/L^2]$ close to the origin, then $\nabla_x u \sim (x/L^2)u$. Thus the ratio of these respective gradients is of order $\lambda z/L^2 \ll 1$.) Hence, taking account only of the gradient of the exponential, we obtain from the integration by parts

$$I_3 \approx -I_2. \qquad (5.202)$$

To determine $I_4$ we use $\nabla_x E_x = -\nabla_y E_y$ in (5.188) (see (5.177)), and then integrate by parts with respect to $y$. This gives

$$I_4 \approx I_2. \qquad (5.203)$$

Thus for an $x$-polarized incident beam the only component of the radiated field (5.181) is

$$\Delta\mathcal{E}_x^{(r)} = i\omega(2\varepsilon_0 c)^{-1}N[a_1 + \eta(a_2 - a_3 + a_4)]\Delta z\mathcal{E}_0 e^{-i\omega(t-z/c)}. \qquad (5.204)$$

Then the resultant field on the axis is

$$\begin{aligned}\mathcal{E}_x + \Delta\mathcal{E}_x^{(r)} &= \left[1 + i\omega(2\varepsilon_0 c)^{-1}Nd_x\Delta z\right]\mathcal{E}_0 e^{-i\omega(t-z/c)} \\ &\approx \mathcal{E}_0 \exp\left[-i\omega\{t - (2\varepsilon_0 c)^{-1}Nd_x\Delta z - z/c\}\right],\end{aligned} \qquad (5.205)$$

where

$$d_x = a_1 + \eta(a_2 - a_3 + a_4). \qquad (5.206)$$

The combined field (5.205) has the same amplitude as the incident wave but its phase is modified in such a way that the wave is retarded: it takes an additional time $(2\varepsilon_0 c)^{-1}Nd_x\Delta z$ to propagate from the lamina at $z = 0$ to the field point at $z$ on the axis. Just as in the scattering theory of the Kerr effect (Section 5.10), this retardation can be interpreted in terms of a refractive index of the gas

$$n_x = 1 + (N/2\varepsilon_0)d_x. \qquad (5.207)$$

Similarly, for a $y$-polarized incident beam, (5.193) implies a refractive index

$$n_y = 1 + (N/2\varepsilon_0)d_y, \qquad (5.208)$$

where
$$d_y = a_1 - \eta(a_2 - a_3 + a_4). \tag{5.209}$$

Hence there is a birefringence

$$n_x - n_y = (N/\varepsilon_0)\eta(a_2 - a_3 + a_4)$$
$$= (N/30\varepsilon_0)\eta\big[3b_{ijij} - b_{iijj} - 3\mathbf{b}_{ijij} + \mathbf{b}_{iijj} - 10\omega^{-1}\varepsilon_{ijk}J'_{ijk}$$
$$+ (kT)^{-1}\big\{3\alpha_{ij}q^{(0)}_{ij} - \alpha_{ii}q^{(0)}_{jj} - p^{(0)}_i(3a_{jji} - a_{ijj} + 10\omega^{-1}\varepsilon_{ijk}G'_{jk})\big\}\big]. \tag{5.210}$$

This is the expression for the birefringence induced in a gas by, and linear in, an applied electrostatic field gradient $\eta$ on the axis of the two-wire capacitor. The temperature-independent and temperature-dependent parts of (5.210) are separately origin independent, as can be verified using results for origin shifts in Section 3.7 and Ref. 50.

The result (5.210) is expressed in terms of the primitive quadrupole moment (1.11). In the far zone, the field radiated by an electric quadrupole can also be expressed in terms of the traceless moment (1.14) (Section 1.15), and therefore so can the birefringence. When this is done the result [51] is the same as that obtained originally by Buckingham and Longuet-Higgins [49].

### 5.11.7   Comparison between theory and experiment

There are three aspects to the Buckingham effect, as the above induced bire-fringence is sometimes called. They are the experiment, the theory on which the interpretation of the experiment is based, and *ab initio* computer calculations which can be compared with experimental results. At first sight, use of (5.210) to extract values of the quadrupole moment from measurements of the birefrin-gence appears to be a formidable task. This task is simplified by noting that, for both non-dipolar and dipolar molecules, the temperature-dependent term in (5.210) can be expressed as

$$(N/30\varepsilon_0 kT)\eta\big(3\alpha_{ij}q^{(0)}_{ij} - \alpha_{ii}q^{(0)}_{jj}\big). \tag{5.211}$$

For non-dipolar molecules this is obvious. For dipolar molecules the validity of (5.211) requires [49] that $q^{(0)}_{ij}$ be taken with respect to a point at which the origin-dependent vector

$$3a_{jji} - a_{ijj} + 10\omega^{-1}\varepsilon_{ijk}G'_{jk} = 0. \tag{5.212}$$

This point, termed the effective quadrupole center (EQC) [49], is therefore a convenient reference point for the origin-dependent quadrupole moment of a dipolar molecule in the Buckingham experiment. In general, the position of the EQC depends on frequency.

We consider the application of (5.211) to a linear molecule. Relative to its main symmetry axis this has only one independent component of its traceless quadrupole moment, which is the commonly quoted value. This can be obtained from (5.211) and measurements of the birefringence at various temperatures and of the relevant component of $\alpha_{ij}$. For the non-dipolar molecule $N_2$ a value $-(4.97 \pm 0.16) \times 10^{-40}\,\mathrm{Cm}^2$ has been found [53], in excellent agreement with the result of an *ab initio* numerical calculation $-(4.93 \pm 0.03) \times 10^{-40}\,\mathrm{Cm}^2$ [54]. Often measurements made at a single temperature are used, together with theoretical values for the temperature-independent term in (5.210), to obtain the quadrupole moment. For $N_2$ this yields $-(5.01 \pm 0.08) \times 10^{-40}\,\mathrm{Cm}^2$ [54,55], in close agreement with the above values.

For the dipolar molecule CO, measurements at fixed temperature and corrected in the above manner give [56,57] $-(8.85 \pm 0.15) \times 10^{-40}\,\mathrm{Cm}^2$ relative to the EQC. The *ab initio* numerical calculation starts by evaluating the moment relative to the centre of mass of the molecule, then locating the EQC relative to the center of mass, and finally correcting the quadrupole moment for this change of origin. This yields [57] $-(8.79 \pm 0.19) \times 10^{-40}\,\mathrm{Cm}^2$.

It is apparent that the theoretical, experimental, and computational aspects of the Buckingham effect have been refined to such an extent that the measured electric quadrupole moments of certain molecules are in impressive agreement with the results of *ab initio* computer calculations. The numerical techniques [57] offer the prospect that they may be used for simple molecules *in lieu* of experiment. The extensive literature on the Buckingham effect presents an impressive body of research which provides strong support for the multipole approach.

## 5.12   Birefringence induced in a gas by an electric field gradient: wave theory

The wave equation (5.4) in Section 5.1 is for propagation of a harmonic plane wave in a homogeneous dielectric. It does not apply to an inhomogeneous medium, like a dielectric subject to a non-uniform electrostatic field. To deduce the wave equation for such a medium, we start with the Maxwell equation (5.1). To electric quadrupole–magnetic dipole order the response fields (1.118) and (1.119) are

$$D_i = \varepsilon_0 (E_i + \mathcal{E}_i) + P_i - \frac{1}{2}\nabla_j Q_{ij} \qquad (5.213)$$

$$H_i = \mu_0^{-1}\mathcal{B}_i - M_i. \qquad (5.214)$$

Here $\mathcal{E}$ and $\mathcal{B}$ are the fields of an electromagnetic wave, $\mathbf{E}$ is an electrostatic field, and

$$P_i = N\bar{p}_i, \qquad Q_{ij} = N\bar{q}_{ij}, \qquad M_i = N\bar{m}_i, \qquad (5.215)$$

where $N$ is the number density of molecules, and $\bar{p}_i$, $\bar{q}_i$, and $\bar{m}_i$ are the average induced moments in a molecule of electric dipole, electric quadrupole, and magnetic dipole, respectively, in the presence of the fields $\mathcal{E}$, $\mathcal{B}$, and $\mathbf{E}$. The moments are defined in (1.10), (1.11), and (1.50), and their average induced values are given by (5.170)–(5.172) for time-harmonic fields $\mathcal{E}$ and $\mathcal{B}$.

From (5.1), (5.213)–(5.215), (5.170)–(5.172), and using the Maxwell equation $\nabla \times \boldsymbol{\mathcal{E}} = -\dot{\boldsymbol{B}}$, one obtains the wave equation [52]

$$\mu_0^{-1}\omega^{-2}(\nabla_i\nabla_j - \delta_{ij}\nabla^2)\mathcal{E}_j - \varepsilon_0\delta_{ij}\mathcal{E}_j - N\langle\alpha_{ij}\rangle\mathcal{E}_j$$
$$+ NE_l(R_{ijkl} - R_{jikl} + S_{ijkl} - S_{jikl})\nabla_k\mathcal{E}_j$$
$$+ N(\nabla_k E_l)(R_{ijkl} + S_{ijkl} - T_{ijkl})\mathcal{E}_j = 0, \quad (5.216)$$

where

$$R_{ijkl} = \omega^{-1}\varepsilon_{ikm}\left(\langle J'_{jml}\rangle + (kT)^{-1}\langle p_l^{(0)} G'_{jm}\rangle\right) \qquad (5.217)$$

$$S_{ijkl} = \frac{1}{2}\langle \mathfrak{b}_{jlik}\rangle + (2kT)^{-1}\langle p_l^{(0)} a_{jik}\rangle \qquad (5.218)$$

$$T_{ijkl} = \frac{1}{2}\langle \mathfrak{b}_{ijkl}\rangle + (2kT)^{-1}\left(\langle\alpha_{ij}q_{kl}^{(0)}\rangle - \langle\alpha_{ij}\rangle\langle q_{kl}^{(0)}\rangle\right). \qquad (5.219)$$

Equation (5.216) is valid to electric quadrupole–magnetic dipole order, and describes a field $\boldsymbol{\mathcal{E}}(\mathbf{r},t)$ with harmonic time dependence $\exp(-i\omega t)$ propagating through a dilute gas in an arbitrary electrostatic field $\mathbf{E}(\mathbf{r})$.

We apply (5.216) to the quadrupole experiment where $E_z = 0$, and $E_x$ and $E_y$ are functions only of $x$ and $y$ (Section 5.11.4). Also $\nabla \times \mathbf{E} = 0$ for an electrostatic field, and $\nabla \cdot \mathbf{E} = 0$. Then the three components of (5.216), obtained by setting $i = x, y, z$ in turn, are [52]

$$[\mu_0^{-1}\omega^{-2}(\nabla_x^2 - \nabla^2) - \varepsilon_0 n_0^2 - NV_1(\nabla_x E_x)]\mathcal{E}_x$$
$$+ [\mu_0^{-1}\omega^{-2}\nabla_x\nabla_y - NV_2(E_x\nabla_y - E_y\nabla_x) - NV_1(\nabla_x E_y)]\mathcal{E}_y$$
$$+ [\mu_0^{-1}\omega^{-2}\nabla_x\nabla_z - NV_2(E_x\nabla_z)]\mathcal{E}_z = 0 \quad (5.220)$$

$$[\mu_0^{-1}\omega^{-2}\nabla_y\nabla_x + NV_2(E_x\nabla_y - E_y\nabla_x) - NV_1(\nabla_x E_y)]\mathcal{E}_x$$
$$+ [\mu_0^{-1}\omega^{-2}(\nabla_y^2 - \nabla^2) - \varepsilon_0 n_0^2 + NV_1(\nabla_x E_x)]\mathcal{E}_y$$
$$+ [\mu_0^{-1}\omega^{-2}\nabla_y\nabla_z - NV_2(E_y\nabla_z)]\mathcal{E}_z = 0 \quad (5.221)$$

$$[\mu_0^{-1}\omega^{-2}\nabla_z\nabla_x + NV_2(E_x\nabla_z)]\mathcal{E}_x + [\mu_0^{-1}\omega^{-2}\nabla_z\nabla_y + NV_2(E_y\nabla_z)]\mathcal{E}_y$$
$$+ [\mu_0^{-1}\omega^{-2}(\nabla_z^2 - \nabla^2) - \varepsilon_0 n_0^2]\mathcal{E}_z = 0. \quad (5.222)$$

Here

$$n_0^2 = 1 + (N/3\varepsilon_0)\alpha_{ii} \qquad (5.223)$$

$$V_1 = \frac{1}{60}\Big[6\mathfrak{b}_{ijij} - 2\mathfrak{b}_{iijj} - \mathfrak{b}_{ijij} - 3\mathfrak{b}_{iijj} - 10\omega^{-1}\varepsilon_{ijk}J'_{ijk}$$
$$+ (kT)^{-1}\{6\alpha_{ij}q_{ij}^{(0)} - 2\alpha_{ii}q_{jj}^{(0)} - p_i^{(0)}(a_{jji} + 3a_{ijj} + 10\omega^{-1}\varepsilon_{ijk}G'_{jk})\}\Big]$$
$$(5.224)$$

$$V_2 = \frac{1}{12}\left[\mathfrak{b}_{iijj} - \mathfrak{b}_{ijij} - 2\omega^{-1}\varepsilon_{ijk}J'_{ijk} - (kT)^{-1}p_i^{(0)}(a_{jji} - a_{ijj} + 2\omega^{-1}\varepsilon_{ijk}G'_{jk})\right]$$
$$(5.225)$$

are independent of $x$, $y$, $z$ and the coordinate origin to which the molecular moments and polarizabilities are referred [49, 50]. In obtaining (5.220)–(5.222) from (5.216) we used intrinsic symmetries of polarizability tensors [52], and the procedure in Ref. 14 to express isotropic averages relative to laboratory axes in terms of molecule-fixed axes. Also, we have omitted terms containing the trace $G'_{ii}$, and thus the theory does not apply to an optically active gas [14, 15].

Approximate solutions to the coupled partial differential equations (5.220)–(5.222) can be obtained as follows. We are interested in time-harmonic solutions which propagate along the $z$-axis (the axis of the cylinder, Section 5.11.4)

$$\boldsymbol{\mathcal{E}} = \mathbf{f}(x, y)e^{-i\omega(t-nz/c)},\tag{5.226}$$

where $\mathbf{f}$ is real and $n$ is a real constant. For this solution the coefficients of $\mathcal{E}_x$ and $\mathcal{E}_y$ in (5.222) are imaginary relative to that of $\mathcal{E}_z$ which is real. Therefore the term involving $\mathcal{E}_z$ in (5.222) is zero

$$(\nabla_x^2 + \nabla_y^2 + K^2)\mathcal{E}_z = 0,\tag{5.227}$$

where $K^2 = \mu_0\varepsilon_0\omega^2 n_0^2$. Because $K \neq 0$ there are no solutions to (5.227) which propagate along the $z$-axis (see the discussion following (5.234)). Thus $\mathcal{E}_z = 0$ and, to electric quadrupole–magnetic dipole order, there are no additional waves [52].

With $\mathcal{E}_z = 0$ we subtract $\nabla_x$ of (5.222) from $\nabla_z$ of (5.220). Because $\mathbf{E}$ depends only on $x$ and $y$, and $n_0$, $V_1$, and $V_2$ are constants, we obtain

$$\left[\mu_0^{-1}\omega^{-2}\nabla^2 + \varepsilon_0 n_0^2 + N(V_1 + V_2)(\nabla_x E_x)\right](\nabla_z\mathcal{E}_x)$$
$$+ NV_2 E_x\nabla_z(\nabla_x\mathcal{E}_x + \nabla_y\mathcal{E}_y) + N(V_1 + V_2)(\nabla_x E_y)(\nabla_z\mathcal{E}_y) = 0.\tag{5.228}$$

We solve (5.228) close to the axis of the cylinder. For $|x|, |y| \ll a$, we have from (5.175)–(5.178) the leading terms

$$E_x = \eta x,\qquad E_y = -\eta y,\qquad \nabla_x E_y = -(6\eta/a^2)xy,\tag{5.229}$$

where $\eta = -2K/a^2$ is the field gradient $\nabla_x E_x$ on the axis. Thus in (5.228) we neglect the terms involving $E_x$ and $\nabla_x E_y$ in comparison with the term containing the constant $\nabla_x E_x = \eta$. Then

$$\left[\mu_0^{-1}\omega^{-2}\nabla^2 + \varepsilon_0 n_0^2 + N\eta(V_1 + V_2)\right](\nabla_z\mathcal{E}_x) = 0.\tag{5.230}$$

Thus for a time-harmonic wave propagating along the $z$-axis

$$\mathcal{E}_x(\mathbf{r}, t) = g(x, y)e^{-i\omega(t-n_xz/c)},\tag{5.231}$$

we have

$$(\nabla_x^2 + \nabla_y^2 + k^2)g(x,y) = 0, \tag{5.232}$$

where

$$k^2 = \mu_0\omega^2\left[\varepsilon_0 n_0^2 + N\eta(V_1 + V_2) - \varepsilon_0 n_x^2\right]. \tag{5.233}$$

Here $n_x$ is the refractive index of the gas for light linearly polarized along the $x$-axis; see the discussion following (5.244). Using the general solution [58] to (5.232) in (5.231) gives

$$\mathcal{E}_x(\mathbf{r},t) = e^{-i\omega(t-n_x z/c)}\int_0^{2\pi} F(u)e^{ik(x\cos u + y\sin u)}\, du. \tag{5.234}$$

This is a superposition of harmonic waves with wave vectors

$$(k\cos u, \; k\sin u, \; \omega n_x/c).$$

For propagation along the $z$-axis we must have $k = 0$, and hence (5.233) becomes

$$\varepsilon_0 n_0^2 + N\eta(V_1 + V_2) - \varepsilon_0 n_x^2 = 0. \tag{5.235}$$

The term involving $\eta$ is small, and so

$$n_x = n_0 + (N/2\varepsilon_0)\eta(V_1 + V_2). \tag{5.236}$$

This is the refractive index for $x$-polarized light (light polarized in the plane of the wires and perpendicular to the axis) propagating along the $z$-axis

$$\mathcal{E}_x(z,t) = \mathcal{E}_0 e^{-i\omega(t-n_x z/c)}, \tag{5.237}$$

where $\mathcal{E}_0$ is a constant; see (5.234) with $k = 0$.

A similar calculation yields the refractive index $n_y$ of light polarized along the $y$-axis. We take $\nabla_z$ of (5.221) minus $\nabla_y$ of (5.222) and repeat the above analysis to obtain

$$\varepsilon_0 n_0^2 - N\eta(V_1 + V_2) - \varepsilon_0 n_y^2 = 0. \tag{5.238}$$

Thus

$$n_y = n_0 - (N/2\varepsilon_0)\eta(V_1 + V_2) \tag{5.239}$$

for light polarized perpendicular to the plane of the wires and propagating along the $z$-axis

$$\mathcal{E}_y(z,t) = \mathcal{E}_o e^{-i\omega(t-n_y z/c)}. \tag{5.240}$$

These results for the refractive indices, and hence the birefringence, are in agreement with those obtained from forward scattering theory in the previous section, as can be shown from (5.207), (5.208), and (5.210).

We emphasize that the wave in the gas is not a plane wave. Consider incident $x$-polarized light. In the gas $\mathcal{E}_x$ is given by (5.237). Then (5.222), with $\mathcal{E}_z = 0$, admits a solution $\mathcal{E}_y$ which propagates along the $z$-axis with the same velocity $c/n_x$ as $\mathcal{E}_x$: for this solution (5.222) is

$$(\mu_0^{-1}\omega^{-2}\nabla_y + NV_2E_y)\mathcal{E}_y = -NV_2E_x\mathcal{E}_x(z,y). \tag{5.241}$$

We solve for $\mathcal{E}_y$ at points close to the axis by taking $\nabla_x$ of (5.241) and using (5.229) to neglect terms in $E_x$ and $\nabla_xE_y$ in comparison with the term containing the constant $\nabla_xE_x = \eta$. Then (5.241) becomes

$$\mu_0^{-1}\omega^{-2}\nabla_x\nabla_y\mathcal{E}_y = -NV_2\eta\mathcal{E}_x(z,t). \tag{5.242}$$

The solution which is a stationary pattern in the $x$ and $y$ directions is

$$\mathcal{E}_y(\mathbf{r},t) = -(\mu_0\omega^2NV_2\eta)xy\mathcal{E}_x(z,t). \tag{5.243}$$

The resultant wave in the gas

$$\mathcal{E}(\mathbf{r},t) = (1, -\mu_0\omega^2NV_2\eta xy, 0)\mathcal{E}_oe^{-i\omega(t-n_xz/c)} \tag{5.244}$$

is clearly not a plane wave.

Thus reference to "$x$-polarized light in the gas" has to be qualified: for incident $x$-polarized light the wave in the gas acquires the small, induced, perpendicular component evident in (5.244). Even though this induced component is zero on the axis, it makes a finite contribution $-NV_2\eta\mathcal{E}_x$ to equation (5.220) through the term $\mu_0^{-1}\omega^{-2}\nabla_x\nabla_y\mathcal{E}_y$ in this equation. When combined with the term $-NV_1\eta\mathcal{E}_x$ in (5.220), the result is a refractive index depending on $V_1 + V_2$, as in (5.236). Thus, if one neglects distortion of the incident wave, as is done in Ref. 50, one obtains an incorrect refractive index $n_x = n_0 + (N/2\varepsilon_0)\eta V_1$. Similar remarks apply to incident $y$-polarized light [52].

## 5.13   Discussion

In the last of the eight papers in which Jones detailed his optical calculus [59], he remarked as follows:"The electromagnetic theory of light propagation in crystals is beset by a fundamental difficulty when one considers the theory to that degree of completeness requisite for the inclusion of optical activity: It is easy to show that the magnetic effects are of the same order of importance as electrical effects in contributing to circular dichroism and circular birefringence. But the simultaneous treatment of both the electrical and magnetic effects in crystals leads to mathematical relations of unmanageable complexity. Accordingly, it is customary in treating crystals to ignore completely the magnetic effects, and to place all of the burden upon the electrical effects for the explanation of experimental results." Jones then proceeds to neglect magnetic effects.

In this regard we emphasize that the multipole treatment of transmission effects in Sections 5.2–5.9 incorporates magnetic contributions alongside the electric ones in a systematic way for both non-magnetic and magnetic systems. The analysis is no more unmanageable with this inclusion than it is without it. In respect of optical activity, in particular, the circular birefringence of a cubic crystal in (5.43) and of a gas (Section 5.3) is expressed entirely in terms of the tensor $G'_{ij}$. (This tensor describes the induction of a magnetic dipole by the time derivative of the electric field of the radiation.) Furthermore, omission of the term involving $G'_{ij}$ in the expression (5.42) for the circular birefringence along the optic axis of a uniaxial crystal would render the expression origin dependent and thus unphysical.

For both crystals and gases under normal conditions, and in the long wavelength regime, a multipole approach provides the theoretical basis for explaining effects in transmission and scattering. Had this approach been formulated at the time, Jones would surely have recognized its applicability to optical activity.

We have seen in the preceding sections that this theory of transmission effects provides physically acceptable (e.g. origin-independent) results despite the fact that it is based on origin-dependent constitutive relations obtained directly from multipole theory in Section 4.1. We have already mentioned that this is associated with the circumstance that the wave equation derived from these constitutive relations is origin independent (Section 5.1), and we comment further on this aspect in Section 9.1. It is now natural to inquire as to the situation in the multipole theory of reflection phenomena: do the multipole constitutive relations lead to physically acceptable results for such phenomena? The answer to this question is one of the central themes of this monograph, and it is considered in the next chapter.

## References

[1] Graham, E. B., Pierrus, J., and Raab, R. E. (1992). Multipole moments and Maxwell's equations. *Journal of Physics* B: *Atomic, Molecular, and Optical Physics*, **25**, 4673–4684.

[2] Birss, R. R. (1963). Macroscopic symmetry in space–time. *Reports on Progress in Physics*, **26**, 307–360.

[3] Birss, R, R. (1966). *Symmetry and magnetism*. North–Holland, Amsterdam.

[4] Graham, C. and Raab, R. E. (1994). Eigenvector approach to the evaluation of the Jones $N$ matrices of non-absorbing crystalline media. *Journal of the Optical Society of America* A, **11**, 2137–2144.

[5] Born, M. and Wolf, E. (1980). *Principles of optics*, p. 673. Pergamon, Oxford.

[6] Gunning, M. J. and Raab, R. E. (1998). Systematic eigenvector approach to crystal optics: an analytic alternative to the geometric ellipsoid model. *Journal of the Optical Society of America* A, **15**, 2199–2207.

[7] Graham, E. B. and Raab, R. E. (1991). Electric dipole effects in magnetic crystals. *Journal of Applied Physics*, **69**, 2549–2551.

[8] Nye, J. F. (1985). *Physical properties of crystals*, Ch. 14. Clarendon Press, Oxford.

[9] Argyres, P. N. (1955). Theory of the Faraday and Kerr effects in ferromagnetics. *Physical Review*, **97**, 334–345.

[10] Suits, J. C., Argyle, B. E., and Freiser, M. J. (1966). Magneto-optical properties of materials containing divalent europium. *Journal of Applied Physics*, **37**, 1391–1397.

[11] Ref. 8, Ch.13.

[12] Hobden, M. V. (1968). Optical activity in a non-enantiomorphous crystal: $AgGaS_2$. *Acta Crystallographica*, A **24**, 676–680.

[13] Kobayashi, J., Uesu, Y., and Sorimachi, H. (1978). Optical activity of some non-enantiomorphous ferroelectrics. *Ferroelectrics*, **21**, 345–346.

[14] Barron, L. D. (1982). *Molecular light scattering and optical activity*, Ch. 4. Cambridge University Press, Cambridge.

[15] Buckingham, A. D. and Dunn, M. B. (1971). Optical activity of oriented molecules. *Journal of the Chemical Society* A, 1988-1991.

[16] Theron, I. P. and Cloete, J. H. (1996). The optical activity of an artificial non-magnetic uniaxial chiral crystal at microwave frequencies. *Journal of Electromagnetic Waves and Applications*, **10**, 539–561.

[17] Theron, I. P. and Cloete, J. H. (1996). The electric quadrupole contribution to the circular birefringence of nonmagnetic anisotropic chiral media: a circular waveguide experiment. *IEEE Transactions on Microwave Theory and Techniques*, **44**, 1451–1459.

[18] Crabbé, P. (1972). *ORD and CD in chemistry and biochemistry.* Academic, New York.

[19] Jones, R. C. (1948). A new calculus for the treatment of optical systems. VII. *Journal of the Optical Society of America*, **38**, 671–685, and references therein to the earlier Jones' papers.

[20] Swindell, W. (ed.) (1975). *Polarized light: Benchmark papers in optics.* Dowden, Hutchinson and Ross, Stroudsberg, Pa. This contains all Jones' papers on the new calculus.

[21] Shurcliff, W. A. (1962). *Polarized light.* Harvard University Press, Cambridge, MA.

[22] Graham, E. B. and Raab, R. E. (1983). On the Jones birefringence. *Proceedings of the Royal Society London* A, **390**, 73–90.

[23] Roth, T. and Rikken, G. L. J. A. (2000). Observation of magnetoelectric Jones birefringence. *Physical Review Letters*, **85**, 4478–4481.

[24] Meintjes, E. M. and Raab, R. E. (1999). A new theory of Pockels birefringence in non-magnetic crystals. *Journal of Optics* A: *Pure and Applied Optics*, **1**, 146–151.

[25] Izdebski, M., Kucharczyk, W., and Raab, R. E. (2001). Effect of beam divergence from the optic axis in an electro-optic experiment to measure an induced Jones birefringence. *Journal of the Optical Society of America* A, **18**, 1393–1398.

[26] Rizzo, A. and Coriani, S. (2003). Jones birefringence in gases: *Ab initio* electron correlated results for atoms and linear moleclues. *Journal of Chemical Physics*, **119**, 11064–11079.

[27] Rikken, G. L. J. A. and Rizzo, C. (2001). Magnetoelectric birefringences of the quantum vacuum. *Physical Review* A, **63**, 0121071–0121074.

[28] Brown, W. F., Shtrikman, S., and Treves, D. (1963). Possibility of visual observation of antiferromagnetic domains. *Journal of Applied Physics*, **34**, 1233–1234.

[29] Birss, R. R. and Shrubsall, R. G. (1967). The propagation of electromagnetic waves in magnetoelectric crystals. *Philosophical Magazine*, **15**, 687–700.

[30] Hornreich, R. M. and Shtrikman, S. (1968). Theory of gyrotropic birefringence. *Physical Review*, **171**, 1065–1074.

[31] Graham, E. B. and Raab, R. E. (1992). Magnetic effects in antiferromagnetic crystals in the electric quadrupole–magnetic dipole approximation. *Philosophical Magazine* B, **66**, 269–284.

[32] Graham, E. B. and Raab, R. E. (1994). The optical properties of antiferromagnetic chromium oxide in transmission. *Journal of Physics: Condensed Matter*, **6**, 6725–6732.

[33] Agranovich, V. M. and Ginzburg, V. L. (1984). *Crystal optics with spatial dispersion, and excitons.* Springer, Berlin.

[34] Pisarev, R. V., Krichevtsov, B. B., and Pavlov, V. V. (1991). Optical study of the antiferromagnetic-paramagnetic phase transition in chromium oxide $Cr_2O_3$. *Phase Transitions* B, **37**, 63–72.

[35] Lorentz, H. A. (1878). Concerning the relation between the velocity of propagation of light and the density and composition of media. In *H. A. Lorentz, collected papers*, (ed. Zeeman, P. and Fokker, A. D.), Vol. 2, pp. 1–119. Nijhoff, The Hague.

[36] Condon, E. U. and Seitz, F. (1932). Lorentz double refraction in the regular system. *Journal of the Optical Society of America*, **22**, 393–401.

[37] Graham, E. B. and Raab, R. E. (1990). Light propagation in cubic and other anisotropic crystals. *Proceedings of the Royal Society London* A, **430**, 593–614.

[38] Ref. 33, p. 6.

[39] Pekar, S. I. (1983). *Crystal optics and additional light waves.* Benjamin, Menlo Park, Ca.

[40] Pastrnak, J. and Vedam, K. (1971). Optical anisotropy of silicon single crystals. *Physical Review* B, **3**, 2567–2571.

[41] Hellwege, K. H. (1951). Optische anisotropie kubischer kristalle bei quadrupolstrahlung. *Zeitschrift für Physik*, **129**, 626–641.

[42] Nettleton, R. E. (1971). Quadrupole exciton transitions and optical anisotropy in cubic crystals. *Journal of the Optical Society of America*, **61**, 1060–1064.

[43] Graham, E. B. and Raab, R. E. (1991). Non-reciprocal rotation in cubic antiferromagnets. *Philosophical Magazine* B, **64**, 267–274.

[44] Buckingham, A. D. and Pople, J. A. (1955). Theoretical studies of the Kerr effect I: deviations from a linear polarization law. *Proceedings of the Physical Society* A, **68**, 905–909.

[45] Buckingham, A. D. (1962). Frequency dependence of the Kerr constant. *Proceedings of the Royal Society London* A, **267**, 271–282.

[46] See Ref. 5, p. 88.

[47] Buckingham, A. D. (1959). Direct method of measuring molecular quadrupole moments. *Journal of Chemical Physics*, **30** 1580–1585.

[48] Buckingham, A. D. and Disch, R. L. (1963). The quadrupole moment of the carbon dioxide molecule. *Proceedings of the Royal Society London* A, **273**, 275–289.

[49] Buckingham, A. D. and Longuet-Higgins, H. C. (1968). The quadrupole moments of dipolar molecules. *Molecular Physics*, **14**, 63–72.

[50] Imrie, D. A. and Raab, R. E. (1991). A new molecular theory of field gradient induced birefringence used for measuring electric quadrupole moments. *Molecular Physics*, **74**, 833–842.

[51] Raab, R. E. and de Lange, O. L. (2003). Forward scattering theory of electric-field-gradient-induced birefringence. *Molecular Physics*, **101**, 3467–3475.

[52] de Lange, O. L. and Raab, R. E. (2004). Reconciliation of the forward scattering and wave theories of electric-field-gradient-induced birefringence. *Molecular Physics*, **102**, 125–130.

[53] Ritchie, G. L. D., Watson, J. N., and Keir, R. I. (2003). Temperature-dependence of electric field-gradient induced birefringence (Buckingham effect) and molecular quadrupole-moment of $N_2$. Comparison of experiment and theory. *Chemical Physics Letters*, **370**, 376, and references therein.

[54] Halkier, A., Coriani, S., and Jorgensen, P. (1998). The molecular electric quadrupole moment of $N_2$. *Chemical Physics Letters*, **294**, 292–296.

[55] Coriani, S., Hättig, C., Jorgensen, P., Rizzo, A., and Ruud, K. (1998). Coupled cluster investigation of the electric-field-gradient-induced birefringence of $H_2$, $N_2$, $C_2H_2$, and $CH_4$. *Journal of Chemical Physics*, **109**, 7176–7184.

[56] Rizzo, A., Coriani, S., Halkier, A., and Hättig, C. (2000). *Ab initio* study of the electric-field-gradient-induced birefringence of a polar molecule: CO. *Journal of Chemical Physics*, **113**, 3077–3087.

[57] Coriani, S., Halkier, A., Jonsson, D., Gauss, J., Rizzo, A., and Christiansen, O. (2003). On the electric field gradient induced birefringence and electric quadrupole moment of CO, $N_2O$, and OCS. *Journal of Chemical Physics*, **118**, 7329–7339.

[58] Morse, P. M. and Feshbach, H. (1953) *Methods of theoretical physics*, Part II. McGraw-Hill, New York, p. 1361.

[59] Jones, R. C. (1956). New calculus for the treatment of optical systems. VIII. Electromagnetic theory. *Journal of the Optical Society of America*, **46**, 126–131.

# 6

## REFLECTION EFFECTS: DIRECT MULTIPOLE RESULTS

*It can't be Nature, for it is not sense.*
Charles Churchill
*(The farewell)*

In this chapter we investigate whether results obtained directly from multipole theory for reflection phenomena satisfy basic requirements of translational invariance and time-reversal invariance (reciprocity). We start with a brief discussion of the reflection matrix (Section 6.1) and the reciprocity principle and its effect on the reflection matrix (Section 6.2). An analysis of equations of continuity in multipole theory (Section 6.3) leads to a determination of matching conditions for electromagnetic fields at the interface between two media (Section 6.4). These matching conditions and multipole theory are used to obtain the reflection matrix for non-magnetic uniaxial and cubic crystals to electric quadrupole–magnetic dipole order (Section 6.5). Solutions of the wave equation (Section 6.6) are used to express the reflection coefficients in a form which displays explicitly their dependence on polarizability tensors of the reflecting medium (Section 6.7). We then test whether these reflection coefficients satisfy the requirements of translational invariance and time-reversal invariance (Section 6.8).

The material in this chapter falls into two categories. That in Sections 6.1 to 6.4 presents general results needed for the description of reflection and refraction. Sections 6.5 to 6.7 contain detailed results obtained from multipole theory for reflection from non-magnetic uniaxial and cubic crystals. The derivation of the latter is algebraically complicated, and the reader may wish to omit these sections and proceed to Sections 6.8 and 6.9, where the unphysical nature of the multipole theory of reflection is discussed.

## 6.1 Reflection and the reflection matrix

We consider the specular reflection of a harmonic, plane electromagnetic wave at the interface between two dielectric media. The interface is taken to be planar and the media are semi-infinite and homogeneous. We suppose that the incident medium is isotropic while the second medium is anisotropic. The plane of incidence is the plane formed by the incident propagation unit vector $\boldsymbol{\sigma}_i$ and the normal at the point of incidence. We assume the following standard kinematic properties [1]: the reflected and transmitted propagation unit vectors $\boldsymbol{\sigma}_r$ and $\boldsymbol{\sigma}_t$ lie in the plane of incidence; the angles of incidence and reflection are equal; and Snell's law applies. These kinematic properties do not depend on the particular form of the matching conditions for the fields at the boundary [1], such as those in Section 6.4.

It is useful to introduce a set of "basis" vectors for the incident and reflected waves. For the incident wave these are the wave vector $\mathbf{k}_i$ and two unit vectors $\hat{\mathbf{p}}_i$ and $\hat{\mathbf{s}}_i$ which are, respectively, parallel to and perpendicular to the plane of incidence, with $\mathbf{p}_i$ perpendicular to $\mathbf{k}_i$. (The notation s originates from the German *senkrecht*.) The vectors $\mathbf{k}_i$, $\hat{\mathbf{p}}_i$, and $\hat{\mathbf{s}}_i$ form a right-hand set, and are depicted in Fig. 6.1 together with the corresponding set $\mathbf{k}_r$, $\hat{\mathbf{p}}_r$, and $\hat{\mathbf{s}}_r$ for the reflected wave. In conformity with this notation, the components of the electric and magnetic fields of the incident and reflected waves are designated with a subscript $p$ if parallel to $\hat{\mathbf{p}}$ and a subscript $s$ if parallel to $\hat{\mathbf{s}}$. (See also Fig. 6.5.)

The reflection matrix is a $2 \times 2$ matrix which relates the two orthogonal components of the electric field of the reflected wave to those of the incident wave. Thus we write

$$\begin{pmatrix} E_{rp} \\ E_{rs} \end{pmatrix} = R \begin{pmatrix} E_{ip} \\ E_{is} \end{pmatrix}, \tag{6.1}$$

where $R$ is the reflection matrix

$$R = \begin{pmatrix} r_{pp} & r_{ps} \\ r_{sp} & r_{ss} \end{pmatrix}. \tag{6.2}$$

The four complex elements of this matrix are known as the Fresnel coefficients of reflection.

In this chapter we shall be concerned with two fundamental requirements which must be satisfied by the above reflection matrix. The first is that the reflected intensity obtained from the Fresnel coefficients should be translationally invariant (independent of the choice of coordinate origin). The second requirement is a consequence of time-reversal invariance, and the associated principle of reciprocity, which imposes a condition on the Fresnel coefficients. Reciprocity and its effect on these coefficients are discussed in the next section.

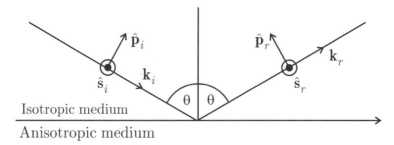

FIG. 6.1. Reflection at an interface between two media. The unit vectors $\hat{\mathbf{p}}_i$ and $\hat{\mathbf{p}}_r$ are parallel to the plane of incidence, whereas $\hat{\mathbf{s}}_i$ and $\hat{\mathbf{s}}_r$ are perpendicular to this plane.

## 6.2   The principle of reciprocity

It is well known that microscopic time-reversal invariance of a system manifests itself in certain macroscopic properties of that system. As an example, consider the thermodynamic equations which relate the response $J_i$ to an appropriate "thermodynamic force" $X_j$; that is, responses such as energy flow and diffusion to forces such as gradients of temperature and chemical potential. For the linear equations

$$J_i = L_{ij}(\mathbf{B})X_j, \tag{6.3}$$

time reversal requires that the coefficients $L_{ij}(\mathbf{B})$ satisfy the famous Onsager reciprocal relations [2]

$$L_{ij}(\mathbf{B}) = L_{ji}(-\mathbf{B}), \tag{6.4}$$

where $\mathbf{B}$ is the magnetic field. These relations connect the results of different experiments, such as measurements of the coefficients of thermal diffusion and thermal effusion, respectively [3]. A particularly clear example is provided by the transport entropy of a moving vortex in a superconductor. This entropy can be measured using either the Nernst effect or the Ettingshausen effect [4], and according to the Onsager relations the results should be the same. This conclusion has been verified experimentally [5].

For the transmission and reflection of electromagnetic radiation by macroscopic matter, time-reversal invariance leads to a result known as the principle of reciprocity. Various statements of this principle have been given [6–8]. Our discussion is based on that of Ref. 7, which can be stated: time reversal of a reflection experiment in respect of (i) its light paths, (ii) the incident and reflected polarization states (obtained by means of suitable polarizer and analyzer respectively), and (iii) the reflecting medium, produces the same analyzed intensity in the reflected beam of the time-reversed experiment as that in the original experiment, provided that the intensities of their respective incident polarizations are the same.

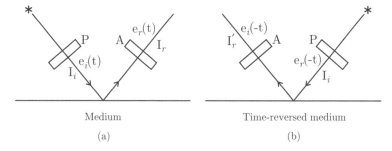

FIG. 6.2. (a) A reflection experiment. $P$ and $A$ denote a polarizer and analyzer, respectively. The other symbols are explained in the text. (b) The time-reversed experiment.

In the reflection experiment of Fig. 6.2(a), the incident beam has polarization $e_i(t)$ and intensity $I_i$, while for the reflected beam after passage through the analyzer these are $e_r(t)$ and $I_r$. For the time-reversed experiment of Fig. 6.2(b) the incident polarization and intensity are $e_r(-t)$ and $I_i$, while the analyzed reflected properties are $e_i(-t)$ and $I_r'$. Then the principle of reciprocity requires

$$I_r' = I_r. \tag{6.5}$$

We now discuss the effect of (6.5) on the reflection matrix (6.2).

Consider first the effect of time reversal on polarization states. A polarization state is judged by an observer looking toward the light source and relative to an outstretched right arm as horizontal. The six basic polarization states transform under time reversal $T$ as follows [7]:

We determine the effect of reciprocity for experiments involving the linear polarization states in I–IV; it is straightforward to present similar analyses for the other polarization states. The reflected field entering the analyzer in Fig. 6.3(a) is

$$\mathbf{E}_r = E_{rp}\hat{\mathbf{p}} + E_{rs}\hat{\mathbf{s}} = (r_{pp}E_{ip} + r_{ps}E_{is})\,\hat{\mathbf{p}} + (r_{sp}E_{ip} + r_{ss}E_{is})\,\hat{\mathbf{s}}. \tag{6.6}$$

This reflected field may be analyzed for any one of the six basic polarization states depicted above. In what follows, a linear polarization along $\hat{\mathbf{p}}$ is referred to as horizontal, and one along $\hat{\mathbf{s}}$ as vertical.

(i) For the experiment shown in Fig. 6.3(a) we choose vertical polarization for the incident beam. Thus in (6.6), $E_{ip} = 0$ and with $E_{is} = E$ we have for the field entering the analyzer $\mathbf{E}_r = (r_{ps}\hat{\mathbf{p}} + r_{ss}\hat{\mathbf{s}})\,E$. The field emerging from a vertical analyzer is then $\mathbf{E}_r = r_{ss}E\hat{\mathbf{s}}$, and the corresponding intensity is

$$I_r = |r_{ss}|^2 I_i, \tag{6.7}$$

where $I_i = |E|^2$ is the incident intensity. For the time-reversed experiment of Fig. 6.3(b) the reflection matrix is

$$R' = \begin{pmatrix} r'_{pp} & r'_{ps} \\ r'_{sp} & r'_{ss} \end{pmatrix}. \tag{6.8}$$

Also, the incident and analyzed polarizations remain vertical (see II above), and the analyzed intensity is

$$I_r' = |r'_{ss}|^2 I_i'. \tag{6.9}$$

Now reciprocity requires that if $I'_i = I_i$, then $I'_r = I_r$. It follows from (6.7) and (6.9) that

$$r'_{ss} = r_{ss}e^{i\alpha_1},\qquad (6.10)$$

where the phase $\alpha_1$ remains to be specified. Similarly, by considering horizontal incident and analyzed polarizations we obtain

$$r'_{pp} = r_{pp}e^{i\alpha_2}.\qquad (6.11)$$

(ii) For horizontal incident and vertical analyzed polarizations the calculations are also similar and yield

$$r'_{ps} = r_{sp}e^{i\alpha_3},\qquad (6.12)$$

while consideration of vertical incident and horizontal analyzed polarizations gives

$$r'_{sp} = r_{ps}e^{i\alpha_4}.\qquad (6.13)$$

(iii) Next we analyze for $+45°$ polarization when the incident polarization is horizontal. The analyzed intensity is

$$I_r = \frac{1}{2}\left(|r_{pp}|^2 + |r_{sp}|^2 + r_{pp}r^*_{sp} + r^*_{pp}r_{sp}\right)I_i.\qquad (6.14)$$

In the time-reversed experiment the incident polarization is $-45°$ (see III), and hence $E'_{ip} = -E'_{is} = E'/\sqrt{2}$. The reflected beam, analyzed for horizontal polarization, has intensity

$$I'_r = \frac{1}{2}\left(|r'_{pp}|^2 + |r'_{ps}|^2 - r'_{pp}r'^*_{ps} - r'^*_{pp}r'_{ps}\right)I'_i.\qquad (6.15)$$

By reciprocity, and using (6.11) and (6.12) in (6.15), we have $\alpha_3 = \alpha_2 + \pi$. Similar calculations involving $\pm45°$ and horizontal polarizations yield $\alpha_2 = \alpha_1$ and $\alpha_4 = \alpha_3$.

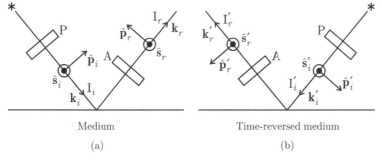

Medium                              Time-reversed medium

(a)                                      (b)

FIG. 6.3. (a) A reflection experiment. (b) The time-reversed experiment. If the incident and reflected polarizations in (a) are $e_i(t)$ and $e_r(t)$, then in (b) the incident polarization is $e_r(-t)$ and the reflected polarization is $e_i(-t)$, see also Fig. 6.2.

The above results show that (6.8) can be written

$$R' = e^{i\alpha_1} \begin{pmatrix} r_{pp} & -r_{sp} \\ -r_{ps} & r_{ss} \end{pmatrix}, \tag{6.16}$$

in which $r_{jk} = r_{jk}(t)$. In (6.16) we can, without loss of generality, choose the over-all phase $\alpha_1$ to be zero [8]. Now $R'$ is the reflection matrix for the time-reversed configuration, and hence we write $r'_{jk} = r_{jk}(-t)$ in (6.8). Then comparing (6.8) and (6.16) we have

$$r_{pp}(-t) = r_{pp}(t) \tag{6.17}$$

$$r_{ss}(-t) = r_{ss}(t) \tag{6.18}$$

$$r_{ps}(-t) = -r_{sp}(t) \tag{6.19}$$

$$r_{sp}(-t) = -r_{ps}(t). \tag{6.20}$$

Equations (6.17)–(6.20) are the desired reciprocity relations for the Fresnel co-efficients of reflection. Note that these are in the $\mathbf{k}$, $\hat{\mathbf{p}}$, $\hat{\mathbf{s}}$ basis of Fig. 6.1, as distinct from the laboratory axes. The connection between the reciprocal relations (6.17)–(6.20) of optics, the Onsager relations (6.4), and microscopic time reversal is discussed in the literature [8, 9].

In this chapter we will use multipole theory to obtain expressions for the above reflection coefficients in terms of polarizability tensors. We will then test whether these expressions satisfy the reciprocity relations (6.17)–(6.20). In this respect we emphasize the following aspects of the time reversal in (6.17)–(6.20): under this reversal time-even and time-odd tensors $A_e$ and $A_o$ are changed according to $A_e \rightarrow A_e$ and $A_o \rightarrow -A_o$.

## 6.3    Equations of continuity

In this section we discuss continuity equations for charge and current densities in the bulk and on interfaces between media. This leads on to a treatment of matching conditions for electromagnetic fields at an interface between two media (Section 6.4). For our purposes it is convenient to adopt a differential approach rather than the more traditional method based on integral relations [10].

This differential approach involves the unit step function and the Dirac delta function [11, 12]. The unit step function is defined by

$$\begin{aligned} u(z) &= 1 \quad (z > 0) & u(-z) &= 0 \quad (z > 0) \\ &= 0 \quad (z < 0) & &= 1 \quad (z < 0). \end{aligned} \tag{6.21}$$

From this it can be shown that

$$\frac{\partial u(\pm z)}{\partial z} = \pm \delta(z), \qquad \frac{\partial^2 u(\pm z)}{\partial z^2} = \pm \delta'(z), \tag{6.22}$$

where $\delta(z)$ is the Dirac delta function defined in Section 1.1, and $\delta'(z)$ is its derivative. In the theory below we obtain equations of the form

$$a\delta(z) + b\delta'(z) = 0, \tag{6.23}$$

where $a$ and $b$ are independent of $z$. By integrating (6.23) over an interval which includes the origin, and with the aid of an integration by parts, we have

$$a = 0. \tag{6.24}$$

From (6.23) and (6.24) it follows that

$$b = 0. \tag{6.25}$$

We choose the Cartesian axes shown in Fig. 6.4, with the $xy$ plane lying in the interface and $+z$-axis as indicated. The properties of the two dielectrics are labelled with subscripts 1 and 2. We work to the order of electric quadrupole–magnetic dipole, and assign the bound charge and current densities (throughout the half-spaces) the trial forms

$$\rho = u(-z)\rho_1(\mathbf{R}) + u(z)\rho_2(\mathbf{R}) + \delta(z)\sigma(\mathbf{r}) + \delta'(z)\hat{\mathbf{z}} \cdot \boldsymbol{\mathcal{P}}(\mathbf{r}) \tag{6.26}$$
$$\mathbf{J} = u(-z)\mathbf{J}_1(\mathbf{R}) + u(z)\mathbf{J}_2(\mathbf{R}) + \delta(z)\mathbf{K}(\mathbf{r}). \tag{6.27}$$

Here $\mathbf{r} = (x, y, 0)$ is the position vector of a macroscopic surface element and $\mathbf{R} = (x, y, z)$. The terms in $\delta(z)$ and $\delta'(z)$ in (6.26) and (6.27) are bound surface contributions, of which those in $\sigma$, $\boldsymbol{\mathcal{P}}$, and $\mathbf{K}$ are the only relevant ones to the order of electric quadrupole and magnetic dipole, as is evident from Section 1.12. (For convenience we have omitted the subscript $b$ on the bound contributions, and in (6.26) and (6.27) we have also omitted the time dependence.)

We now use the trial forms (6.26) and (6.27) in the continuity equation

$$\boldsymbol{\nabla} \cdot \mathbf{J} = -\dot{\rho}. \tag{6.28}$$

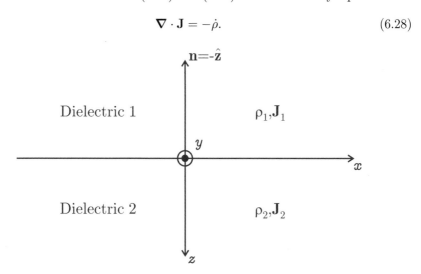

FIG. 6.4. Semi-infinite dielectrics and the coordinate system used in the text.

Noting that $\mathbf{\nabla} \cdot \mathbf{J}_1 = -\dot{\rho}_1$ and $\mathbf{\nabla} \cdot \mathbf{J}_2 = -\dot{\rho}_2$, and using (6.23)–(6.25), we obtain

$$\hat{\mathbf{z}} \cdot (\mathbf{J}_1 - \mathbf{J}_2) + \mathbf{\nabla} \cdot \mathbf{K} + \dot{\sigma} = 0 \tag{6.29}$$

and

$$\hat{\mathbf{z}} \cdot (\mathbf{K} + \dot{\boldsymbol{\mathcal{P}}}) = 0. \tag{6.30}$$

In these equations $\mathbf{K} = \mathbf{K}(x, y)$ is a surface property and

$$\mathbf{\nabla} \cdot \mathbf{K} = \nabla_x K_x + \nabla_y K_y \tag{6.31}$$

is the two-dimensional divergence of $\mathbf{K}$. Equation (6.29) is the equation of continuity on a surface; it has also been obtained from integral forms by Magid [13]. Equation (6.30) is an additional continuity condition that is essential for consistency to electric quadrupole–magnetic dipole order.

Next we investigate whether the multipole forms for $\sigma$, $\boldsymbol{\mathcal{P}}$, $\mathbf{K}$, and $\mathbf{J}$ satisfy these continuity equations. To electric quadrupole–magnetic dipole order, we have from (1.102), (1.103), and (1.106)

$$\sigma = \left( P_i - \frac{1}{2} \nabla_j Q_{ij} \right) n_i \tag{6.32}$$

$$\mathcal{P}_i = -\frac{1}{2} Q_{ij} n_j \tag{6.33}$$

$$K_i = \left( \frac{1}{2} \dot{Q}_{ij} - \varepsilon_{ijk} M_k \right) n_j. \tag{6.34}$$

Also $\mathbf{n} = -\hat{\mathbf{z}}$ and $\mathbf{J}_2 - \mathbf{J}_1 = \mathbf{J}$ where, from (1.105), $\mathbf{J}$ has Cartesian components

$$J_i = \dot{P}_i - \nabla_j \left( \frac{1}{2} \dot{Q}_{ij} - \varepsilon_{ijk} M_k \right). \tag{6.35}$$

When these are substituted in (6.29) and (6.30), we find that (6.30) is satisfied but (6.29) is not.

The explanation of this rather surprising result is contained in Ref. 14 where a modified form of (6.32) is introduced, namely

$$\sigma = \left( P_i - \frac{1}{2} \nabla_j Q_{ij} \right) n_i + \mathbf{\nabla} \cdot \boldsymbol{\mathcal{P}}. \tag{6.36}$$

Its use ensures that the continuity equation (6.29) is satisfied. The need for such a modification to the surface charge density has also been noted by Langreth [15] and discussed by Graham and Raab [16]. The origin of the term $\mathbf{\nabla} \cdot \boldsymbol{\mathcal{P}}$ in (6.36) is associated with a certain arbitrariness in the expression (1.103) for the bound surface electric dipole moment density obtained from the scalar potential (1.99).

Specifically, the surface term $-R_j Q_{ij}/2R^3$ in (1.99) can be transformed using the identity

$$\frac{R_j}{R^3} Q_{ij} n_i = \frac{1}{R} \nabla_j Q_{ij} n_i - \nabla_j \left( \frac{1}{R} Q_{ij} \right) n_i. \qquad (6.37)$$

The last term in (6.37) makes no contribution to a surface integral over a closed surface, such as those in (1.99). Thus the surface integral of the left-hand side of (6.37) leads to an additional contribution

$$-\frac{1}{2} \nabla_j Q_{ij} n_i = \boldsymbol{\nabla} \cdot \boldsymbol{\mathcal{P}}$$

to the bound surface charge density in (1.99) and (1.102). We note that, to electric quadrupole–magnetic dipole order, no such arbitrariness exists in the interpretation of the densities $\mathbf{J}$ and $\mathbf{K}$ obtained from the expansion (1.100) of the vector potential.

Thus, the surface continuity equations are satisfied to electric quadrupole–magnetic dipole order by the multipole expressions (6.33)–(6.35) and a modified surface charge density (6.36). One can extend this discussion of continuity equations to higher multipole orders by generalizing the trial functions (6.26) and (6.27), and the surface charge density (6.36). This has been done to electric octopole-magnetic quadrupole order in Ref. 14.

## 6.4   Matching conditions in multipole theory

In the derivation of matching conditions in this section, it is more suited to our purpose to use the two inhomogeneous Maxwell equations (1.112) and (1.115), which contain the bound source densities $\rho_b$ and $\mathbf{J}_b$ [17], rather than the traditional equations involving $\mathbf{D}$ and $\mathbf{H}$. Consequently, and contrary to usual practice, these two response fields do not enter the matching conditions below, which, instead, are couched entirely in terms of the fields $\mathbf{E}$ and $\mathbf{B}$ and induced multipole contributions. In the previous section we have shown that the trial forms for the bound source densities in (6.26) and(6.27), together with their multipole surface contributions, satisfy the equations of continuity. We now proceed to use them to derive the matching conditions for electromagnetic fields at an interface between two media, to electric quadrupole and magnetic dipole order.

On the basis of the above approach we assume the $\mathbf{E}$ and $\mathbf{B}$ fields in both half-spaces to have the forms

$$\mathbf{E} = u(-z)\mathbf{E}_1(\mathbf{R}) + u(z)\mathbf{E}_2(\mathbf{R}) + \delta(z)\boldsymbol{\mathcal{E}}(\mathbf{r}) \qquad (6.38)$$

$$\mathbf{B} = u(-z)\mathbf{B}_1(\mathbf{R}) + u(z)\mathbf{B}_2(\mathbf{R}) + \delta(z)\boldsymbol{\mathcal{B}}(\mathbf{r}), \qquad (6.39)$$

where $\boldsymbol{\mathcal{E}}$ and $\boldsymbol{\mathcal{B}}$, if they exist at all, are surface contributions with the dimensions of their respective fields $\times$ length. The fields (6.38) and (6.39) are now used in the Maxwell equations (1.112)–(1.115), from which the free source contributions $\rho_f$ and $\mathbf{J}_f$ are omitted for a dielectric. When these fields and the source densities

(6.26) and (6.27) are substituted in Maxwell's equations, we obtain equations of the form (6.23). These yield, via (6.24) and (6.25), the desired matching conditions. We discuss this for each Maxwell equation in turn.

(i) In

$$\nabla \cdot \mathbf{B} = 0 \tag{6.40}$$

we substitute (6.39). Noting that $\mathbf{B}_1$ and $\mathbf{B}_2$ each satisfy (6.40), we obtain an equation of the form (6.23). Hence, using (6.24) and (6.25), we find

$$B_{2z} - B_{1z} + \nabla \cdot \boldsymbol{\mathcal{B}} = 0 \tag{6.41}$$

$$\mathcal{B}_z = 0. \tag{6.42}$$

(ii)

$$\nabla \times \mathbf{E} = -\dot{\mathbf{B}}. \tag{6.43}$$

Consider first the $x$ component

$$\nabla_y E_z - \nabla_z E_y = -\dot{B}_x. \tag{6.44}$$

We substitute (6.38) and (6.39) in (6.44) and note that the fields $\mathbf{E}_1$ and $\mathbf{B}_1$ satisfy (6.43), as do $\mathbf{E}_2$ and $\mathbf{B}_2$. Then using (6.23)–(6.25) we obtain

$$E_{1y} - E_{2y} + \nabla_y \mathcal{E}_z + \dot{\mathcal{B}}_x = 0 \tag{6.45}$$

$$\mathcal{E}_y = 0. \tag{6.46}$$

Similarly, the $y$ component of (6.43) yields

$$E_{2x} - E_{1x} - \nabla_x \mathcal{E}_z + \dot{\mathcal{B}}_y = 0 \tag{6.47}$$

$$\mathcal{E}_x = 0. \tag{6.48}$$

The $z$ component of (6.43) provides no additional information.

(iii)

$$\nabla \cdot \mathbf{E} = \varepsilon_0^{-1} \rho, \tag{6.49}$$

where we continue to omit the subscript $b$ on bound source densities. It is helpful to note that from (6.46), (6.48), and the surface nature of $\boldsymbol{\mathcal{E}} = \boldsymbol{\mathcal{E}}(x, y)$, we have

$$\nabla \cdot \boldsymbol{\mathcal{E}} = 0. \tag{6.50}$$

Substituting (6.38) in (6.49), and using (6.50), we find

$$E_{2z} - E_{1z} - \varepsilon_0^{-1} \sigma = 0 \tag{6.51}$$

$$\mathcal{E}_z - \varepsilon_0^{-1} \mathcal{P}_z = 0. \tag{6.52}$$

(iv)

$$\boldsymbol{\nabla} \times \mathbf{B} = \mu_0(\varepsilon_0 \dot{\mathbf{E}} + \mathbf{J}).\tag{6.53}$$

In this $\mathbf{J}$ is given to electric quadrupole–magnetic dipole order by (6.27). The calculations are similar to the above, and we simply state the results. From the $x$ component of (6.53) we have

$$B_{1y} - B_{2y} - \mu_0 K_x = 0 \tag{6.54}$$

$$\mathcal{B}_y = 0. \tag{6.55}$$

The $y$ and $z$ components yield

$$B_{2x} - B_{1x} - \mu_0 K_y = 0 \tag{6.56}$$
$$\mathcal{B}_x = 0 \tag{6.57}$$
$$\varepsilon_0 \dot{\mathcal{E}}_z + K_z = 0. \tag{6.58}$$

This completes our analysis based on Maxwell's equations, and we now summarize the results. First, from (6.42), (6.55), and (6.57)

$$\mathcal{B} = 0, \tag{6.59}$$

while (6.46), (6.48), and (6.52) show that the only non-vanishing component of $\mathcal{E}$ is

$$\mathcal{E}_z = \varepsilon_0^{-1}\mathcal{P}_z. \tag{6.60}$$

From (6.45), (6.47), (6.51), (6.59), and (6.60) the matching conditions on the components of the electric field at the interface are, to electric quadrupole–magnetic dipole order,

$$E_{2x} = E_{1x} + \varepsilon_0^{-1}\nabla_x \mathcal{P}_z \tag{6.61}$$
$$E_{2y} = E_{1y} + \varepsilon_0^{-1}\nabla_y \mathcal{P}_z \tag{6.62}$$
$$E_{2z} = E_{1z} + \varepsilon_0^{-1}\sigma. \tag{6.63}$$

The corresponding conditions on $\mathbf{B}$ are given by (6.41), (6.54), (6.56), and (6.59), namely

$$B_{2x} = B_{1x} + \mu_0 K_y \tag{6.64}$$
$$B_{2y} = B_{1y} - \mu_0 K_x \tag{6.65}$$
$$B_{2z} = B_{1z}. \tag{6.66}$$

Equations (6.61)–(6.66) are the desired matching conditions expressed in terms of $\mathbf{E}$ and $\mathbf{B}$. They are valid for dielectric media which are linear, homogeneous, anisotropic, and dissipative. The extension of these matching conditions has been carried out to electric octopole-magnetic quadrupole order [14].

We remark that $\mathcal{P}$ and $\mathbf{K}$ are of electric quadrupole–magnetic dipole order, and it therefore follows from the above results that to electric dipole order

$$E_{2x} = E_{1x}, \qquad E_{2y} = E_{1y}, \qquad E_{2z} = E_{1z} - \varepsilon_0^{-1}P_z, \qquad (6.67)$$
$$B_{2x} = B_{1x}, \qquad B_{2y} = B_{1y}, \qquad B_{2z} = B_{1z}. \qquad (6.68)$$

These agree with the standard Maxwell results for a dielectric [17] when $\mathbf{D}$ and $\mathbf{H}$ are expressed to electric dipole order as $\mathbf{D} = \varepsilon_0\mathbf{E} + \mathbf{P}$, $\mathbf{H} = \mu_0^{-1}\mathbf{B}$. Thus the Maxwell boundary conditions are an approximation, valid to electric dipole order.

For the applications presented in the next section it is convenient to express (6.61)–(6.66) in terms of the multipole moment densities in (6.33), (6.34), and (6.36). In doing so we remind the reader that $\sigma$, $\mathcal{P}$, and $\mathbf{K}$ are surface properties (see (6.26) and (6.27)), as are $P_i$, $Q_{ij}$, and $M_i$ in (6.33), (6.34), and (6.36); that is, they are functions of $x$ and $y$ only. Thus we obtain

$$E_{2x} = E_{1x} + \frac{1}{2}\varepsilon_0^{-1}\nabla_x Q_{zz} \qquad (6.69)$$

$$E_{2y} = E_{1y} + \frac{1}{2}\varepsilon_0^{-1}\nabla_y Q_{zz} \qquad (6.70)$$

$$E_{2z} = E_{1z} - \varepsilon_0^{-1}(P_z - \nabla_x Q_{xz} - \nabla_y Q_{yz}) \qquad (6.71)$$

$$B_{2x} = B_{1x} - \mu_0 \left(\frac{1}{2}\dot{Q}_{yz} - M_x\right) \qquad (6.72)$$

$$B_{2y} = B_{1y} + \mu_0 \left(\frac{1}{2}\dot{Q}_{xz} + M_y\right) \qquad (6.73)$$

$$B_{2z} = B_{1z}. \qquad (6.74)$$

## 6.5   The reflection matrix for non-magnetic uniaxial and cubic crystals

Our purpose in this section is to derive expressions containing the dependence of the reflection matrix on polarizability tensors; that is, equations (6.138)–(6.141) below. As in the preceding sections, we work to electric quadrupole–magnetic dipole order. In principle, the calculations are straightforward but the algebraic details are rather involved. To assist the reader we present an outline of these.

For a ray incident on the surface of an anisotropic medium there are, in general, two refracted rays travelling in different directions with different velocities, and each satisfying Snell's law [18]. For simplicity we make the following assumptions: the incident medium (medium 1 in Fig. 6.4) is a vacuum; the other half-space (medium 2) comprises a non-magnetic uniaxial or cubic crystal; the refracted rays in the crystal do not involve an additional (longitudinal) wave (Section 5.7). We also suppose that the crystal is so oriented that its crystallographic axes coincide with the corresponding laboratory axes of Figs 6.4 and 6.5. Thus the laboratory $z$-axis is the main symmetry axis of the crystal, which for a uniaxial is its optic axis. The $xz$ plane is the plane of incidence and the $xy$ plane that of reflection.

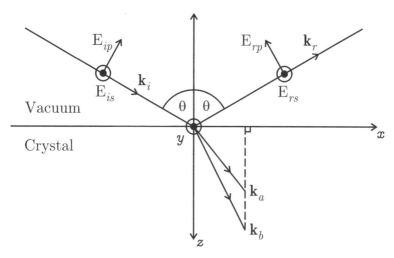

FIG. 6.5. Reflection and transmission at a crystal–vacuum interface. The symbols are defined in the text.

We distinguish the properties of the two refracted waves by subscripts $a$ and $b$. For harmonic plane waves their wave vectors are

$$\mathbf{k}_a = \frac{\omega n_a}{c}\boldsymbol{\sigma}_a, \qquad \mathbf{k}_b = \frac{\omega n_b}{c}\boldsymbol{\sigma}_b, \tag{6.75}$$

where $n$ denotes a refractive index and $\boldsymbol{\sigma}$ a wavefront normal. Since $\mathbf{k}_a$ and $\mathbf{k}_b$ lie in the plane of incidence,

$$k_{iy} = k_{ry} = k_{ay} = k_{by} = 0. \tag{6.76}$$

From Fig. 6.5

$$k_{iz} = -k_{rz} = \frac{\omega}{c}\sigma_z = \frac{\omega}{c}\cos\theta = e\cos\theta \tag{6.77}$$

$$k_{ix} = k_{rx} = \frac{\omega}{c}\sigma_x = \frac{\omega}{c}\sin\theta = e\sin\theta, \tag{6.78}$$

where $e = \omega c^{-1}$. In the following we will also use the notation $\mathbf{k}_1 = \mathbf{k}_i$ or $\mathbf{k}_r$ and $\mathbf{k}_2 = \mathbf{k}_a$ or $\mathbf{k}_b$. Then Snell's law can be written

$$k_{1x} = k_{ix} = k_{ax} = k_{bx} = k_{2x}. \tag{6.79}$$

As outlined in Section 2.11, the multipole moment densities that are induced in a non-magnetic dielectric by the fields of a harmonic plane wave are given, to electric quadrupole–magnetic dipole order, by

$$P_i = \alpha_{ij}E_j + \frac{1}{2}a_{ijk}\nabla_k E_j + \omega^{-1}G'_{ij}\dot{B}_j \tag{6.80}$$

$$Q_{ij} = a_{kij}E_k \tag{6.81}$$

$$M_i = -\omega^{-1}G'_{ji}\dot{E}_j. \tag{6.82}$$

Tables show that for all non-magnetic uniaxial and cubic symmetries the following tensor components vanish relative to crystallographic axes, as do those with permutations of the same subscripts [19],

$$\alpha_{xy}, \quad \alpha_{xz}, \quad \alpha_{yz}, \quad a_{xzz}, \quad a_{yzz}, \quad G'_{xz}, \quad G'_{yz}. \tag{6.83}$$

From this and (6.80)–(6.82) the multipole moment densities that enter the matching conditions (6.69)–(6.74) are

$$P_z = \alpha_{zz} E_z + \frac{i}{2}(k_x a_{zxx} E_x + k_x a_{zyx} E_y + k_z a_{zzz} E_z) - \frac{i}{\omega}k_x G'_{zz} E_y \tag{6.84}$$

$$Q_{xz} = a_{xxz} E_x + a_{yxz} E_y \tag{6.85}$$

$$Q_{yz} = a_{xyz} E_x + a_{yyz} E_y \tag{6.86}$$

$$Q_{zz} = a_{zzz} E_z \tag{6.87}$$

$$M_x = i(G'_{xx} E_x + G'_{yx} E_y) \tag{6.88}$$

$$M_y = i(G'_{xy} E_x + G'_{yy} E_y), \tag{6.89}$$

where we have used (6.43) and the substitutions $\nabla_j \to ik_j$ and $\partial/\partial t \to -i\omega$ for harmonic plane fields (4.1).

In (6.84)–(6.89), $\mathbf{E} = \mathbf{E}_a + \mathbf{E}_b$ and $\mathbf{B} = \mathbf{B}_a + \mathbf{B}_b$ are the total fields in the crystal due to waves $a$ and $b$. For brevity we denote them in what follows by $\mathbf{E}_2$ and $\mathbf{B}_2$. Similarly, the total vacuum fields are $\mathbf{E}_1 = \mathbf{E}_i + \mathbf{E}_r$ and $\mathbf{B}_1 = \mathbf{B}_i + \mathbf{B}_r$. Then the matching conditions (6.69)–(6.74) become

$$E_{2x} = E_{1x} + \frac{i}{2}\varepsilon_0^{-1}k_{2x}a_{zzz}E_{2z} \tag{6.90}$$

$$E_{2y} = E_{1y} \tag{6.91}$$

$$E_{2z} = E_{1z} - \varepsilon_0^{-1}\left[\alpha_{zz}E_{2z} + ik_{2x}\left\{\left(\frac{1}{2}a_{zxx} - a_{xxz}\right)E_{2x}\right.\right.$$
$$\left.\left. + \left(\frac{1}{2}a_{zyx} - a_{yxz}\right)E_{2y} - \omega^{-1}G'_{zz}E_{2y}\right\} + \frac{i}{2}k_{2z}a_{zzz}E_{2z}\right] \tag{6.92}$$

$$B_{2x} = B_{1x} + i\mu_0\left[\left(G'_{xx} + \frac{1}{2}\omega a_{xyz}\right)E_{2x} + \left(G'_{yx} + \frac{1}{2}\omega a_{yyz}\right)E_{2y}\right] \tag{6.93}$$

$$B_{2y} = B_{1y} + i\mu_0\left[\left(G'_{xy} - \frac{1}{2}\omega a_{xxz}\right)E_{2x} + \left(G'_{yy} - \frac{1}{2}\omega a_{yxz}\right)E_{2y}\right] \tag{6.94}$$

$$B_{2z} = B_{1z}. \tag{6.95}$$

According to the above notation, a quantity like $k_{2x}E_{2x}$ in (6.92) represents $k_{ax}E_{ax} + k_{bx}E_{bx}$. The fields $\mathbf{E}$ and $\mathbf{B}$ at the interface are given by (4.1) with $z = 0$. Then use of (6.76) and (6.79) shows that we can cancel the phase factor in (4.1). Thus the fields in (6.90)–(6.95) are amplitudes.

For normal incidence $k_{2x} = 0$ (see (6.78) and (6.79)) and therefore, to electric quadrupole–magnetic dipole order, the tangential components of $\mathbf{E}$ are continuous across the interface. This is not the case for the next multipole order, for which the continuity (6.95) of the normal component of $\mathbf{B}$ also fails [14].

The reflection matrix in (6.1) relates components of electric fields. We therefore use (6.43) to eliminate $\mathbf{B}_1$ and $\mathbf{B}_2$ from (6.93)–(6.95), which become respectively

$$i\mu_0\omega\left(G'_{xx} + \frac{1}{2}\omega a_{xyz}\right)E_{2x} + \left[k_{2z} + i\mu_0\omega\left(G'_{yx} + i\omega a_{yyz}\right)\right]E_{2y} = k_{1z}E_{1y} \quad (6.96)$$

$$\left[k_{2z} - i\mu_0\omega\left(G'_{xy} - \frac{1}{2}\omega a_{xxz}\right)\right]E_{2x} - i\mu_0\omega\left(G'_{yy} - \frac{1}{2}\omega a_{yxz}\right)E_{2y} - k_{2x}E_{2z}$$
$$= k_{1z}E_{1x} - k_{1x}E_{1z} \quad (6.97)$$

$$k_{2x}E_{2y} = k_{1x}E_{1y}. \quad (6.98)$$

Because of (6.79), equations (6.98) and (6.91) are identical.

Equation (6.92) yields for $E_{2z}$, to first order in $a_{ijk}$,

$$E_{2z} = n_E^{-2}\left(1 - \frac{i}{2}\varepsilon_0^{-1}n_E^{-2}k_{2z}a_{zzz}\right)E_{1z} - n_E^{-2}i\varepsilon_0 k_{2x}\left[\left(\frac{1}{2}a_{zxx} - a_{xxz}\right)E_{2x}\right.$$
$$\left. + \left(\frac{1}{2}a_{zyx} - a_{yxz} - \omega^{-1}G'_{zz}\right)E_{2y}\right], \quad (6.99)$$

where from (5.53), $n_E^2 = 1 + \varepsilon_0^{-1}\alpha_{zz}$ for a uniaxial crystal. For a cubic crystal $n_E = n_O$. We use (6.99) to eliminate $E_{2z}$ in (6.90). Thus

$$E_{2x} = E_{1x} + i(2\varepsilon_0 n_E^2)^{-1}k_{1x}a_{zzz}E_{1z}, \quad (6.100)$$

where we have used (6.79) and neglected terms quadratic in polarizability densities $a_{ijk}$ and $G'_{ij}$. Similarly, (6.97) and (6.99) yield

$$\left[k_{2z} + i\varepsilon_0^{-1}n_E^{-2}k_{1x}^2\left(\frac{1}{2}a_{zxx} - a_{xxz}\right) - i\mu_0\omega\left(G'_{xy} - \frac{1}{2}\omega a_{xxz}\right)\right]E_{2x}$$
$$+ i\left[\varepsilon_0^{-1}n_E^{-2}k_{1x}^2\left(\frac{1}{2}a_{zyx} - a_{yxz} - \omega^{-1}G'_{zz}\right) - \mu_0\omega\left(G'_{yy} - \frac{1}{2}\omega a_{yxz}\right)\right]E_{2y}$$
$$= k_{1z}E_{1x} + k_{1x}\left[n_E^{-2}\left(1 - \frac{i}{2}\varepsilon_0^{-1}n_E^{-2}k_{2z}a_{zzz}\right) - 1\right]E_{1z}. \quad (6.101)$$

Thus we have the four matching conditions (6.100), (6.91), (6.96), and (6.101), which we write respectively as

$$E_{2x} = E_{1x} + ik_{1x}F_1 E_{1z} \quad (6.102)$$
$$E_{2y} = E_{1y} \quad (6.103)$$

$$iF_2 E_{2x} + (k_{2z} + iF_3)E_{2y} = k_{1z}E_{1y} \quad (6.104)$$
$$(k_{2z} + iF_4 + iF_5)E_{2x} + i(F_6 + F_7)E_{2y} = k_{1z}E_{1x} + k_{1x}F_8 E_{1z}. \quad (6.105)$$

Here

$$F_1 = (2\varepsilon_0 n_E^2)^{-1} a_{zzz} \tag{6.106}$$

$$F_2 = \mu_0 \omega \left( G'_{xx} + \frac{1}{2}\omega a_{xyz} \right) \tag{6.107}$$

$$F_3 = \mu_0 \omega \left( G'_{yx} + \frac{1}{2}\omega a_{yyz} \right) \tag{6.108}$$

$$F_4 = (\varepsilon_0 n_E^2)^{-1} k_{1x}^2 \left( \frac{1}{2}a_{zxx} - a_{xxz} \right) \tag{6.109}$$

$$F_5 = -\mu_0 \omega \left( G'_{xy} - \frac{1}{2}\omega a_{xxz} \right) \tag{6.110}$$

$$F_6 = (\varepsilon_0 n_E^2)^{-1} k_{1x}^2 \left( \frac{1}{2}a_{zyx} - a_{yxz} - \omega^{-1}G'_{zz} \right) \tag{6.111}$$

$$F_7 = -\mu_0 \omega \left( G'_{yy} - \frac{1}{2}\omega a_{yxz} \right) \tag{6.112}$$

$$F_8 = n_E^{-2} \left( 1 - \frac{i}{2}\varepsilon_0^{-1} n_E^{-2} k_{2z} a_{zzz} \right) - 1. \tag{6.113}$$

Before proceeding, we note the following: (i) Recall that in the above $\mathbf{E}_2$ denotes the superposition $\mathbf{E}_a + \mathbf{E}_b$, and $k_{2z}\mathbf{E}_2$ denotes $k_{az}\mathbf{E}_a + k_{bz}\mathbf{E}_b$. Similar superpositions apply to the field $\mathbf{E}_1$ in the vacuum. (ii) From Fig. 6.5 we see that

$$E_{ix} = E_{ip}\cos\theta, \qquad E_{rx} = -E_{rp}\cos\theta, \tag{6.114}$$
$$E_{iy} = E_{is}, \qquad E_{ry} = E_{rs}, \tag{6.115}$$
$$E_{iz} = -E_{ip}\sin\theta, \qquad E_{rz} = -E_{rp}\sin\theta. \tag{6.116}$$

(iii) The fields $\mathbf{E}_a$ and $\mathbf{E}_b$ will be obtained as unnormalized solutions of the wave equation (Section 6.6), and thus require inclusion of scaling factors $S_a$ and $S_b$.

Taking account of these three points and using (6.77) and (6.78), we obtain the matching conditions (6.102)–(6.105) in the desired forms

$$E_{ax}S_a + E_{bx}S_b = N_1 \tag{6.117}$$
$$E_{ay}S_a + E_{by}S_b = N_2 \tag{6.118}$$
$$\alpha_a S_a + \alpha_b S_b = N_3 \tag{6.119}$$
$$\beta_a S_a + \beta_b S_b = N_4. \tag{6.120}$$

Here

$$N_1 = (E_{ip} - E_{rp})\cos\theta - ieF_1(E_{ip} + E_{rp})\sin^2\theta \qquad (6.121)$$

$$N_2 = E_{is} + E_{rs} \qquad (6.122)$$

$$N_3 = e(E_{is} - E_{rs})\cos\theta \qquad (6.123)$$

$$N_4 = e(E_{ip} + E_{rp})[1 - (1 + F_8)\sin^2\theta] \qquad (6.124)$$

$$\alpha_m = iF_2 E_{mx} + (k_{mz} + iF_3)E_{my} \qquad (6.125)$$

$$\beta_m = [k_{mz} + i(F_4 + F_5)]E_{mx} + i(F_6 + F_7)E_{my}, \qquad (6.126)$$

where in (6.125) and (6.126), $m$ is $a$ or $b$.

We now solve (6.117) and (6.118) for $S_a$ and $S_b$ and then substitute these into (6.119) and (6.120) to obtain

$$c_1 E_{rp} + c_2 E_{rs} + c_3 E_{ip} + c_4 E_{is} = 0 \qquad (6.127)$$

$$d_1 E_{rp} + d_2 E_{rs} + d_3 E_{ip} + d_4 E_{is} = 0, \qquad (6.128)$$

where

$$c_1 = (\cos\theta + ieF_1\sin^2\theta)(\alpha_a E_{by} - \alpha_b E_{ay}) \qquad (6.129)$$

$$c_2 = \alpha_a E_{bx} - \alpha_b E_{ax} - e(E_{ax}E_{by} - E_{bx}E_{ay})\cos\theta \qquad (6.130)$$

$$c_3 = (\cos\theta - ieF_1\sin^2\theta)(\alpha_b E_{ay} - \alpha_a E_{by}) \qquad (6.131)$$

$$c_4 = \alpha_a E_{bx} - \alpha_b E_{ax} + e(E_{ax}E_{by} - E_{bx}E_{ay})\cos\theta \qquad (6.132)$$

$$d_1 = (\cos\theta + ieF_1\sin^2\theta)(\beta_a E_{by} - \beta_b E_{ay})$$
$$+ e(E_{ax}E_{by} - E_{bx}E_{ay})[1 - (1 + F_8)\sin^2\theta] \qquad (6.133)$$

$$d_2 = d_4 = \beta_a E_{bx} - \beta_b E_{ax} \qquad (6.134)$$

$$d_3 = (\cos\theta - ieF_1\sin^2\theta)(\beta_b E_{ay} - \beta_a E_{by})$$
$$+ e(E_{ax}E_{by} - E_{bx}E_{ay})[1 - (1 + F_8)\sin^2\theta]. \qquad (6.135)$$

Finally, from (6.127) and (6.128), the reflected amplitudes $E_{rp}$ and $E_{rs}$ are given in terms of the incident amplitudes $E_{ip}$ and $E_{is}$ by

$$E_{rp} = r_{pp}E_{ip} + r_{ps}E_{is} \qquad (6.136)$$

$$E_{rs} = r_{sp}E_{ip} + r_{ss}E_{is}, \qquad (6.137)$$

where

$$r_{pp} = (c_3 d_2 - c_2 d_3)/\Delta \qquad (6.138)$$

$$r_{ps} = (c_4 d_2 - c_2 d_4)/\Delta \qquad (6.139)$$

$$r_{sp} = (c_1 d_3 - c_3 d_1)/\Delta \qquad (6.140)$$

$$r_{ss} = (c_1 d_4 - c_4 d_1)/\Delta \qquad (6.141)$$

and

$$\Delta = c_2 d_1 - c_1 d_2. \qquad (6.142)$$

Thus (6.138)–(6.141) are the four Fresnel reflection coefficients in (6.2). Specifically, they are reflection coefficients for non-magnetic uniaxial and cubic crystals, in the configuration mentioned at the beginning of this section, given by multipole theory to electric quadrupole–magnetic dipole order.

To the order of electric dipole $a_{ijk} = G'_{ij} = 0$, so that from (6.106) to (6.112) $F_i = 0$ ($i = 1, \ldots, 7$). Then from (6.125), (6.126), and (6.129)–(6.135) the quantities in the Fresnel coefficients (6.138)–(6.141) are given by

$$c_1 = -c_3 = (k_{az} - k_{bz})E_{ay}E_{by}\cos\theta \tag{6.143}$$

$$c_2 = (k_{az} + e\cos\theta)E_{ay}E_{bx} - (k_{bz} + e\cos\theta)E_{ax}E_{by} \tag{6.144}$$

$$c_4 = (k_{az} - e\cos\theta)E_{ay}E_{bx} - (k_{bz} - e\cos\theta)E_{ax}E_{by} \tag{6.145}$$

$$d_1 = [k_{az}\cos\theta + e(1 - n_E^{-2}\sin^2\theta)]E_{ax}E_{by}$$
$$\quad - [k_{bz}\cos\theta + e(1 - n_E^{-2}\sin^2\theta)]E_{ay}E_{bx} \tag{6.146}$$

$$d_2 = d_4 = (k_{az} - k_{bz})E_{ax}E_{bx} \tag{6.147}$$

$$d_3 = -[k_{az}\cos\theta - e(1 - n_E^{-2}\sin^2\theta)]E_{ax}E_{by}$$
$$\quad + [k_{bz}\cos\theta - e(1 - n_E^{-2}\sin^2\theta)]E_{ay}E_{bx}. \tag{6.148}$$

We also require for later use the Fresnel coefficients (6.138)–(6.141) for normal incidence ($\theta = 0$). From (6.78) and (6.79), $k_{1x} = 0$ and hence $F_4 = F_6 = 0$. From (6.125) and (6.126), $\alpha_a$ and $\alpha_b$ are unaffected, while

$$\beta_m = (k_{mz} + iF_5)E_{mx} + iF_7 E_{my} \tag{6.149}$$

($m = a$ or $b$). Equations (6.129)–(6.135) become

$$c_1 = -c_3 = \alpha_a E_{by} - \alpha_b E_{ay} \tag{6.150}$$

$$c_2 = \alpha_a E_{bx} - \alpha_b E_{ax} - e(E_{ax}E_{by} - E_{bx}E_{ay}) \tag{6.151}$$

$$c_4 = \alpha_a E_{bx} - \alpha_b E_{ax} + e(E_{ax}E_{by} - E_{bx}E_{ay}) \tag{6.152}$$

$$d_1 = \beta_a E_{by} - \beta_b E_{ay} + e(E_{ax}E_{by} - E_{bx}E_{ay}) \tag{6.153}$$

$$d_2 = d_4 = \beta_a E_{bx} - \beta_b E_{ax} \tag{6.154}$$

$$d_3 = \beta_b E_{ay} - \beta_a E_{by} + e(E_{ax}E_{by} - E_{bx}E_{ay}). \tag{6.155}$$

Consequently, (6.139) and (6.140) become

$$r_{ps} = d_2(c_4 - c_2)/\Delta \tag{6.156}$$

$$r_{sp} = c_1(c_4 - c_2)/\Delta, \tag{6.157}$$

while the expressions (6.138) and (6.141) for $r_{pp}$ and $r_{ss}$ remain the same.

## 6.6   Solutions of the wave equation

The polarizability tensors of the medium enter the reflection coefficients (6.138)–(6.141) both through the matching conditions (Section 6.5) and the eigenpolarizations of the waves $a$ and $b$ propagating in the medium. We now determine the latter by solving the wave equation (5.4), in which for non-magnetic crystals and to electric quadrupole–magnetic dipole order we retain (5.6) and (5.9) but omit the other multipole terms. Compared with the wave equations in Sections 5.2–5.9, where simple propagation directions were chosen inside the crystal, the solution for the two refracted waves is now more complicated. For an arbitrary angle of incidence, the propagation directions of these two rays, as determined from Snell's law, are not in general along crystallographic axes. One simplification is that $k_{ay} = k_{by} = 0$ (see (6.76)).

The crystal is placed in the configuration relative to laboratory axes that is described in Section 6.5, so that (6.83) applies. Furthermore, the odd-rank polar tensor $a_{ijk}$ and the even-rank tensor $G'_{ij}$ vanish relative to crystallographic axes for all non-magnetic centrosymmetric crystals (Section 3.6). We therefore consider the remaining uniaxial and cubic symmetry classes, which are

$$\text{uniaxials:} \quad 4,\ \bar{4},\ 422,\ 4mm,\ \bar{4}2m,\ 3,\ 32,\ 3m,\ 6,\ \bar{6},\ 622,\ 6mm,\ \bar{6}m2 \quad (6.158)$$

$$\text{cubics:} \quad 23,\ 432,\ \bar{4}3m. \quad (6.159)$$

The wave equation in the form of the secular equation in (5.12) and (5.13) can be shown to be

$$\begin{vmatrix} -j_z^2 + n_O^2 & i\varepsilon_0^{-1}c^{-1}j_z A_{xyz} & j_x j_z + i\varepsilon_0^{-1}c^{-1}j_x A_{xzx} \\ -i\varepsilon_0^{-1}c^{-1}j_z A_{xyz} & -j_x^2 - j_z^2 + n_O^2 & i\varepsilon_0^{-1}c^{-1}j_x A_{yzx} \\ j_x j_z - i\varepsilon_0^{-1}c^{-1}j_x A_{xzx} & -i\varepsilon_0^{-1}c^{-1}j_x A_{yzx} & -j_x^2 + n_E^2 \end{vmatrix} = 0, \quad (6.160)$$

where

$$\mathbf{j} = \frac{c}{\omega}\mathbf{k} = n\boldsymbol{\sigma} \quad (6.161)$$

and

$$A_{ijk} = \varepsilon_{jkl}G'_{il} - \varepsilon_{ikl}G'_{jl} + \frac{1}{2}\omega(a_{ijk} - a_{jki}) = -A_{jik}. \quad (6.162)$$

For a uniaxial crystal $n_O$ and $n_E$ are the ordinary and extraordinary refractive indices, given by (5.52) and (5.53), respectively. For a cubic crystal $n_E = n_O$. Table 6.1 lists the components $A_{xyz} = A_1$, $A_{xzx} = A_2$, and $A_{yzx} = A_3$ for the classes in (6.158) and (6.159).

We consider separately the solutions to (6.160) for oblique incidence and for normal incidence. For oblique incidence ($j_x \neq 0$), and to first order in the polarizability tensors $a_{ijk}$ and $G'_{jk}$, the solutions to (6.160) are given by

$$j_{az}^2 = n_O^2 - j_x^2 \quad (6.163)$$

$$\frac{E_{ax}}{E_{ay}} = \frac{i(n_O^2 - j_x^2)^{1/2}[(n_E^2 - j_x^2)A_1 + j_x^2 A_3]}{\varepsilon_0 c j_x^2(n_O^2 - n_E^2)} \quad (6.164)$$

$$\frac{E_{az}}{E_{ay}} = -\frac{i[(n_O^2 - j_x^2)A_1 + j_x^2 A_3]}{\varepsilon_0 c j_x(n_O^2 - n_E^2)} \quad (6.165)$$

and

$$j_{bz}^2 = n_E^{-2}n_O^2(n_E^2 - j_x^2) \tag{6.166}$$

$$\frac{E_{by}}{E_{bx}} = \frac{in_O n_E[(n_E^2 - j_x^2)A_1 + j_x^2 A_3]}{\varepsilon_0 c j_x^2(n_O^2 - n_E^2)(n_E^2 - j_x^2)^{1/2}} \tag{6.167}$$

$$\frac{E_{bz}}{E_{bx}} = \frac{j_x[in_E A_2 - \varepsilon_0 c n_O(n_E^2 - j_x^2)^{1/2}]}{\varepsilon_0 c n_E(n_E^2 - j_x^2)}. \tag{6.168}$$

In the fields above, the symbol $E$ denotes an amplitude (Section 5.1).

The solutions (6.163)–(6.168) apply to oblique incidence on uniaxial crystals but not on cubics because of the singularity in (6.164), (6.165), and (6.167) when $n_E = n_O$. For cubics one must return to the wave equation (6.160) and set $n_E = n_O$ and, for the classes 23 and 432, $A_{xyz} = A_{yzx} = A_1$ and $A_{xzx} = 0$, as in Table 6.1. Then, to first order in the polarizability tensor, the solutions are given by

$$j_{az}^2 = n_O^2 - j_x^2 + \varepsilon_0^{-1}c^{-1}n_O A_1 \tag{6.169}$$

$$\mathbf{E}_a = [in_O^{-2}j_{az}(n_O - \varepsilon_0^{-1}c^{-1}A_1), 1, -in_O^{-2}j_x(n_O - \varepsilon_0^{-1}c^{-1}A_1)] \tag{6.170}$$

$$j_{bz}^2 = n_O^2 - j_x^2 - \varepsilon_0^{-1}c^{-1}n_O A_1 \tag{6.171}$$

$$\mathbf{E}_b = [-in_O^{-2}j_{bz}(n_O + \varepsilon_0^{-1}c^{-1}A_1), 1, in_O^{-2}j_x(n_O + \varepsilon_0^{-1}c^{-1}A_1)]. \tag{6.172}$$

**Table 6.1** *Components of the tensor $A_{ijk}$ in (6.162) for non-centrosymmetric non-magnetic uniaxial and cubic crystals.*

| Class | $A_{xyz} = A_1$ | $A_{xzx} = A_2$ | $A_{yzx} = A_3$ |
|---|---|---|---|
| Uniaxial | | | |
| 4, 3, 6 | $L_1$ | $L_3$ | $L_4$ |
| 422, 32, 622 | $L_1$ | 0 | $L_4$ |
| $\bar{4}$ | 0 | $L_3$ | $L_5$ |
| $\bar{4}2m$ | 0 | 0 | $L_5$ |
| $4mm$, $3m$, $6mm$ | 0 | $L_3$ | 0 |
| $\bar{6}$, $\bar{6}m2$ | 0 | 0 | 0 |
| Cubic | | | |
| 23 | $L_2$ | 0 | $L_2$ |
| 432 | $L_2$ | 0 | $L_2$ |
| $\bar{4}3m$ | 0 | 0 | 0 |

$$L_1 = 2G'_{xx} + \omega a_{xyz}, \quad L_2 = 2G'_{xx}, \quad L_3 = G'_{xy} + \frac{1}{2}\omega(a_{xxz} - a_{zxx}),$$

$$L_4 = G'_{xx} + G'_{zz} - \frac{1}{2}\omega a_{xyz}, \quad L_5 = -G'_{xx} + \frac{1}{2}\omega(a_{xyz} - a_{zxy})$$

Table 6.1 shows that $A_1 = 0$ for the cubic class $\bar{4}3m$, and its solutions then follow from (6.169) to (6.172).

It is useful to note from (6.164), (6.165), (6.167), (6.168), (6.170), (6.172), and Table 6.1 that all three amplitude components of both refracted waves are non-zero except for the uniaxial crystal classes

$$4mm, \quad 3m, \quad 6mm \tag{6.173}$$

and

$$\bar{6}, \quad \bar{6}m2. \tag{6.174}$$

For these the amplitudes are

$$\mathbf{E}_a = (0, E_{ay}, 0), \qquad \mathbf{E}_b = (E_{bx}, 0, E_{bz}). \tag{6.175}$$

Next we solve the wave equation for normal incidence. From (6.78) and (6.161), $j_x = 0$, so that (6.160) becomes

$$\begin{vmatrix} -j_z^2 + n_O^2 & i\varepsilon_0^{-1}c^{-1}j_z A_{xyz} & 0 \\ -i\varepsilon_0^{-1}c^{-1}j_z A_{xyz} & -j_z^2 + n_O^2 & 0 \\ 0 & 0 & n_E^2 \end{vmatrix} = 0. \tag{6.176}$$

Of the uniaxial and cubic classes in Table 6.1, $A_{xyz}$ in (6.176) exists only for

$$4, 3, 6; \quad 422, 32, 622; \quad 23, 432, \tag{6.177}$$

which are the only optically active ones listed (see Section 5.3). For them the solutions of (6.176) to first order in $A_{xyz}$ are

$$j_{az} = n_O + (2\varepsilon_0 c)^{-1} A_{xyz}, \qquad \mathbf{E}_a = (1, -i, 0) \tag{6.178}$$
$$j_{bz} = n_O - (2\varepsilon_0 c)^{-1} A_{xyz}, \qquad \mathbf{E}_b = (1, i, 0). \tag{6.179}$$

From (5.27) and (5.28) the eigenpolarizations in (6.178) and (6.179) are circular, as expected.

For the remaining classes in Table 6.1 the solutions for normal incidence are exact, namely

$$j_{az} = n_O, \qquad \mathbf{E}_a = (0, 1, 0) \tag{6.180}$$
$$j_{bz} = n_O, \qquad \mathbf{E}_b = (1, 0, 0). \tag{6.181}$$

Equations (6.178)–(6.181) include the case of normal incidence for cubics.

## 6.7  Reflection coefficients

To evaluate the reflection coefficients (6.138)–(6.141) we require the eight quantities $F_i$ defined in (6.106)–(6.113). These are given in Table 6.2 in terms of polarizability tensors for various symmetry classes of non-magnetic uniaxial and cubic crystals. Consider first oblique incidence. Because of the complicated algebra involved in calculating the reflection coefficients when all amplitude components exist, we restrict ourselves to the classes in (6.173) and (6.174). We begin with those in (6.173), for which, from (6.125), (6.126), (6.175), and Table 6.2, we find

$$\alpha_a = (k_{az} + iF_3)E_{ay}, \qquad \alpha_b = 0, \qquad\qquad (6.182)$$
$$\beta_a = 0, \qquad\qquad \beta_b = [k_{bz} + i(F_4 + F_5)]E_{bx}. \qquad (6.183)$$

Using these and (6.175) in (6.129)–(6.135), we obtain

**Table 6.2** *Expressions for $F_i$ in (6.106)–(6.113) for non-centrosymmetric non-magnetic uniaxial and cubic crystals.*

| Class | $F_1$ | $F_2$ | $F_3$ | $F_4$ | $F_5$ | $F_6$ | $F_7$ | $F_8$ |
|---|---|---|---|---|---|---|---|---|
| Uniaxial | | | | | | | | |
| 4, 3, 6 | $I_1$ | $I_2$ | $I_5$ | $I_6$ | $I_5$ | $I_7$ | $-I_2$ | $I_{13}$ |
| 422, 32, 622 | 0 | $I_2$ | 0 | 0 | 0 | $I_7$ | $-I_2$ | $I_{14}$ |
| $\bar{4}$ | 0 | $I_2$ | $-I_5$ | $I_6$ | $I_5$ | $I_8$ | $I_2$ | $I_{14}$ |
| $\bar{4}2m$ | 0 | $I_2$ | 0 | 0 | 0 | $I_8$ | $I_2$ | $I_{14}$ |
| 4mm, 3m, 6mm | $I_1$ | 0 | $I_5$ | $I_6$ | $I_5$ | 0 | 0 | $I_{13}$ |
| $\bar{6}$, $\bar{6}m2$ | 0 | 0 | 0 | 0 | 0 | 0 | 0 | $I_{14}$ |
| Cubic | | | | | | | | |
| 23 | 0 | $I_2$ | 0 | 0 | 0 | $I_9$ | $I_{12}$ | $I_{14}$ |
| 432 | 0 | $I_3$ | 0 | 0 | 0 | $I_{10}$ | $-I_3$ | $I_{14}$ |
| $\bar{4}3m$ | 0 | $I_4$ | 0 | 0 | 0 | $I_{11}$ | $I_4$ | $I_{14}$ |

$$I_1 = (2\varepsilon_0 n_E^2)^{-1}a_{zzz}, \quad I_2 = \mu_0\omega\left(G'_{xx} + \frac{1}{2}\omega a_{xyz}\right), \quad I_3 = \mu_0\omega G'_{xx},$$

$$I_4 = \frac{1}{2}\mu_0\omega^2 a_{xyz}, \quad I_5 = -\mu_0\omega\left(G'_{xy} - \frac{1}{2}\omega a_{xxz}\right),$$

$$I_6 = (\varepsilon_0 n_E^2)^{-1}k_{1x}^2\left(\frac{1}{2}a_{zxx} - a_{xxz}\right), \quad I_7 = -(\varepsilon_0 n_E^2)^{-1}k_{1x}^2(\omega^{-1}G'_{zz} + a_{yzx}),$$

$$I_8 = (\varepsilon_0 n_E^2)^{-1}k_{1x}^2\left(\frac{1}{2}a_{zzy} - a_{xyz}\right), \quad I_9 = -(\omega n_E)^{-2}c^2 k_{1x}^2 I_2,$$

$$I_{10} = -(\omega n_E)^{-2}c^2 k_{1x}^2 I_3, \quad I_{11} = -(\omega n_E)^{-2}c^2 k_{1x}^2 I_4,$$

$$I_{12} = -\mu_0\omega\left(G'_{xx} - \frac{1}{2}\omega a_{xyz}\right), \quad I_{13} = n_E^{-2}\left(1 - \frac{i}{2}\varepsilon_0^{-1}n_E^{-2}k_{2z}a_{zzz}\right) - 1, \quad I_{14} = n_E^{-2} - 1$$

$$c_1 = c_3 = d_2 = d_4 = 0 \tag{6.184}$$

$$c_2 = (k_{az} + iF_3 + e\cos\theta)E_{bx}E_{ay} \tag{6.185}$$

$$c_4 = (k_{az} + iF_3 - e\cos\theta)E_{bx}E_{ay} \tag{6.186}$$

$$d_1 = -\big[ieF_1 k_{bz}\sin^2\theta + \cos\theta\{k_{bz} + i(F_4 + F_5)\} \\ + e\{1 - (1 + F_8)\sin^2\theta\}\big]E_{bx}E_{ay} \tag{6.187}$$

$$d_3 = \big[-ieF_1 k_{bz}\sin^2\theta + \cos\theta\{k_{bz} + i(F_4 + F_5)\} \\ - e\{1 - (1 + F_8)\sin^2\theta\}\big]E_{bx}E_{ay}. \tag{6.188}$$

Finally, according to (6.138)–(6.142) and (6.184) we have

$$r_{pp} = -d_3/d_1 \tag{6.189}$$

$$r_{ss} = -c_4/c_2 \tag{6.190}$$

$$r_{ps} = r_{sp} = 0. \tag{6.191}$$

Equations (6.189)–(6.191) give the reflection coefficients for oblique incidence on a non-magnetic crystal belonging to one of the uniaxial classes in (6.173).

For the classes in (6.174), Table 6.2 shows that the only non-zero $F_i$ is $F_8$. Thus (6.189)–(6.191) apply, where from (6.185) to (6.188),

$$c_2 = (k_{az} + e\cos\theta)E_{bx}E_{ay} \tag{6.192}$$

$$c_4 = (k_{az} - e\cos\theta)E_{bx}E_{ay} \tag{6.193}$$

$$d_1 = -\big[k_{bz}\cos\theta + e(1 - n_E^{-2}\sin^2\theta)\big]E_{bx}E_{ay} \tag{6.194}$$

$$d_3 = \big[k_{bz}\cos\theta - e(1 - n_E^{-2}\sin^2\theta)\big]E_{bx}E_{ay}. \tag{6.195}$$

Note that, unlike (6.185)–(6.188), these are independent of polarizability tensors of electric quadrupole–magnetic dipole order.

For the other symmetry classes in Table 6.1 all three amplitude components of the refracted waves are non-zero for oblique incidence (Section 6.6). The corresponding reflection coefficients are more complicated than those given above. We therefore consider normal incidence and present results for a few of these classes:

(i) For $\bar{4}$

$$r_{pp} = \frac{e^2(n_O^2 - 1) + 2ieI_5}{e^2(n_O + 1)^2} \tag{6.196}$$

$$r_{ss} = -\frac{e^2(n_O^2 - 1) - 2ieI_5}{e^2(n_O + 1)^2} \tag{6.197}$$

$$r_{ps} = \frac{2ieI_2}{e^2(n_O + 1)^2} = -r_{sp}, \tag{6.198}$$

where $I_2$ and $I_5$ are given in Table 6.2.

(ii) For 4, 3, 6

$$r_{pp} = \frac{e^2(n_O^2 - 1) + 2ien_O I_5}{e^2(n_O^2 + 1)^2 + ie(n_O + 2)I_5} = -r_{ss} \tag{6.199}$$

and $r_{ps} = r_{sp} = 0$. Here $I_5$ is given in Table 6.2. In (6.196)–(6.199) we have neglected contributions which are quadratic in the polarizability tensors.

## 6.8 Tests of translational and time-reversal invariance

We now examine whether the reflection coefficients obtained in the previous section are consistent with the requirements of translational invariance and time-reversal invariance.

1. *Translational invariance.* Consider uniaxial crystals of the classes in (6.173), and a beam with amplitude $E_{is}$ (i. e. $E_{ip} = 0$) incident obliquely on the crystal. Then from (6.1), (6.2), and (6.191) the reflected amplitude is $r_{ss}E_{is}$ and the reflected intensity is $I = |r_{ss}E_{is}|^2$. Thus if $\delta$ denotes a change due to a shift in the origin of coordinates, the corresponding change in the reflected intensity is

$$\delta I = (r_{ss}\delta r_{ss}^* + r_{ss}^*\delta r_{ss})|E_{is}|^2. \tag{6.200}$$

The change in $r_{ss}$ due to an arbitrary infinitesimal shift $\mathbf{d} = (d_x, d_y, d_z)$ of the origin is found from (6.185), (6.186), and (6.190), and is

$$\delta r_{ss} = -\frac{c_2\delta c_4 - c_4\delta c_2}{c_2^2} = -\frac{2ie\cos\theta\delta F_3}{(k_{az} + e\cos\theta + iF_3)^2}. \tag{6.201}$$

From Table 6.2 we have

$$F_3 = -\mu_0\omega\left(G'_{xy} - \frac{1}{2}\omega a_{xxz}\right), \tag{6.202}$$

and use of (3.66) and (3.69) shows that

$$\delta F_3 = -\mu_0\omega^2\alpha_{xx}d_z. \tag{6.203}$$

Equations (6.200), (6.201), and (6.203) yield

$$\delta I = -\frac{8e\mu_0\omega^2 k_{az}\cos^2\theta F_3\alpha_{xx}|E_{is}|^2}{[(k_{az} + e\cos\theta)^2 + F_3^2]^2}d_z, \tag{6.204}$$

where $k_{az} = e(n_O^2 - e^2\sin^2\theta)^{1/2}$. Equation (6.204) applies to uniaxial crystals of the classes in (6.173), that are used in the configuration described in Section 6.5, and is valid for arbitrary angle of incidence $\theta$. Clearly the calculated reflected intensity has the unphysical property that it depends on the choice of coordinate origin.

A similar calculation for the classes in (6.173) can be performed for an incident beam with amplitude $E_{ip}$ (i.e. $E_{is} = 0$). Then

$$\delta I = (r_{pp} \delta r_{pp}^* + r_{pp}^* \delta r_{pp}) |E_{ip}|^2, \qquad (6.205)$$

where $r_{pp}$ is given by (6.187)–(6.189). For arbitrary incidence, $\delta I$ is again nonzero and proportional to $d_z$, but the result is more complicated than (6.204). We present it for normal incidence ($\theta = 0$) only. Then $k_{1x} = 0$ and from Table 6.2, $F_4 = I_6 = 0$. Thus

$$\delta I = -\frac{8e^2 \mu_0 n_O \omega^2 F_5 \alpha_{xx} |E_{ip}|^2}{[e^2 (n_O + 1)^2 + F_5^2]^2} d_z, \qquad (6.206)$$

where $F_5 = F_3$ is given by (6.202). Again, the calculated intensity does not satisfy the physical requirement of translational invariance.

Additional examples are provided by the classes $\bar{4}$ and 4, 3, 6. For normal incidence the reflection coefficients are given by (6.196)–(6.199), and these also yield reflected intensities which are origin dependent.

We remark that for certain crystals possessing a high degree of symmetry, the reflected intensity is translationally invariant, at least to electric quadrupole–magnetic dipole order. The symmetry classes in (6.174) illustrate this. For these only one of the eight quantities $F_i$ is non-zero (Table 6.2), and the reflection coefficients given by (6.189)–(6.195) are all independent of the choice of origin. Therefore so are the reflected intensities. We discuss this further in Section 9.3.

2. *Time-reversal invariance (reciprocity)*. For non-magnetic crystals all polarizability tensors are time even. Thus the test for reciprocity (Section 6.2) is particularly simple. For the reflection coefficients calculated in Section 6.7 one has, by inspection, that $r_{pp}(-t) = r_{pp}(t)$ and $r_{ss}(-t) = r_{ss}(t)$. Also, from (6.198), $r_{ps}(-t) = r_{ps}(t) = -r_{sp}(t)$. Thus all the reflection coefficients in Section 6.7 satisfy the reciprocity relations (6.17)–(6.20).

## 6.9   Discussion

In this chapter we have demonstrated that multipole theory, when taken to electric quadrupole–magnetic dipole order, yields unphysical results for reflection of harmonic plane waves from crystal surfaces. In general, the algebra required for this purpose is rather involved, and we therefore concentrated on non-magnetic uniaxial crystals. We showed that for these the simplest crystal classes to consider are $4mm$, $3m$, $6mm$, and $\bar{6}$, $\bar{6}m2$ (Section 6.6). For the former group we found that the calculated reflection coefficients (Section 6.7) yield reflected intensities which are not translationally invariant (independent of choice of origin) (Section 6.8). The reflection coefficients do, however, satisfy the reciprocity relations required by time-reversal invariance.

These conclusions apply also to other symmetry classes. There are, however, exceptions to this pattern. Examples are the classes $\bar{6}$ and $\bar{6}m2$: for such crystals there are no contributions to the reflection coefficients coming from terms of electric quadrupole–magnetic dipole order, and consequently the requirements of both translational and time-reversal invariance are satisfied (Section 6.8).

The analysis can be extended to magnetic crystals by including the contributions of the time-odd polarizability tensors $G_{ij}$ and $a'_{ijk}$ (Sections 2.5 and 3.5). Then, as outlined in Section 2.11, equations (6.80)–(6.82) are replaced by

$$P_i = \alpha_{ij}E_j + \omega^{-1}\alpha'_{ij}\dot{E}_j + \frac{1}{2}a_{ijk}\nabla_k E_j + \frac{1}{2}\omega^{-1}a'_{ijk}\nabla_k \dot{E}_j$$
$$+ G_{ij}B_j + \omega^{-1}G'_{ij}\dot{B}_j \tag{6.207}$$

$$Q_{ij} = a_{kij}E_k - \omega^{-1}a'_{kij}\dot{E}_k \tag{6.208}$$

$$M_i = G_{ji}E_j - \omega^{-1}G'_{ji}\dot{E}_j. \tag{6.209}$$

The calculated reflection coefficients yield origin-dependent reflected intensities. Reflection from magnetic crystals is discussed further in Chapter 9.

We remind the reader of the discussion in Chapter 4 of linear constitutive relations for the response fields $\mathbf{D}$ and $\mathbf{H}$ in multipole theory. There we obtained multipole contributions to the material constants (permittivity, inverse permeability, and magnetoelectric coefficients), and showed the unphysical (origin dependent) nature of the contributions beyond electric dipole order. This, together with the results of the present chapter, indicates that existing multipole theory is unphysical when taken beyond electric dipole order. For certain phenomena, notably transmission effects such as those discussed in Chapter 5, the results obtained from such multipole theory are nevertheless physically acceptable. This has to do with the circumstance that origin-dependent constitutive relations yield an origin-independent wave equation (Sections 5.1 and 9.1).

The above remarks lead one to inquire whether it is possible to reformulate multipole theory in a manner which is free from these defects. Such a formulation should (i) yield origin-independent material constants, (ii) give results for reflection phenomena which are consistent with translational invariance, (iii) maintain the time-reversal invariance of the theory, and (iv) leave the multipole theory of transmission effects unaltered. This task is undertaken in the next two chapters.

## References

[1] Jackson, J. D. (1999). *Classical electrodynamics*, Sect. 7.3. Wiley, New York.

[2] See, for example, Landau, L. D. and Lifshitz, E. M. (1968). *Statistical physics*. Pergamon, Oxford.

[3] Sommerfeld, A. (1956). *Thermodynamics and statistical mechanics*. Academic Press, New York.

[4] Solomon, P. R. and Otter, F. A. (1967). Thermomagnetic effects in superconductors. *Physical Review*, **164**, 608–618.

[5] de Lange, O. L. and Otter, F. A. (1975). Flux flow effects in a nearly reversible Type II superconductor. *Journal of Low Temperature Physics*, **18**, 31–42.

[6] Perrin, F. (1942). Polarization of light scattered by isotropic opalescent media. *Journal of Chemical Physics*, **10**, 415–427. This paper also contains references to earlier work on the reciprocity principle.

[7] de Figueiredo, I. M. B. and Raab, R. E. (1980). A pictorial approach to macroscopic space-time symmetry, with particular reference to light scattering. *Proceedings of the Royal Society London* A, **369**, 501–516.

[8] Shelankov, A. L. and Pikus, G. E. (1992). Reciprocity in reflection and transmission of light. *Physical Review* B, **46**, 3326–3336.

[9] Canright, G. S. and Rojo, A. G. (1992). Ellipsometry and broken time-reversal symmetry in the high-temperature superconductors. *Physical Review* B, **46**, 14078–14088.

[10] See Ref. 1, Sect. 1.5.

[11] Born, M. and Wolf, E. (1980). *Principles of optics*, p. 763, Pergamon, Oxford.

[12] Hall, D. G. (1995). A few remarks on the matching conditions at interfaces in electromagnetic theory. *American Journal of Physics*, **63**, 508–512.

[13] Magid, L. M. (1972). *Electromagnetic fields, energy and waves*, Sect. 4.3. Wiley, New York.

[14] Graham, E. B. and Raab, R. E. (2000). Multipole solution for the macroscopic electromagnetic boundary conditions at a vacuum-dielectric interface. *Proceedings of the Royal Society London* A, **456**, 1193–1215.

[15] Langreth, D. C. (1989). Macroscopic approach to the theory of reflectivity. *Physical Review* B, **39**, 10020–10027.

[16] Graham, E. B. and Raab, R. E. (2001). The role of the macroscopic surface density of bound electric dipole moment in reflection. *Proceedings of the Royal Society London* A, **457**, 471–484.

[17] Wangsness, R. K. (1986). *Electromagnetic fields*, p. 354. Wiley, New York.

[18] See Ref. 11, Sect. 14.2 and 14.3.

[19] Birss, R. R. (1966). *Symmetry and magnetism*. North–Holland, Amsterdam.

# 7

## TRANSFORMATIONS OF THE RESPONSE FIELDS AND THE CONSTITUTIVE TENSOR

*There will be time to audit*
*The accounts later, there will be sunlight later*
*And the equation will come out at last.*
Louis Macneice
(*Autumn Journal*)

*Listen: there's a hell*
*of a good universe next door; let's go*
E. E. Cummings
(*Pity this busy monster, manunkind*)

The fundamental role of transformations in the theory of the electromagnetic field has been known for a long time. The possibility of transformations arises because certain quantities are not uniquely specified in electromagnetism. In this chapter we discuss three such transformations: (i)The well-known gauge transformation of the 4-vector potential which leaves the fields $\mathbf{E}$ and $\mathbf{B}$ unchanged (Section 7.1). (ii) Transformations of the response fields $\mathbf{D}$ and $\mathbf{H}$ of macroscopic electromagnetism which leave the inhomogeneous Maxwell equations unchanged (Section 7.2). (iii) Transformations which leave $\mathbf{D}$ and $\mathbf{H}$ unchanged but alter the relative contributions of $\mathbf{E}$ and $\mathbf{B}$ to $\mathbf{D}$ and $\mathbf{H}$ (Section 7.3). The latter two transformations are then formulated in a manner which can be applied to the linear constitutive relations of multipole theory (Section 7.4).

### 7.1   Gauge transformations of the 4-vector potential

It is clear from the relation $\mathbf{B} = \boldsymbol{\nabla} \times \mathbf{A}$ that $\mathbf{B}$ is unchanged by the substitution $\mathbf{A} \to \mathbf{A} + \boldsymbol{\nabla}\chi$, or $\mathbf{B} \to \mathbf{B} + \boldsymbol{\nabla} \times (\boldsymbol{\nabla}\chi)$, where $\chi(\mathbf{r}, t)$ is an (almost) arbitrary function. A similar invariance applies to $\mathbf{E} = -\boldsymbol{\nabla}\Phi - \dot{\mathbf{A}}$ if we also let $\Phi \to \Phi - \dot{\chi}$. In general, the electromagnetic field tensor

$$F^{\nu\mu} = \partial^{\nu} A^{\mu} - \partial^{\mu} A^{\nu} \tag{7.1}$$

is unchanged by the gauge transformation

$$A^{\mu} \to A^{\mu} + \partial_{\mu}\chi, \tag{7.2}$$

provided $\partial^{\nu}\partial_{\mu}\chi = \partial^{\mu}\partial_{\nu}\chi$. (For an explanation of the notation used above, see Section 4.3.) Thus the 4-vector potential $A^{\mu}$ is not uniquely specified by the electric and magnetic fields $\mathbf{E}$ and $\mathbf{B}$: one can change $A^\mu$ in the manner (7.2) without changing $\mathbf{E}$ and $\mathbf{B}$ in (4.27). This gauge invariance of the electromagnetic field is (when extended to include non-Abelian fields) the basis for the gauge principle and a theory of fundamental interactions [1, 2].

## 7.2   "Gauge transformations" of response fields

It is well known that the inhomogeneous macroscopic Maxwell equations of Section 1.13,

$$\nabla \cdot \mathbf{D} = \rho_f \qquad (7.3)$$

and

$$\nabla \times \mathbf{H} = \mathbf{J}_f + \dot{\mathbf{D}}, \qquad (7.4)$$

do not specify the response fields $\mathbf{D}$ and $\mathbf{H}$ uniquely. This follows from Helmholtz's theorem which requires that both the divergence and curl of a vector field, together with a suitable boundary condition, be known for the field to be specified completely [3]. To discuss the non-uniqueness of $\mathbf{D}$ and $\mathbf{H}$ it is convenient to consider the Fourier transforms of (7.3) and (7.4). Using the Fourier expansion

$$\mathbf{D}(\mathbf{r}, t) = \iint_{-\infty}^{\infty} d^3k \, d\omega \, e^{i(\mathbf{k}\cdot\mathbf{r}-\omega t)} \mathbf{D}(\mathbf{k}, \omega) \qquad (7.5)$$

and similar expansions for $\mathbf{H}$, $\mathbf{J}_f$, and $\rho_f$ in (7.3) and (7.4), we obtain

$$\mathbf{k} \cdot \mathbf{D}(\mathbf{k}, \omega) = -i\rho_f(\mathbf{k}, \omega) \qquad (7.6)$$

and

$$\mathbf{k} \times \mathbf{H}(\mathbf{k}, \omega) = -i\mathbf{J}_f(\mathbf{k}, \omega) - \omega \mathbf{D}(\mathbf{k}, \omega). \qquad (7.7)$$

Thus (7.6) and (7.7) are invariant under the transformations

$$\mathbf{H}(\mathbf{k}, \omega) \to \mathbf{H}(\mathbf{k}, \omega) + \mathbf{H}^G(\mathbf{k}, \omega), \quad \mathbf{D}(\mathbf{k}, \omega) \to \mathbf{D}(\mathbf{k}, \omega) - \frac{1}{\omega} \mathbf{k} \times \mathbf{H}^G(\mathbf{k}, \omega), \quad (7.8)$$

for arbitrary $\mathbf{H}^G$. We adopt a terminology that is sometimes used [4] and refer to (7.8) as "gauge transformations" (although they are distinct from (7.2)).

The application we have in mind is simpler than the above in that we are concerned with fields represented by complex harmonic plane waves

$$\mathbf{D}(\mathbf{r}, t) = \mathbf{D}_0 \, e^{i(\mathbf{k}\cdot\mathbf{r}-\omega t)}, \qquad (7.9)$$

and similarly for $\mathbf{H}$. For these the gauge property follows directly from (7.3) and (7.4), and it can be stated in terms of the fields rather than their Fourier transforms: if $\mathbf{H}^G(\mathbf{r}, t)$ is a complex harmonic field like (7.9) then the gauge transformations

$$H_i \to H_i + H_i^G, \qquad D_i \to D_i - \frac{1}{\omega} \varepsilon_{ijk} k_j H_k^G \qquad (7.10)$$

of the fields $\mathbf{H}(\mathbf{r}, t)$ and $\mathbf{D}(\mathbf{r}, t)$ will leave (7.3) and (7.4) unchanged. (In (7.10) we have, for later convenience, expressed a vector product in terms of the Levi–Civita tensor $\varepsilon_{ijk}$ defined in (1.32).)

## 7.3    Faraday transformations

For complex harmonic fields like (7.9), Faraday's law $\nabla \times \mathbf{E} + \dot{\mathbf{B}} = 0$ can be written

$$\varepsilon_{jmn} k_m E_n - \omega B_j = 0. \tag{7.11}$$

Therefore $\mathbf{D}$ and $\mathbf{H}$ are unchanged by the substitutions

$$H_i \rightarrow H_i + H_i^F, \qquad D_i \rightarrow D_i + D_i^F, \tag{7.12}$$

where

$$H_i^F = Y_{ij}^F \left( B_j - \frac{1}{\omega} \varepsilon_{jmn} k_m E_n \right) \tag{7.13}$$

$$D_i^F = Z_{ij}^F \left( B_j - \frac{1}{\omega} \varepsilon_{jmn} k_m E_n \right) \tag{7.14}$$

and $Y_{ij}^F$ and $Z_{ij}^F$ are any second-rank tensors. The properties of these two tensors, in applications to multipole theory, are discussed in the next section. We refer to (7.12)–(7.14) as "Faraday transformations" to distinguish them from the gauge transformations (7.10). Note that, *by construction, Faraday transformations cannot change the fields* $\mathbf{D}$ *and* $\mathbf{H}$. They can, however, play a useful role in transforming linear constitutive relations because the linear combinations of $\mathbf{E}$ and $\mathbf{B}$ in (7.13) and (7.14) will alter the relative contributions of $\mathbf{E}$ and $\mathbf{B}$ to $\mathbf{D}$ and $\mathbf{H}$, and hence the material constants (see below).

## 7.4    Transformations of linear constitutive relations in multipole theory

We wish to develop a transformation theory [5] for the linear constitutive relations of multipole theory, based on the gauge and Faraday transformations of Sections 7.2 and 7.3. To do so, we first recall certain features of multipole theory which should be built into our transformation theory: (i) The constitutive relations (4.2) and (4.3) are characterized by a linear homogeneous dependence of $\mathbf{D}$ and $\mathbf{H}$ on the fields $\mathbf{E}$ and $\mathbf{B}$, and also on the macroscopic polarizability tensors of each order (Section 4.1). (ii) These fields and polarizability tensors possess certain properties under space inversion and time reversal (Section 3.5). (iii) Each order of the multipole expansion is characterized by the highest power of the wave vector $k_i$ that appears in the corresponding material constant. For example, in $A_{ij}^M$ to electric dipole order it is $k_i^0$; to electric quadrupole–magnetic dipole order it is $k_i$; to electric octopole–magnetic quadrupole order it is $k_i^2$, and so on (Section 4.5). These powers of $k_i$ are, of course, associated with the various space derivatives that enter each order of the multipole expansion (Section 4.1). (iv) The discussion in Chapters 4, 5, and 6 also motivates us to impose as a goal of the transformation theory that it should yield origin-independent material constants and preserve the symmetries (4.40)–(4.42) for non-dissipative media.

We therefore consider the most general transformations of the type (7.10) and (7.12)–(7.14) which satisfy the following requirements [5]:

1. Preserve the linear homogeneous dependence of **D** and **H** on the fields **E** and **B**.
2. Preserve the linear homogeneous dependence of **D** and **H** on the appropriate macroscopic polarizability tensors.
3. Maintain the order of the multipole expansion.
4. Be consistent with requirements of space inversion and time reversal for the fields **E**, **B**, **D**, **H** and the polarizability tensors.
5. Impose the conditions (4.9) for origin independence of the material constants.
6. Preserve the symmetries (4.40)–(4.42) for non-dissipative media.

(The reader may wonder why we do not also impose the Post constraint (4.53) as a requirement. The answer is that it is unnecessary to do so; see Chapter 8.)

The requirement 1 applied to the transformations (7.10) and (7.12)–(7.14) means first that in (7.10) we should take

$$H_i^G = U_{ij}^G E_j + X_{ij}^G B_j, \tag{7.15}$$

and second that the tensors $U_{ij}^G$ and $X_{ij}^G$, like the tensors $Y_{ij}^F$ and $Z_{ij}^F$ in (7.13) and (7.14), *should be independent of the fields* **E** *and* **B**. The **D** and **H** fields (4.2) and (4.3) transformed according to (7.10) and (7.12)–(7.14) also satisfy linear constitutive relations, but with transformed material constants

$$A_{ij} = A_{ij}^M - \frac{1}{\omega} \varepsilon_{ikl} k_k U_{lj}^G + \frac{1}{\omega} \varepsilon_{jkl} k_k Z_{il}^F \tag{7.16}$$

$$T_{ij} = T_{ij}^M - \frac{1}{\omega} \varepsilon_{ikl} k_k X_{lj}^G + Z_{ij}^F \tag{7.17}$$

$$U_{ij} = U_{ij}^M + U_{ij}^G + \frac{1}{\omega} \varepsilon_{jkl} k_k Y_{il}^F \tag{7.18}$$

$$X_{ij} = X_{ij}^M + X_{ij}^G + Y_{ij}^F, \tag{7.19}$$

where, as before, the superscript $M$ refers to the tensors given directly by multipole theory (Section 4.5). The field-independent tensors $U_{ij}^G$, $X_{ij}^G$, $Y_{ij}^F$, and $Z_{ij}^F$ in (7.16)–(7.19) remain to be determined.

According to (7.16)–(7.19), the transformed constitutive tensor is

$$\begin{pmatrix} A_{ij}^M - \omega^{-1}\varepsilon_{ikl}k_k U_{lj}^G + \omega^{-1}\varepsilon_{jkl}k_k Z_{il}^F & T_{ij}^M - \omega^{-1}\varepsilon_{ikl}k_k X_{lj}^G + Z_{ij}^F \\ U_{ij}^M + U_{ij}^G + \omega^{-1}\varepsilon_{jkl}k_k Y_{il}^F & X_{ij}^M + X_{ij}^G + Y_{ij}^F \end{pmatrix}. \tag{7.20}$$

Thus the effects of the gauge $(G)$ and Faraday $(F)$ transformations on the constitutive tensor given directly by multipole theory can be indicated schematically by

$$\begin{pmatrix} A_{ij}^M & \overset{F}{\longleftrightarrow} & T_{ij}^M \\ G\uparrow & & \uparrow G \\ U_{ij}^M & \underset{F}{\longleftrightarrow} & X_{ij}^M \end{pmatrix}. \tag{7.21}$$

The gauge transformations change the constitutive tensor and the fields $\mathbf{D}$ and $\mathbf{H}$. In its effect on $\mathbf{D}$, the first term in (7.15) transforms between the permittivity tensor $A_{ij}^M$ and the magnetoelectric tensor $U_{ij}^M$, while the second term transforms between the magnetoelectric tensor $T_{ij}^M$ and the inverse permeability tensor $X_{ij}^M$ (see (7.21)). By contrast, the Faraday transformations change only the constitutive tensor (structural field): they cannot, as we have emphasized above, change the fields $\mathbf{D}$ and $\mathbf{H}$. (In this regard they are analogous to the gauge transformation (7.2) which changes the 4-vector potential without changing the fields $\mathbf{E}$ and $\mathbf{B}$.) Specifically, the transformation (7.14) "shuffles" between the tensors $A_{ij}^M$ and $T_{ij}^M$, whereas (7.13) "shuffles" between $U_{ij}^M$ and $X_{ij}^M$ (see (7.21)). In our transformations of multipole theory, both gauge and Faraday transformations are required (Chapter 8).

The transformations leading to the material constants (7.16)–(7.19) preserve the linear homogeneous dependence of $\mathbf{D}$ and $\mathbf{H}$ on $\mathbf{E}$ and $\mathbf{B}$ in (4.2) and (4.3) (requirement 1). The application of (7.16)–(7.19) to multipole theory involves determining first, for a given multipole order, the most general form of the tensors $U_{ij}^G$, $X_{ij}^G$, $Y_{ij}^F$, and $Z_{ij}^F$: according to requirement 2, they should be found by taking *linear combinations of all the second-rank tensors that are constructed to be linear and homogeneous in the macroscopic polarizability tensors of a particular multipole order.* Thus, in constructing these second-rank tensors, one should use the polarizability tensors $\alpha_{ij}$ and $\alpha'_{ij}$ for transforming electric dipole order; $G'_{ij}$, $a_{ijk}$, $G_{ij}$, and $a'_{ijk}$ for transforming electric quadrupole–magnetic dipole order, and so on (Section 4.5). Additional building blocks for this purpose are $\delta_{ij}$, $\varepsilon_{ijk}$, and the wave vector $k_i$. (See the examples (8.3), (8.4), (8.13), and (8.14) in Chapter 8.) With regard to the wave vector, its use should be consistent with the order of the multipole expansion (requirement 3). We will see that it cannot play a role in transformations of the first two orders of the multipole expansion, and starts to contribute at electric octopole–magnetic quadrupole order (Chapter 8).

Next, one imposes the requirements of space inversion and time reversal (requirement 4). There are two general properties which follow from this: first, because $\mathbf{D}$ and $\mathbf{E}$ are polar vectors while $\mathbf{H}$ and $\mathbf{B}$ are axial vectors (Section 3.5), it follows from (7.13) to (7.15), and the fact that $\mathbf{k}$ is polar while $\varepsilon_{ijk}$ is axial, that $U_{ij}^G$ and $Z_{ij}^F$ are axial tensors, and $X_{ij}^G$ and $Y_{ij}^F$ are polar tensors. Similarly, because $\mathbf{D}$ and $\mathbf{E}$ are time even whereas $\mathbf{H}$ and $\mathbf{B}$ are time odd, it follows that $U_{ij}^G$ and $Z_{ij}^F$ are time odd, and $X_{ij}^G$ and $Y_{ij}^F$ are time even.

Starting with a given pair of harmonic response fields, such as those obtained directly from multipole theory, the transformation theory creates an infinite number of fields which are consistent with requirements 1–4 (see Chapter 8). The final step involves imposing requirements 5 (origin independence) and 6 (symmetries), with the goal of obtaining response fields which are (i) unique and (ii) physically acceptable. The extent to which these two goals are achieved is discussed in Chapters 8 and 9.

# References

[1] See references in Cheng, T. P. and Li, L. (1988). Resource Letter: Gl–1 Gauge invariance. *American Journal of Physics*, **56**, 586–600.

[2] O'Raifeartaigh, L. (1997). *The dawning of gauge theory*. Princeton University Press, Princeton.

[3] Morse, P. M. and Feshbach, H. (1953). *Methods of theoretical physics*, Part 1, p. 53. McGraw-Hill, New York. See also Baierlein, R. (1995). Representing a vector field: Helmholtz's theorem derived from a Fourier identity. *American Journal of Physics*, **63**, 180–182.

[4] Smith, C. R. and Inguva, R. (1984).Electrodynamics in a dispersive medium: $E$, $B$, $D$, and $H$. *American Journal of Physics*, **52**, 27–34.

[5] de Lange, O. L. and Raab, R. E. (2003). Completion of multipole theory for the electromagnetic response fields $D$ and $H$. *Proceedings of the Royal Society London* A, **459**, 1325–1341.

8

# APPLICATIONS OF THE GAUGE AND FARADAY TRANSFORMATIONS

*It was a room in which you had no
reason for sitting in one place rather
than another.*
George Eliot
(*Middlemarch*)

*Let's all move one place on.*
Lewis Caroll
(*Alice's adventures in wonderland*)

We apply the transformation theory of Section 7.4 to the first two orders of multipole theory, namely electric dipole order and electric quadrupole–magnetic dipole order. Both non-magnetic and magnetic media are considered [1]. The purpose of these calculations is to obtain transformed material constants, to electric quadrupole–magnetic dipole order, which satisfy the physical requirements given in Sections 4.2–4.4; that is equations (8.65)–(8.68) below. We also obtain, to this multipole order, polarizability densities which are invariant (origin independent), see (8.73) and (8.74).

## 8.1  Electric dipole order

We have seen that to this order multipole theory produces physically acceptable results for linear constitutive relations, in that the material constants are origin independent and satisfy the necessary symmetries (Section 4.2). We therefore expect, on grounds of uniqueness, that one cannot transform these material constants into different, yet physically acceptable, material constants. Thus, to electric dipole order the transformation theory of Section 7.4 should yield a null result: the details are straightforward, nevertheless it is useful to consider this case before discussing more complicated higher-order transformations.

Consider first a non-magnetic medium. There is just one polarizability tensor, the symmetric tensor $\alpha_{ij}$ (Section 4.5). Hence the available second-rank tensors for constructing the unknown tensors $U_{ij}^G$, $X_{ij}^G$, $Y_{ij}^F$, and $Z_{ij}^F$ in (7.16)–(7.19) are

$$\alpha_{ij}, \quad \alpha_{ll}\delta_{ij}. \tag{8.1}$$

All other possibilities in (8.1) involve powers of the wave vector $k_i$, such as $\alpha_{lk}k_k\varepsilon_{ijl}$ and $\alpha_{ll}k_ik_j$. These must be excluded because it is clear that, when

178

used in (7.16)–(7.19), they produce higher powers of $k_i$ than is appropriate for this order (cf. (4.56)) and so violate requirement 3 of Section 7.4. (Even those in (8.1) violate this condition and should be omitted, see (8.6) and (8.7). However, as mentioned above, we continue with them to illustrate the procedure at the simple level of electric dipole.)

The tensors (8.1) are polar (Section 3.5) and therefore so are the linear combinations $U_{ij}^G$, $X_{ij}^G$, $Y_{ij}^F$, and $Z_{ij}^F$ constructed from them. However, $U_{ij}^G$ and $Z_{ij}^F$ are necessarily axial (Section 7.4) and hence

$$U_{ij}^G = Z_{ij}^F = 0. \tag{8.2}$$

Based on (8.1) and the discussion in Section 7.4, we write

$$X_{ij}^G = \beta_1 \alpha_{ij} + \beta_2 \alpha_{ll} \delta_{ij} \tag{8.3}$$

$$Y_{ij}^F = \gamma_1 \alpha_{ij} + \gamma_2 \alpha_{ll} \delta_{ij}. \tag{8.4}$$

The coefficients $\beta_i$ and $\gamma_i$ remain to be determined: they are, by definition, independent of the fields $\mathbf{E}$ and $\mathbf{B}$, and the wave vector $k_i$. All the tensors in (8.3) and (8.4) are time-even (Sections 3.5 and 7.4) and therefore $\beta_i$ and $\gamma_i$ are real.

Substituting (8.2)–(8.4) in (7.16)–(7.19), and using the direct multipole results (4.56), we have the transformed material constants

$$A_{ij} = \alpha_{ij} \tag{8.5}$$

$$T_{ij} = -\frac{1}{\omega} \varepsilon_{ikl} k_k (\beta_1 \alpha_{lj} + \beta_2 \alpha_{mm} \delta_{lj}) \tag{8.6}$$

$$U_{ij} = \frac{1}{\omega} \varepsilon_{jkl} k_k (\gamma_1 \alpha_{il} + \gamma_2 \alpha_{mm} \delta_{il}) \tag{8.7}$$

$$X_{ij} = (\beta_1 + \gamma_1) \alpha_{ij} + (\beta_2 + \gamma_2) \alpha_{mm} \delta_{ij} \tag{8.8}$$

for the electric dipole contribution in a non-magnetic medium. Equations (8.5)–(8.8) have been constructed to satisfy requirements $1, 2$, and $4$ of Section 7.4. We now consider their compatibility with requirements $3, 5$, and $6$. Because $\alpha_{ij}$ is origin independent (Section 3.7), so are (8.5)–(8.8) (requirement 5). With regard to requirement 6, the symmetries (4.40) and (4.42) are satisfied because $\alpha_{ij}$ is symmetric and $\beta_i$ and $\gamma_i$ are real. The symmetry (4.41) requires $\gamma_i = \beta_i$. Hence, using (8.5) and (8.6) in (4.2) we have

$$D_i = (\epsilon_0 \delta_{ij} + \alpha_{ij}) E_j - \frac{1}{\omega} \varepsilon_{ikl} k_k (\gamma_1 \alpha_{lj} + \gamma_2 \alpha_{mm} \delta_{lj}) B_j. \tag{8.9}$$

From (8.9) it is apparent that the transformed field $\mathbf{D}$ depends on the coefficients in the Faraday transformation (see (8.4)). But, as emphasized in Section 7.4, the Faraday transformation cannot change $\mathbf{D}$. Therefore

$$\gamma_i = \beta_i = 0. \tag{8.10}$$

The same result follows by noting that the terms linear in $k_i$ in (8.6) and (8.7) violate requirement 3. We conclude that, as expected, at electric dipole order no transformations of the type (7.16)–(7.19) are possible which produce physically acceptable constitutive relations. This result can readily be extended to a magnetic medium by including the (real, antisymmetric, time-odd) polarizability tensor $\alpha'_{ij}$ (Section 4.5), and it confirms the uniqueness of the direct multipole result to electric dipole order (Section 4.5).

## 8.2   Electric quadrupole–magnetic dipole order, non-magnetic medium

For a non-magnetic medium and to the above order, the relevant polarizability tensors are $G'_{ij}$ and $a_{ijk}$ (Section 4.5). These provide the independent second-rank tensors

$$G'_{ij}, \quad G'_{ji}, \quad G'_{ll}\delta_{ij}, \quad \varepsilon_{ikl}a_{klj}, \quad \varepsilon_{jkl}a_{kli}, \quad \varepsilon_{ijk}a_{kll}, \quad \varepsilon_{ijk}a_{llk}. \tag{8.11}$$

In constructing this list we have taken account of the symmetry (2.83) of $a_{ijk}$. Tensors constructed with the aid of the wave vector, such as $a_{ijk}k_k$ and $\varepsilon_{ijl}G_{lk}k_k$, are not included in (8.11) because according to (7.16)–(7.19) they produce terms quadratic in $k_i$ in the material constants, and therefore not of electric quadrupole–magnetic dipole order (Section 7.4).

The tensors (8.11) are axial (Section 3.5) and therefore so are the linear combinations $U^G_{ij}$, $X^G_{ij}$, $Y^F_{ij}$, and $Z^F_{ij}$ constructed from them. Because $X^G_{ij}$ and $Y^F_{ij}$ are necessarily polar (Section 7.4), it follows that

$$X^G_{ij} = Y^F_{ij} = 0. \tag{8.12}$$

Using (8.11) and the discussion in Section 7.4, we write

$$U^G_{ij} = \beta_1 G'_{ij} + \beta_2 G'_{ji} + \beta_3 G'_{ll}\delta_{ij} + \beta_4 \varepsilon_{ikl}a_{klj} + \beta_5 \varepsilon_{jkl}a_{kli} + \beta_6 \varepsilon_{ijk}a_{kll} + \beta_7 \varepsilon_{ijk}a_{llk} \tag{8.13}$$

$$Z^F_{ij} = \gamma_1 G'_{ij} + \gamma_2 G'_{ji} + \gamma_3 G'_{ll}\delta_{ij} + \gamma_4 \varepsilon_{ikl}a_{klj} + \gamma_5 \varepsilon_{jkl}a_{kli} + \gamma_6 \varepsilon_{ijk}a_{kll} + \gamma_7 \varepsilon_{ijk}a_{llk}, \tag{8.14}$$

where the 14 coefficients $\beta_i$ and $\gamma_i$ are to be determined. The tensors on the right-hand sides of (8.13) and (8.14) are all time even (Section 3.5) while $U^G_{ij}$ and $Z^F_{ij}$ are time odd (Section 7.4). It follows that $\beta_i$ and $\gamma_i$ are either imaginary or zero.

Substituting (8.12)–(8.14) in (7.16)–(7.19) we obtain the electric quadrupole–magnetic dipole contribution to the transformed material constants. Before writing these down we note that, according to (7.16), (8.13), and (8.14), $A_{ij}E_j$ contains the terms

$$-\frac{1}{\omega}(\varepsilon_{ikl}\beta_3\delta_{lj} - \varepsilon_{jkl}\gamma_3\delta_{il})G'_{mm}k_k E_j = -(\beta_3 + \gamma_3)G'_{mm}\delta_{ij}B_j. \tag{8.15}$$

Thus these terms can be incorporated in $T_{ij}B_j$ in (4.2). The above leads to

$$A_{ij} = \frac{i}{2}(a_{ijk} - a_{jik})k_k - \frac{1}{\omega}\varepsilon_{ikl}k_k(\beta_1 G'_{lj} + \beta_2 G'_{jl} + \beta_4\varepsilon_{lmn}a_{mnj} + \beta_5\varepsilon_{jmn}a_{mnl}$$

$$+ \beta_6\varepsilon_{ljm}a_{mnn} + \beta_7\varepsilon_{ljm}a_{nnm}) + \frac{1}{\omega}\varepsilon_{jkl}k_k(\gamma_1 G'_{il} + \gamma_2 G'_{li}$$

$$+ \gamma_4\varepsilon_{imn}a_{mnl} + \gamma_5\varepsilon_{lmn}a_{mni} + \gamma_6\varepsilon_{ilm}a_{mnn} + \gamma_7\varepsilon_{ilm}a_{nnm}) \tag{8.16}$$

$$T_{ij} = -iG'_{ij} + \gamma_1 G'_{ij} + \gamma_2 G'_{ji} + \gamma_4\varepsilon_{ikl}a_{klj} + \gamma_5\varepsilon_{jkl}a_{kli} + \gamma_6\varepsilon_{ijk}a_{kll}$$

$$+ \gamma_7\varepsilon_{ijk}a_{llk} - \beta_3 G'_{ll}\delta_{ij} \tag{8.17}$$

$$U_{ij} = -iG'_{ji} + \beta_1 G'_{ij} + \beta_2 G'_{ji} + \beta_3 G'_{ll}\delta_{ij} + \beta_4\varepsilon_{ikl}a_{klj} + \beta_5\varepsilon_{jkl}a_{kli}$$

$$+ \beta_6\varepsilon_{ijk}a_{kll} + \beta_7\varepsilon_{ijk}a_{llk} \tag{8.18}$$

$$X_{ij} = 0, \tag{8.19}$$

where we have used the multipole forms (4.57). Note that a consequence of using (8.15) is to eliminate a term involving $\gamma_3$ which would otherwise appear in (8.17); thus there are 13 coefficients $\beta_i$ and $\gamma_i$ to be determined.

Equations (8.16)–(8.19) have been constructed by imposing requirements 1–4 of Section 7.4. Next, we impose origin independence on the transformed material constants (requirement 5). Consider first the invariance of $U_{ij}$ for an arbitrary shift of origin $\mathbf{d} = (d_x, d_y, d_z)$: using (8.18), (3.66), and (3.69) we have for the corresponding change in $U_{ij}$

$$\Delta U_{ij} = \frac{1}{2}\omega\beta_1\varepsilon_{jkl}d_k\alpha_{il} + \frac{1}{2}\omega(-i + \beta_2)\varepsilon_{ikl}d_k\alpha_{jl} - \beta_4\varepsilon_{ikl}(d_j\alpha_{kl} + d_l\alpha_{kj})$$

$$- \beta_5\varepsilon_{jkl}(d_i\alpha_{kl} + d_l\alpha_{ki}) - 2\beta_6\varepsilon_{ijk}d_l\alpha_{kl} - \beta_7\varepsilon_{ijk}(d_k\alpha_{ll} + d_l\alpha_{lk}). \tag{8.20}$$

(In (8.20) we have used the invariance $\Delta G'_{ll} = 0$; see (3.69) and (2.73).) Thus, in particular, for $\mathbf{d} = (0, 0, d_z)$

$$\Delta U_{xy} = d_z\left[\left(\frac{1}{2}\omega\beta_1 + \beta_5 - \beta_7\right)\alpha_{xx} + \left(\frac{i}{2}\omega - \frac{1}{2}\omega\beta_2 - \beta_4 - \beta_7\right)\alpha_{yy}\right.$$

$$\left. - 2(\beta_6 + \beta_7)\alpha_{zz}\right]. \tag{8.21}$$

Because $\alpha_{xx}$, $\alpha_{yy}$, and $\alpha_{zz}$ are independent components of $\alpha_{ij}$, it follows from the invariance $\Delta U_{xy} = 0$ that

$$\frac{1}{2}\omega\beta_1 + \beta_5 - \beta_7 = 0 \tag{8.22}$$

$$\frac{i}{2}\omega - \frac{1}{2}\omega\beta_2 - \beta_4 - \beta_7 = 0 \tag{8.23}$$

$$\beta_6 + \beta_7 = 0. \tag{8.24}$$

For the invariance of $T_{ij}$ in (8.17), a similar calculation to the above leads to

$$\Delta T_{ij} = \frac{1}{2}\omega(\gamma_1 - i)\varepsilon_{jkl}d_k\alpha_{il} + \frac{1}{2}\omega\gamma_2\varepsilon_{ikl}d_k\alpha_{jl} - \gamma_4\varepsilon_{ikl}(d_j\alpha_{kl} + d_l\alpha_{kj})$$

$$- \gamma_5\varepsilon_{jkl}(d_i\alpha_{kl} + d_l\alpha_{ki}) - 2\gamma_6\varepsilon_{ijk}d_l\alpha_{kl} - \gamma_7\varepsilon_{ijk}(d_k\alpha_{ll} + d_l\alpha_{lk}). \tag{8.25}$$

For $\mathbf{d} = (0, 0, d_z)$

$$\Delta T_{xy} = d_z \left[ \left( -\frac{i}{2}\omega + \frac{1}{2}\omega\gamma_1 + \gamma_5 - \gamma_7 \right) \alpha_{xx} \right.$$
$$\left. - \left( \frac{1}{2}\omega\gamma_2 + \gamma_4 + \gamma_7 \right) \alpha_{yy} - 2(\gamma_6 + \gamma_7)\alpha_{zz} \right]. \qquad (8.26)$$

The invariance $\Delta T_{xy} = 0$ requires that

$$-\frac{i}{2}\omega + \frac{1}{2}\omega\gamma_1 + \gamma_5 - \gamma_7 = 0 \qquad (8.27)$$

$$\frac{1}{2}\omega\gamma_2 + \gamma_4 + \gamma_7 = 0 \qquad (8.28)$$

$$\gamma_6 + \gamma_7 = 0. \qquad (8.29)$$

Origin independence of the other components of (8.16)–(8.18) does not lead to additional independent equations for the $\beta_i$ and $\gamma_i$. Using (8.22)–(8.24) we can eliminate $\beta_5$, $\beta_6$, and $\beta_7$ from (8.18). Also, the identity (A.7) shows that the resulting terms in $\beta_4$ cancel in (8.18). Equations (8.27)–(8.29) are used in (8.17) in a similar manner. Thus we obtain

$$T_{ij} = (\gamma_1 - i)V_{ij} + \gamma_2 V_{ji} - \beta_3 G'_{ll}\delta_{ij} \qquad (8.30)$$
$$U_{ij} = \beta_1 V_{ij} + (\beta_2 - i)V_{ji} + \beta_3 G'_{ll}\delta_{ij}, \qquad (8.31)$$

where

$$V_{ij} = G'_{ij} - \frac{1}{2}\omega\varepsilon_{jkl}a_{kli} \qquad (8.32)$$

is origin independent (see Appendix G).

Equations (8.30) and (8.31) are the most general origin-independent forms of the transformations (8.17) and (8.18). We note that, as assumed in Ref. 1, the trace terms $a_{kll}$ and $a_{llk}$ in the list (8.11) play no role in these transformations. To proceed further we consider a non-dissipative medium and impose the symmetry (4.41), $U_{ij} = -T^*_{ji}$. (The restriction to a non-dissipative medium is lifted later, see Section 8.4.) It then follows from (8.30) and (8.31) that

$$\beta_1 = \gamma_2, \qquad \beta_2 = \gamma_1, \qquad \beta_3 = 0. \qquad (8.33)$$

We recall that the terms in $\beta_i$ in (8.16)–(8.18) are associated with a gauge transformation, while the terms in $\gamma_i$ are associated with a Faraday transformation (see (8.13) and (8.14)). Therefore, (8.31), (8.33), (4.3), and (8.19) mean that the transformed field $\mathbf{H}$ depends on the coefficients $\gamma_i$ of the Faraday transformation. However, by construction the Faraday transformation cannot change $\mathbf{H}$ (Section 7.4). We thus conclude that

$$\gamma_1 = \gamma_2 = 0 \qquad (8.34)$$

and hence from (8.33)

$$\beta_1 = \beta_2 = 0. \tag{8.35}$$

Substituting (8.22)–(8.24), (8.27)–(8.29), (8.34), and (8.35) into (8.16), we find that $A_{ij} = 0$.

From the above we obtain the desired transformations of (4.57)

$$A_{ij} = 0 \tag{8.36}$$

$$T_{ij} = -i\left(G'_{ij} - \frac{1}{2}\omega\varepsilon_{jkl}a_{kli}\right) \tag{8.37}$$

$$U_{ij} = -i\left(G'_{ji} - \frac{1}{2}\omega\varepsilon_{ikl}a_{klj}\right) \tag{8.38}$$

$$X_{ij} = 0. \tag{8.39}$$

Comparing (8.16)–(8.18) and (8.36)–(8.38) we see that of the 13 coefficients $\beta_i$ and $\gamma_i$, the non-zero ones are $\beta_4 = \frac{1}{2}i\omega$ and $\gamma_5 = \frac{1}{2}i\omega$. Equations (8.36)–(8.39) are the electric quadrupole–magnetic dipole contributions to the transformed material constants for a non-magnetic medium. They are origin independent and they satisfy the symmetry (4.41) and the Post constraint (4.53). The transformation theory has achieved these properties by shifting the terms involving $a_{ijk}$ in the permittivity tensor $A_{ij}$ into the magnetoelectric tensors $T_{ij}$ and $U_{ij}$, see (4.57) and (8.36)–(8.39).

## 8.3   Electric quadrupole–magnetic dipole order, magnetic medium

For a magnetic medium, and to the above order, one has in addition to $G'_{ij}$ and $a_{ijk}$ the polarizability tensors $G_{ij}$ and $a'_{ijk}$ (Section 4.5). The latter provide the independent second-rank tensors

$$G_{ij}, \quad G_{ji}, \quad G_{ll}\delta_{ij}, \quad \varepsilon_{ikl}a'_{klj}, \quad \varepsilon_{jkl}a'_{kli}, \quad \varepsilon_{ijk}a'_{kll}, \quad \varepsilon_{ijk}a'_{llk}, \tag{8.40}$$

where we have taken account of the symmetry (2.84) of $a'_{ijk}$. The tensors (8.40) are all axial (Section 3.5) and therefore (8.12) applies here as well. For $U^G_{ij}$ and $Z^F_{ij}$ we have (8.13) and (8.14) with $G'$ and $a$ replaced by $G$ and $a'$. The tensors (8.40) are time-odd (Section 3.5), and therefore time reversal requires that the coefficients $\beta_i$ and $\gamma_i$ are real. The transformed material constants have the additional contributions due to the magnetic property of the medium

$$A_{ij} = \frac{1}{2}(a'_{ijk} + a'_{jik})k_k - \frac{1}{\omega}\varepsilon_{ikl}k_k(\beta_1 G_{lj} + \beta_2 G_{jl} + \beta_4\varepsilon_{lmn}a'_{mnj} + \beta_5\varepsilon_{jmn}a'_{mnl}$$

$$+ \beta_6\varepsilon_{ljm}a'_{mnn} + \beta_7\varepsilon_{ljm}a'_{nnm}) + \frac{1}{\omega}\varepsilon_{jkl}k_k(\gamma_1 G_{il} + \gamma_2 G_{li}$$

$$+ \gamma_4\varepsilon_{imn}a'_{mnl} + \gamma_5\varepsilon_{lmn}a'_{mni} + \gamma_6\varepsilon_{ilm}a'_{mnn} + \gamma_7\varepsilon_{ilm}a'_{nnm}) \tag{8.41}$$

$$T_{ij} = G_{ij} + \gamma_1 G_{ij} + \gamma_2 G_{ji} + \gamma_4\varepsilon_{ikl}a'_{klj} + \gamma_5\varepsilon_{jkl}a'_{kli} + \gamma_6\varepsilon_{ijk}a'_{kll}$$

$$+ \gamma_7\varepsilon_{ijk}a'_{llk} - \beta_3 G_{ll}\delta_{ij} \tag{8.42}$$

$$U_{ij} = -G_{ji} + \beta_1 G_{ij} + \beta_2 G_{ji} + \beta_3 G_{ll}\delta_{ij} + \beta_4 \varepsilon_{ikl}a'_{klj} + \beta_5 \varepsilon_{jkl}a'_{lki}$$
$$+ \beta_6 \varepsilon_{ijk}a'_{kll} + \beta_7 \varepsilon_{ijk}a'_{llk} \tag{8.43}$$
$$X_{ij} = 0, \tag{8.44}$$

where we have used (7.16)–(7.19) and the multipole forms (4.61). Also, we have used a relation like (8.15) to eliminate a term involving $\gamma_3$ in (8.42). The 13 coefficients $\beta_i$ and $\gamma_i$ in these expressions are to be determined.

Next, we impose origin independence on (8.41)–(8.43). The details are similar to those for a non-magnetic medium (Section 8.2) but with two differences. For an arbitrary shift of origin $\mathbf{d} = (d_x, d_y, d_z)$ the corresponding change in $U_{ij}$ obtained from (8.43), (3.67), and (3.68) is

$$\Delta U_{ij} = -\frac{1}{2}\omega\beta_1\varepsilon_{jkl}d_k\alpha'_{il} + \frac{1}{2}\omega(1-\beta_2)\varepsilon_{ikl}d_k\alpha'_{jl} - \frac{1}{2}\omega\beta_3\varepsilon_{klm}d_l\alpha'_{km}\delta_{ij}$$
$$- \beta_4\varepsilon_{ikl}(d_j\alpha'_{kl} + d_l\alpha'_{kj}) - \beta_5\varepsilon_{jkl}(d_i\alpha'_{kl} + d_l\alpha'_{ki})$$
$$- 2\beta_6\varepsilon_{ijk}d_l\alpha'_{kl} - \beta_7\varepsilon_{ijk}d_l\alpha'_{lk}. \tag{8.45}$$

Here we note the first difference when compared with the corresponding expression (8.20) for a non-magnetic medium. It is that the trace $G_{ll}$ is not invariant (Section 3.7) whereas $G'_{ll}$ in (8.18) is: consequently, a term in $\beta_3$ appears in (8.45) but not in (8.20). This comparison also shows that while (8.20) has a term in $\beta_7\alpha_{ll}$, the corresponding term in $\beta_7\alpha'_{ll}$ is absent from (8.45). This is due to the second difference in the two calculations, namely, the antisymmetry of $\alpha'_{ij}$ compared with the symmetry of $\alpha_{ij}$.

For an arbitrary shift of origin $\mathbf{d} = (d_x, d_y, d_z)$ we have from (8.45)

$$\Delta U_{xy} = \left(\frac{1}{2}\omega\beta_1 + 3\beta_5 + 2\beta_6 - \beta_7\right)d_x\alpha'_{xz}$$
$$+ \left(\frac{1}{2}\omega - \frac{1}{2}\omega\beta_2 - 3\beta_4 + 2\beta_6 - \beta_7\right)d_y\alpha'_{yz}. \tag{8.46}$$

Thus the invariance $\Delta U_{xy} = 0$ requires that

$$\frac{1}{2}\omega\beta_1 + 3\beta_5 + 2\beta_6 - \beta_7 = 0 \tag{8.47}$$

$$\frac{1}{2}\omega - \frac{1}{2}\omega\beta_2 - 3\beta_4 + 2\beta_6 - \beta_7 = 0. \tag{8.48}$$

From the invariance $\Delta U_{xx} = 0$ for an origin shift $\mathbf{d} = (d_x, 0, 0)$ we obtain

$$\omega\beta_3 - 2\beta_4 - 2\beta_5 = 0. \tag{8.49}$$

The invariance of $T_{ij}$ yields additional relations between coefficients. From (8.42), (3.67), and (3.68)

$$\Delta T_{ij} = -\frac{1}{2}\omega(1+\gamma_1)\varepsilon_{jkl}d_k\alpha'_{il} - \frac{1}{2}\omega\gamma_2\varepsilon_{ikl}d_k\alpha'_{jl} - \gamma_4\varepsilon_{ikl}(d_j\alpha'_{kl} + d_l\alpha'_{kj})$$
$$- \gamma_5\varepsilon_{jkl}(d_i\alpha'_{kl} + d_l\alpha'_{ki}) - 2\gamma_6\varepsilon_{ijk}d_l\alpha'_{kl} - \gamma_7\varepsilon_{ijk}d_l\alpha'_{lk}$$
$$+ \frac{1}{2}\omega\beta_3\varepsilon_{klm}d_l\alpha'_{km}\delta_{ij}. \tag{8.50}$$

Thus $\Delta T_{xy} = 0$ for a shift of origin $\mathbf{d} = (d_x,\, d_y,\, d_z)$ requires

$$\frac{1}{2}\omega + \frac{1}{2}\omega\gamma_1 + 3\gamma_5 + 2\gamma_6 - \gamma_7 = 0 \tag{8.51}$$

$$\frac{1}{2}\omega\gamma_2 + 3\gamma_4 - 2\gamma_6 + \gamma_7 = 0, \tag{8.52}$$

while $\Delta T_{xx} = 0$ for $\mathbf{d} = (d_x,\, 0,\, 0)$ gives

$$\omega\beta_3 + 2\gamma_4 + 2\gamma_5 = 0. \tag{8.53}$$

Invariance of other components of the material constants, such as $\Delta A_{ij} = 0$, provides no further independent equations for the $\beta_i$ and $\gamma_i$.

We now use (8.51)–(8.53) to eliminate $\beta_3$, $\gamma_5$, and $\gamma_7$ from $T_{ij}$ in (8.42) and (8.47)–(8.49) to eliminate $\beta_3$, $\beta_5$, and $\beta_7$ from $U_{ij}$ in (8.43). We then obtain

$$T_{ij} = (1+\gamma_1)W_{ij} + \gamma_2 W_{ji} - \left(\frac{1}{6}\omega\gamma_2 + \gamma_4 - \gamma_6\right)\varepsilon_{ijk}(a'_{kll} + 2a'_{llk}) \tag{8.54}$$

$$U_{ij} = \beta_1 W_{ij} + (\beta_2 - 1)W_{ji} - \left[\frac{1}{6}\omega(\beta_2 - 1) + \beta_4 - \beta_6\right]\varepsilon_{ijk}(a'_{kll} + 2a'_{llk}). \tag{8.55}$$

In these

$$W_{ij} = G_{ij} - \frac{1}{3}G_{ll}\delta_{ij} - \frac{1}{6}\omega\varepsilon_{jkl}a'_{kli}, \tag{8.56}$$

which is shown in Appendix G to be origin independent.

Imposing the symmetry $U_{ij} = -T^*_{ji}$ (see (4.41)) and recalling that $\beta_i$ and $\gamma_i$ are real, we find from (8.54) and (8.55) that

$$\beta_1 = -\gamma_2, \qquad \beta_2 = -\gamma_1, \qquad \frac{1}{6}\omega(\beta_2 - 1) + \beta_4 - \beta_6 = \frac{1}{6}\omega\gamma_2 + \gamma_4 - \gamma_6. \tag{8.57}$$

Next, we use (8.57) in (8.55) to express $U_{ij}$ in terms of the $\gamma_i$, and then apply the reasoning that led to (8.34). This gives

$$\gamma_1 = \gamma_2 = \beta_1 = \beta_2 = 0, \qquad \gamma_6 = \gamma_4, \qquad \beta_6 = \beta_4 - \frac{1}{6}\omega. \tag{8.58}$$

Finally, we substitute (8.58) into (8.47)–(8.49) and (8.51)–(8.53) and then use these in (8.41) to obtain

$$A_{ij} = S_{ijk}k_k, \tag{8.59}$$

where

$$S_{ijk} = \frac{1}{3}(a'_{ijk} + a'_{jki} + a'_{kij}). \tag{8.60}$$

This is origin independent (see Appendix G) and its symmetry ensures that of $A_{ij}$ (see (4.40)).

The above calculations yield the desired transformations of (4.61)

$$A_{ij} = \frac{1}{3}\left(a'_{ijk} + a'_{jki} + a'_{kij}\right)k_k \tag{8.61}$$

$$T_{ij} = G_{ij} - \frac{1}{3}G_{ll}\delta_{ij} - \frac{1}{6}\omega\varepsilon_{jkl}a'_{kli} \tag{8.62}$$

$$U_{ij} = -G_{ji} + \frac{1}{3}G_{ll}\delta_{ij} + \frac{1}{6}\omega\varepsilon_{ikl}a'_{klj} \tag{8.63}$$

$$X_{ij} = 0. \tag{8.64}$$

Comparing (8.41)–(8.43) and (8.61)–(8.63) we see that of the 13 coefficients $\beta_i$ and $\gamma_i$, the non-zero ones are $\beta_3 = \frac{1}{3}$, $\beta_4 = \frac{1}{6}\omega$, and $\gamma_5 = -\frac{1}{6}\omega$. Equations (8.61)–(8.64) are the additional transformed material constants, to order electric quadrupole–magnetic dipole, due to the magnetic nature of the medium. We note from (8.56) that $W_{ii} = -\omega\varepsilon_{ikl}a'_{kli}/6 = 0$ because of the symmetry (2.84). Thus $W_{ij}$ is traceless, a property which is essential for restoring the validity of the Post constraint in multipole theory (see Section 8.4).

## 8.4   Discussion

(i) Starting from a given pair of response fields **D** and **H**, such as the direct multipole solutions of Section 4.1, the transformation equations (7.16)–(7.19) produce an infinite number of solutions for these fields. The examples of Sections 8.2 and 8.3 show how, by imposing certain of the invariances (4.9) and the symmetries (4.40)–(4.42), one can determine a unique solution from this infinite set. Note that it is not necessary to impose all the desired invariances and symmetries: thus for the example of a magnetic medium in Section 8.3 we required only the changes $\Delta U_{xx}$, $\Delta U_{xy}$, $\Delta T_{xx}$, and $\Delta T_{xy}$ to be zero, and that the symmetry $U_{ij} = -T^*_{ji}$ (i.e. (4.41)) should hold. The resulting transformed material constants satisfy the remaining invariances and symmetries.

(ii) For a magnetic medium and to electric quadrupole–magnetic dipole order, the total transformed material constants are obtained by adding the transformed contributions (8.36)–(8.39) and (8.61)–(8.64) to the vacuum and electric dipole terms, which are evident in (4.4)–(4.7). This yields

$$A_{ij} = \varepsilon_0\delta_{ij} + \alpha_{ij} - i\alpha'_{ij} + \frac{1}{3}\left(a'_{ijk} + a'_{jki} + a'_{kij}\right)k_k \tag{8.65}$$

$$T_{ij} = -i\left(G'_{ij} - \frac{1}{2}\omega\varepsilon_{jkl}a_{kli}\right) + G_{ij} - \frac{1}{3}G_{ll}\delta_{ij} - \frac{1}{6}\omega\varepsilon_{jkl}a'_{kli} \tag{8.66}$$

$$U_{ij} = -i\left(G'_{ji} - \frac{1}{2}\omega\varepsilon_{ikl}a_{klj}\right) - G_{ji} + \frac{1}{3}G_{ll}\delta_{ij} + \frac{1}{6}\omega\varepsilon_{ikl}a'_{klj} \tag{8.67}$$

$$X_{ij} = \mu_0^{-1}\delta_{ij}. \tag{8.68}$$

Equations (8.65)–(8.68) are the main results of this chapter.

(iii) Comparing (8.65)–(8.68) with the expressions given directly by multipole theory (Section 4.5), we see that the effect of the transformation theory of Section 7.4 has been four-fold: 1. In the magnetoelectric coefficients $T_{ij}$ and $U_{ij}$, the origin-dependent tensors $G'_{ij}$ and $G_{ij}$ in (4.57) and (4.61) have been replaced by the invariant tensors $V_{ij}$ (see (8.32)) and $W_{ij}$ (see (8.56)), respectively. 2. In $A_{ij}$ the origin-dependent combinations of $a_{ijk}$ in (4.57) and $a'_{ijk}$ in (4.61) have been replaced in (8.65) by the invariant, symmetric tensor $S_{ijk}$ (see (8.60)). 3. The transformed magnetoelectric tensors (8.66) and (8.67) have traces $T_{ii} = -iG'_{ii} = U_{ii}$, and therefore satisfy the Post constraint (4.53). By contrast, the magnetoelectric coefficients given directly by multipole theory do not satisfy this constraint (Section 4.5). 4. The above changes to the constitutive tensor are carried out without affecting the necessary symmetries (4.40)–(4.42). Also, it is clear from Sections 8.2 and 8.3 that, for the application of the transformation theory to electric quadrupole–magnetic dipole order of multipole theory, the results (8.65)–(8.68) are unique.

(iv) We emphasize that origin independence of the magnetoelectric tensors $T_{ij}$ and $U_{ij}$ is achieved by forming linear combinations of the magnetic dipole polarizability tensors $G_{ij}$ and $G'_{ij}$ with the electric quadrupole contributions $a_{ijk}$ and $a'_{ijk}$ (see (8.66) and (8.67)). It is easily seen from the transformation theory of Section 7.4 that non-zero invariant tensors $T_{ij}$ and $U_{ij}$ cannot be obtained if the electric quadrupole terms are neglected, that is, by using only $G_{ij}$ and $G'_{ij}$. This need to use both the quadrupole and dipole terms is, of course, also evident in the hierarchy (1.122). It is therefore clear that theories of the magnetoelectric effect based only on the magnetic dipole contribution [2] cannot yield physically acceptable results.

(v) The constitutive relations (4.2) and (4.3) with the transformed (invariant) material constants (8.65)–(8.68) can be used in the expressions (1.118) and (1.119) to obtain transformed multipole moment densities [3]. The results in the absence of dissipation are

$$P_i = \tilde{\alpha}_{ij}E_j + \frac{1}{2}\tilde{a}_{ijk}\nabla_k E_j + \tilde{G}_{ij}B_j \tag{8.69}$$

$$Q_{ij} = \tilde{a}^*_{kij}E_k \tag{8.70}$$

$$M_i = \tilde{G}^*_{ji}E_j, \tag{8.71}$$

where

$$\tilde{\alpha}_{ij} = \alpha_{ij} - i\alpha'_{ij} \tag{8.72}$$

$$\tilde{a}_{ijk} = -iS_{ijk} \tag{8.73}$$

$$\tilde{G}_{ij} = W_{ij} - iV_{ij}. \tag{8.74}$$

Here $\tilde{a}_{ijk}$ and $\tilde{G}_{ij}$ are transformed polarizability densities given in terms of the invariant tensors $V_{ij}$, $W_{ij}$, and $S_{ijk}$. Thus these polarizability densities are origin independent, and we refer to them as the invariant polarizability densities: they are the unique invariant densities given by the above transformation theory, and should be contrasted with the non-invariant polarizability densities given directly by multipole theory. From (6.207) to (6.209) the latter are

$$\tilde{a}_{ijk} = a_{ijk} - ia'_{ijk} \tag{8.75}$$

$$\tilde{G}_{ij} = G_{ij} - iG'_{ij}. \tag{8.76}$$

Comparing (8.73) and (8.74) with (8.75) and (8.76) we see that origin independence has been achieved by replacing the origin-dependent combination of $a_{ijk}$ and $a'_{ijk}$ with the invariant tensor $S_{ijk}$, and the origin-dependent tensors $G'_{ij}$ and $G_{ij}$ with the invariant tensors $V_{ij}$ and $W_{ij}$. Note that the invariant polarizability density $S_{ijk}$ in (8.73) does not involve the non-magnetic quadrupole polarizability $a_{ijk}$; instead this tensor appears in $V_{ij}$ in (8.74). We have already seen (Section 6.8) that the above non-invariant polarizability densities yield unphysical results for reflected intensities. In Chapter 9 we use the invariant polarizability densities in the multipole theory of reflection.

(vi) It is interesting to note that a fully symmetric time-odd polar tensor of rank 3, like $S_{ijk}$, was identified in a phenomenological theory, with no reference to multipoles, to play a role in the boundary conditions [4]. The significance of this tensor for certain reflection phenomena is discussed in the next chapter.

(vii) The symmetric part of the tensor $V_{ij}$ and its origin dependence were first discussed in a theory of optical activity of aligned molecules [5]. (See also Section 5.3.)

(viii) The transformed material constants (8.65)–(8.68) agree with those obtained previously by trial-and-error methods [3,6].

(ix) In the treatment presented in Ref. 1, no account was taken of the last two entries in (8.11) and (8.40). This affects the details of the analysis but not the results (8.36)–(8.39) and (8.61)–(8.64).

(x) The structure of the material constants (8.65)–(8.68) has previously been attributed to the requirement of covariance [3, 7]; this was done on the basis of Post's covariant formulation [8] of the symmetries (4.40)–(4.42) possessed by these material constants. However, the multipole expansions of **D** and **H** in terms of space and time derivatives of the fields **E** and **B** (Section 4.1) are not, in general, covariant. Furthermore, the direct multipole expressions for the material constants also possess the symmetries (4.40)–(4.42) (see Section 4.5). The calculations presented in Sections 8.2 and 8.3 show rather that the material constants (8.65)–(8.68) are a consequence of the transformation theory of Section 7.4. In this theory it is translational invariance, rather than covariance, that plays a crucial role.

(xi) The above derivation of the transformed material constants (8.65)–(8.68) is for non-dissipative media. The results can be extended to dissipative media by replacing the real polarizability tensors in (8.65)–(8.67) with the appropriate complex forms which apply in absorption (Section 2.8) and by making the wave vector $\mathbf{k}$ in (8.65) complex to allow for attenuation of the wave. The origin dependences of the complex polarizability tensors are the same as those for non-dissipative media (Section 3.7), and it therefore follows that the transformed material constants in dissipative media are also origin independent. However, these transformed material constants violate the symmetries (4.40)–(4.42), as one expects for dissipative media (Sections 4.3 and 4.5).

(xii) Thus the symmetries (4.40)–(4.42) play an intermediate role in applications of the transformation theory of Section 7.4. By considering a non-dissipative medium one can use these symmetries to obtain unique solutions from the infinite transformed sets such as (8.30), (8.31) and (8.54), (8.55); see the calculations leading to (8.36)–(8.38) and (8.61)–(8.63). Having done so, one can remove the restriction to non-dissipative media, as explained above.

(xiii) Translationally invariant combinations of polarizability tensors also occur in the wave equation; see (5.8)–(5.11). Those of electric quadrupole-magnetic dipole order consist of linear combinations of the invariant tensors $V_{ij}, W_{ij}$, and $S_{ijk}$ mentioned in (iii) and (v) above for the transformed material constants and transformed polarizability denisties. Specifically, (5.8) and (5.9) can be expressed as

$$\beta_{ij}^{s} = \sigma_k \left[ -\epsilon_{ikl} W_{jl} - \epsilon_{jkl} W_{il} + \omega S_{ijk} \right] \tag{8.77}$$

$$\beta_{ij}^{a} = \sigma_k \left[ \epsilon_{ikl} V_{jl} - \epsilon_{jkl} V_{il} \right]. \tag{8.78}$$

(xiv) The transformation theory can also be applied to electric octopole–magnetic quadrupole order. These calculations are considerably more involved than those presented above for electric quadrupole–magnetic dipole order because of the large number of second-rank tensors which can be constructed from the relevant polarizability tensors, the wave vector, and $\varepsilon_{ijk}$ and $\delta_{ij}$. Also, the polarizability tensors have complicated translational properties; see (3.70)–(3.79). The calculations are probably at the limit of what is feasible, and for details the reader is referred to the literature [9].

## References

[1] de Lange, O. L. and Raab, R. E. (2003). Completion of multipole theory for the electromagnetic response fields $\mathbf{D}$ and $\mathbf{H}$. *Proceedings of the Royal Society London* A, **459**, 1325–1341.

[2] Muto, M., Tanabe, Y., Iizuka-Sakano, T., and Hanamura, E. (1998). Magnetoelectric and second-harmonic spectra in antiferromagnetic $Cr_2O_3$. *Physical Review* B, **57**, 9586–9607.

[3] Graham, E. B. and Raab, R. E. (2000). Multipole solution for the macroscopic electromagnetic boundary conditions at a vacuum–dielectric interface. *Proceedings of the Royal Society London* A, **456**, 1193–1215.

[4] Krichevtsov, B. B., Pavlov, V. V., Pisarev, R. V., and Gridnev, V. N. (1993). Spontaneous non-reciprocal reflection of light from antiferromagnetic $Cr_2O_3$. *Journal of Physics: Condensed Matter*, **5**, 8233–8244.

[5] Buckingham, A. D. and Dunn, M. B. (1971). Optical activity of oriented molecules. *Journal of the Chemical Society* A, 1988–1991.

[6] Raab, R. E. and de Lange, O. L. (2001). Symmetry contraints for electromagnetic constitutive relations. *Journal of Optics* A: *Pure and Applied Optics*, **3**, 446–451.

[7] Graham, E. B. and Raab, R. E. (1977). Covariant **D** and **H** fields for reflection from a magnetic anisotropic chiral medium. *Journal of the Optical Society of America* A, **14**, 131–134.

[8] Post, E. J. (1962). *Formal structure of electromagnetics*. North–Holland, Amsterdam. (Reprinted 1997. Dover, New York.)

[9] Raab, R. E. and de Lange, O. L. (2004). Transformed multipole theory of the response fields **D** and **H** to electric octopole–magnetic quadrupole order. *Proceedings of the Royal Society London* A, in press.

# 9

# TRANSMISSION AND REFLECTION EFFECTS:
# TRANSFORMED MULTIPOLE RESULTS

*Good reasons must, of force,*
*give place to better.*
William Shakespeare
(*Julius Caesar*)

In Chapters 5 and 6 we applied the standard formulation of multipole theory to transmission and reflection phenomena. While this produced a satisfactory description of the former, the theory of reflection effects was found to be unphysical. In this chapter we apply the transformed multipole theory of Chapters 7 and 8 to transmission and reflection phenomena. For transmission the outcome is particularly simple: for the phenomena considered in Chapter 5, transformed multipole theory yields precisely the same results as the standard theory (Section 9.1). For reflection the results are different. In Sections 9.2–9.5 we consider reflection from non-magnetic uniaxial and cubic crystals. Reflection from a magnetic crystal ($Cr_2O_3$) is treated in Sections 9.6 and 9.7, and the results compared with experiment in Section 9.8. In each case, detailed expressions are derived for the Fresnel reflection coefficients, and their behaviour under space translation and time reversal is discussed.

## 9.1 The wave equation and transmission

In Section 5.1 we obtained a wave equation from multipole theory. This was done by using the constitutive relations (4.2)–(4.7), given directly by multipole theory, in the inhomogeneous Maxwell equation $\nabla \times \mathbf{H} = \dot{\mathbf{D}}$. We worked to electric octopole–magnetic quadrupole order, and the resulting wave equation is (5.4).

We now ask, what is the wave equation given by the constitutive relations obtained using the transformation theory of Section 7.4? The answer is simple: because this transformation theory leaves the inhomogeneous Maxwell equations unchanged (Sections 7.2–7.4), it follows that the wave equation is invariant under these transformations. Thus, to all multipole orders, one will obtain the same wave equation from the direct multipole constitutive relations and the transformed constitutive relations. In particular, to electric octopole–magnetic quadrupole order the wave equation is given by (5.4), whether one starts from direct multipole constitutive relations or from transformed relations. (This can be verified explicitly, to electric quadrupole–magnetic dipole order, using the transformed constitutive relations with material constants (8.65)–(8.68).) We conclude that all theoretical results for transmission effects based on the multi-pole wave equation, such as those in Sections 5.2–5.8, are unaffected by use of the transformation theory of Section 7.4. To put it differently, for the description of transmission effects using a wave equation, it is unnecessary to transform direct multipole constitutive relations such as (4.2)–(4.7).

In Section 5.1 we remarked that it is somewhat surprising that constitutive relations with origin-dependent material constants, such as (4.2)–(4.7), lead to an origin-independent wave equation (5.4). The following comment may shed some light on this: suppose one had been able, *a priori*, to start with origin-independent constitutive relations (like those of Chapter 8). One would then be satisfied to see these yielding an origin-independent wave equation. Now transform (using the theory of Chapter 7) these origin-independent constitutive relations given by multipole theory. (That is, the inverse of the procedure in Chapter 8.) Because the wave equation is invariant under this transformation, it may then be less surprising that

$$\left.\begin{array}{l}\text{Origin-dependent multipole}\\ \text{constitutive relations}\end{array}\right\} \longrightarrow \left.\begin{array}{l}\text{Origin-independent}\\ \text{wave equation}\end{array}\right\}. \qquad (9.1)$$

## 9.2 Reflection from non-magnetic uniaxial and cubic crystals

Compared with our previous analysis of this problem based on results of direct multipole theory (Section 6.5), a major difference occurs at the start of the calculation. Instead of using the multipole moment densities (6.80)–(6.82) involving non-invariant polarizability densities, we use the transformed multipole moment densities (8.69)–(8.74) which contain invariant polarizability densities. For a non-magnetic dielectric and to electric quadrupole–magnetic dipole order $\alpha'_{ij} = a'_{ijk} = G_{ij} = 0$ (Table 3.2). Thus we have from (8.69)–(8.74) and (8.32)

$$P_i = \alpha_{ij} E_j - i\left(G'_{ij} - \frac{1}{2}\omega\varepsilon_{jkl}a_{kli}\right)B_j \qquad (9.2)$$

$$Q_{ij} = 0 \qquad (9.3)$$

$$M_i = i\left(G'_{ji} - \frac{1}{2}\omega\varepsilon_{ikl}a_{klj}\right)E_j. \qquad (9.4)$$

The subsequent calculation of the reflection matrix is similar to that in Section 6.5, and we present an outline.

The orientation of the crystal is specified in Section 6.5. Thus the plane of reflection is the crystallographic $xy$ plane (Fig. 6.5), and the crystal is oriented so that its $xz$ plane is that of incidence. From (6.69)–(6.74) and (9.2)–(9.4) the matching conditions are

$$E_{2x} = E_{1x} \qquad (9.5)$$

$$E_{2y} = E_{1y} \qquad (9.6)$$

$$E_{2z} = E_{1z} - \varepsilon_0^{-1} \left[ \alpha_{zj} E_{2j} - i \left( G'_{zj} - \frac{1}{2} \omega \varepsilon_{jkl} a_{klz} \right) B_{2j} \right] \qquad (9.7)$$

$$B_{2x} = B_{1x} + i\mu_0 \left( G'_{jx} - \frac{1}{2} \omega \varepsilon_{xkl} a_{klj} \right) E_{2j} \qquad (9.8)$$

$$B_{2y} = B_{1y} + i\mu_0 \left( G'_{jy} - \frac{1}{2} \omega \varepsilon_{ykl} a_{klj} \right) E_{2j} \qquad (9.9)$$

$$B_{2z} = B_{1z}. \qquad (9.10)$$

Next, we use the following. First, for non-magnetic uniaxials and cubics the components (6.83) of $\alpha_{ij}$, $a_{ijk}$, and $G'_{ij}$ vanish. Second, we use the Maxwell equation (6.43), applied to harmonic plane waves, to eliminate $\mathbf{B}_1$ and $\mathbf{B}_2$ from (9.7) to (9.10). (In doing so, we note that (9.10) and (9.6) are identical, see Section 6.5.) Third, we use (9.7) to eliminate the component $E_{2z}$ from (9.9). As a result, (9.8) and (9.9) can be written

$$iF_1 E_{2x} + (k_{2z} + iF_2) E_{2y} = k_{1z} E_{1y} \qquad (9.11)$$

$$(k_{2z} - iF_3) E_{2x} - i(F_4 + F_5) E_{2y} = k_{1z} E_{1x} - k_{1x}(1 - n_E^{-2}) E_{1z}, \qquad (9.12)$$

where $n_E^2 = 1 + \varepsilon_0^{-1} \alpha_{zz}$ for a uniaxial crystal, and $n_E = n_O$ for a cubic crystal. In (9.11) and (9.12)

$$F_1 = \mu_0 \omega \left( G'_{xx} - \frac{1}{2} \omega a_{yzx} + \frac{1}{2} \omega a_{zxy} \right) \qquad (9.13)$$

$$F_2 = \mu_0 \omega \left( G'_{yx} - \frac{1}{2} \omega a_{yyz} + \frac{1}{2} \omega a_{zyy} \right) \qquad (9.14)$$

$$F_3 = \mu_0 \omega \left( G'_{xy} + \frac{1}{2} \omega a_{xxz} - \frac{1}{2} \omega a_{zxx} \right) \qquad (9.15)$$

$$F_4 = \mu_0 \omega \left( G'_{yy} + \frac{1}{2} \omega a_{xyz} - \frac{1}{2} \omega a_{zxy} \right) \qquad (9.16)$$

$$F_5 = \varepsilon_0^{-1} \omega^{-1} n_E^{-2} \left( G'_{zz} - \frac{1}{2} \omega a_{xyz} + \frac{1}{2} \omega a_{yzx} \right) k_{1x}^2. \qquad (9.17)$$

Equations (9.5), (9.6), (9.11), and (9.12) are the matching conditions in the transformed multipole theory; they may be compared with the corresponding conditions (6.102)–(6.105) in direct multipole theory. By taking account of the three points listed after (6.113), we can write (9.5), (9.6), (9.11), and (9.12) in the form

$$E_{ax} S_a + E_{bx} S_b = N_1 \qquad (9.18)$$

$$E_{ay} S_a + E_{by} S_b = N_2 \qquad (9.19)$$

$$\alpha_a S_a + \alpha_b S_b = N_3 \qquad (9.20)$$

$$\beta_a S_a + \beta_b S_b = N_4. \qquad (9.21)$$

Here $S_a$ and $S_b$ are scaling factors for the transmitted amplitudes $E_a$ and $E_b$, and

$$N_1 = (E_{ip} - E_{rp})\cos\theta \tag{9.22}$$

$$N_2 = E_{is} + E_{rs} \tag{9.23}$$

$$N_3 = e(E_{is} - E_{rs})\cos\theta \tag{9.24}$$

$$N_4 = e(E_{ip} + E_{rp})(1 - n_E^{-2}\sin^2\theta) \tag{9.25}$$

$$\alpha_m = iF_1 E_{mx} + (k_{mz} + iF_2)E_{my} \tag{9.26}$$

$$\beta_m = (k_{mz} - iF_3)E_{mx} - i(F_4 + F_5)E_{my}, \tag{9.27}$$

where $m$ is $a$ or $b$, and $e = \omega c^{-1}$. The components $E_{ip}$, $E_{is}$, $E_{rp}$, $E_{rs}$ of the incident and reflected fields are those depicted in Fig. 6.5.

We solve (9.18) and (9.19) for $S_a$ and $S_b$, and substitute the results in (9.20) and (9.21) to obtain

$$c_1 E_{rp} + c_2 E_{rs} + c_3 E_{ip} + c_4 E_{is} = 0 \tag{9.28}$$

$$d_1 E_{rp} + d_2 E_{rs} + d_3 E_{ip} + d_4 E_{is} = 0, \tag{9.29}$$

where

$$c_1 = -c_3 = -(\alpha_a E_{by} - \alpha_b E_{ay})\cos\theta \tag{9.30}$$

$$c_2 = -\alpha_a E_{bx} + \alpha_b E_{ax} + e(E_{ax}E_{by} - E_{bx}E_{ay})\cos\theta \tag{9.31}$$

$$c_4 = -\alpha_a E_{bx} + \alpha_b E_{ax} - e(E_{ax}E_{by} - E_{bx}E_{ay})\cos\theta \tag{9.32}$$

$$d_1 = -(\beta_a E_{by} - \beta_b E_{ay})\cos\theta - e(1 - n_E^{-2}\sin^2\theta)(E_{ax}E_{by} - E_{bx}E_{ay}) \tag{9.33}$$

$$d_2 = d_4 = -\beta_a E_{bx} + \beta_b E_{ax} \tag{9.34}$$

$$d_3 = (\beta_a E_{by} - \beta_b E_{ay})\cos\theta - e(1 - n_E^{-2}\sin^2\theta)(E_{ax}E_{by} - E_{bx}E_{ay}). \tag{9.35}$$

We remind the reader that in the above, $E$ denotes a field amplitude.

According to (9.28) and (9.29), the reflected amplitudes are given in terms of the incident amplitudes by

$$E_{rp} = r_{pp}E_{ip} + r_{ps}E_{is} \tag{9.36}$$

$$E_{rs} = r_{sp}E_{ip} + r_{ss}E_{is}, \tag{9.37}$$

where the four Fresnel coefficients are

$$r_{pp} = (c_3 d_2 - c_2 d_3)/\Delta \tag{9.38}$$

$$r_{ps} = d_2(c_4 - c_2)/\Delta \tag{9.39}$$

$$r_{sp} = c_1(d_1 + d_3)/\Delta \tag{9.40}$$

$$r_{ss} = (c_1 d_4 - c_4 d_1)/\Delta, \tag{9.41}$$

with

$$\Delta = c_2 d_1 - c_1 d_2. \tag{9.42}$$

Equations (9.38)–(9.41) are expressions for the reflection coefficients of non-magnetic uniaxial and cubic crystals given, to electric quadrupole–magnetic dipole order, by the transformed multipole theory of Chapters 7 and 8. The orientation of the crystal is specified at the beginning of Section 6.5. These results of transformed multipole theory may be compared with the corresponding reflection coefficients obtained directly from multipole theory, see (6.129)–(6.135) and (6.138)–(6.141).

To proceed further, we require the eigenpolarizations deduced in Section 6.6 for the waves $a$ and $b$ propagating in the crystal. These enable us to obtain explicit expressions for the above reflection coefficients in terms of polarizability tensors of the crystal. We present these results separately for uniaxial and cubic crystals in the next two sections.

## 9.3 Explicit results for non-magnetic uniaxial crystals

We wish to obtain explicit expressions for the reflection coefficients to lowest order in the polarizability tensors $G'_{ij}$ and $a_{ijk}$; equivalently, to lowest order in the quantities $F_i$ defined in (9.13)–(9.17). For brevity we adopt the following terminology. A quantity which is independent of the $F_i$ is of order $F^0$, written $O(F^0)$; a quantity linear in the $F_i$ is of order $F$, written $O(F)$; etc.

The amplitudes $E_{ax}$ and $E_{by}$ which appear in (9.26), (9.27), and (9.30)–(9.35) are given in terms of $E_{ay}$ and $E_{bx}$ by (6.164) and (6.167). The tensors $A_1$ and $A_3$ in the latter equations can easily be expressed in terms of the $F_i$ using (6.162) and (9.13), (9.16), and (9.17). One finds

$$A_1 = (\mu_0\omega)^{-1}(F_1 + F_4) \tag{9.43}$$

$$A_3 = (\mu_0\omega)^{-1}(F_4 + n_E^2 \sin^{-2}\theta\, F_5). \tag{9.44}$$

Thus, from (6.164), (6.167), (9.43), (9.44), and our use of scaling factors, $E_{ax}$ and $E_{by}$ are $O(F)$, while $E_{ay}$ and $E_{bx}$ are $O(F^0)$. Consequently, neglecting terms $O(F^2)$ in (9.26) and (9.27), we obtain

$$\alpha_a = k_{az}E_{ay} + iF_2 E_{ay} \tag{9.45}$$

$$\alpha_b = k_{bz}E_{by} + iF_1 E_{bx} \tag{9.46}$$

$$\beta_a = k_{az}E_{ax} - i(F_4 + F_5)E_{ay} \tag{9.47}$$

$$\beta_b = k_{bz}E_{bx} - iF_3 E_{bx}. \tag{9.48}$$

From (9.30)–(9.35) and (9.45)–(9.48), and neglecting terms $O(F^2)$, we have

$$c_1 = -c_3 = -[(k_{az} - k_{bz})E_{ay}E_{by} - iF_1 E_{bx}E_{ay}]\cos\theta \tag{9.49}$$

$$c_2 = -(k_{az} + e\cos\theta)E_{bx}E_{ay} - iF_2 E_{bx}E_{ay} \tag{9.50}$$

$$c_4 = -(k_{az} - e\cos\theta)E_{bx}E_{ay} - iF_2 E_{bx}E_{ay} \tag{9.51}$$

$$d_1 = [k_{bz}\cos\theta + e(1 - n_E^{-2}\sin^2\theta)]E_{bx}E_{ay} - iF_3 E_{bx}E_{ay}\cos\theta \tag{9.52}$$

$$d_2 = d_4 = -(k_{az} - k_{bz})E_{ax}E_{bx} + i(F_4 + F_5)E_{bx}E_{ay} \tag{9.53}$$

$$d_3 = -[k_{bz}\cos\theta - e(1 - n_E^{-2}\sin^2\theta)]E_{bx}E_{ay} + iF_3 E_{bx}E_{ay}\cos\theta. \tag{9.54}$$

It follows from these equations that $c_2$, $c_4$, $d_1$, and $d_3$ are $O(F^0)$, while $c_1$, $c_3$, $d_2$, and $d_4$ are $O(F)$. Thus in (9.42) we can approximate $\Delta = c_2 d_1$, and (9.38)–(9.41) approximate to

$$r_{pp} = -d_3/d_1 \tag{9.55}$$

$$r_{ss} = -c_4/c_2 \tag{9.56}$$

$$r_{ps} = d_2(c_4 - c_2)/c_2 d_1 \tag{9.57}$$

$$r_{sp} = c_1(d_1 + d_3)/c_2 d_1. \tag{9.58}$$

In (9.49)–(9.54) we use

$$k_{az} = e j_{az} = e(n_O^2 - \sin^2 \theta)^{1/2} \tag{9.59}$$

and

$$k_{bz} = e j_{bz} = e(n_O/n_E)(n_E^2 - \sin^2 \theta)^{1/2} \tag{9.60}$$

(see Section 6.6). Then (9.50)–(9.52) and (9.54)–(9.56) yield

$$r_{pp} = \frac{n_O \cos \theta - (1 - n_E^{-2} \sin^2 \theta)^{1/2}}{n_O \cos \theta + (1 - n_E^{-2} \sin^2 \theta)^{1/2}} \tag{9.61}$$

$$r_{ss} = \frac{\cos \theta - (n_O^2 - \sin^2 \theta)^{1/2}}{\cos \theta + (n_O^2 - \sin^2 \theta)^{1/2}}, \tag{9.62}$$

where terms $O(F)$ have been neglected. Similarly, from (9.57), (9.58), (9.49)–(9.54), (6.164), and (6.167), and neglecting terms $O(F^2)$, we obtain

$$r_{ps} = g\left[ \frac{(n_O^2 - \sin^2 \theta)^{1/2}}{n_E(n_O^2 - n_E^2) \sin^2 \theta} \left\{ n_E(n_O^2 - \sin^2 \theta)^{1/2} - n_O(n_E^2 - \sin^2 \theta)^{1/2} \right\} \right.$$
$$\left. \times \left\{ (n_E^2 - \sin^2 \theta)F_1 + n_E^2(F_4 + F_5) \right\} - (F_4 + F_5) \right] \cos \theta \tag{9.63}$$

and

$$r_{sp} = g(1 - n_E^{-2} \sin^2 \theta)\left[ \frac{n_O n_E}{(n_O^2 - n_E^2)(n_E^2 - \sin^2 \theta)^{1/2} \sin^2 \theta} \left\{ (n_O^2 - \sin^2 \theta)^{1/2} \right. \right.$$
$$\left. - \frac{n_O}{n_E}(n_E^2 - \sin^2 \theta)^{1/2} \right\} \left\{ (n_E^2 - \sin^2 \theta)F_1 + n_E^2(F_4 + F_5) \right\} - F_1 \right] \cos \theta, \tag{9.64}$$

where

$$g = (2i/e)\left[ \left\{ (n_O^2 - \sin^2 \theta)^{1/2} + \cos \theta \right\} \right.$$
$$\left. \times \left\{ n_O(1 - n_E^{-2} \sin^2 \theta)^{1/2} \cos \theta + (1 - n_E^{-2} \sin^2 \theta) \right\} \right]^{-1}. \tag{9.65}$$

Equations (9.61)–(9.64) are the desired reflection coefficients, for arbitrary angle of incidence $\theta$, given explicitly in terms of polarizability tensors. Note that

$r_{pp}$ and $r_{ss}$ in (9.61) and (9.62) are $O(F^0)$, while $r_{ps}$ and $r_{sp}$ in (9.63) and (9.64) are $O(F)$. To complete the presentation we list in Table 9.1 explicit expressions for the quantities $F_i$ for various symmetry classes of uniaxials [1].

For normal incidence ($\theta = 0$) the above results for reflection coefficients simplify considerably. For all classes of uniaxials listed in Table 9.1

$$r_{pp} = -r_{ss} = \frac{n_O - 1}{n_O + 1}. \tag{9.66}$$

Also, for all classes except $\bar{4}2m$ and $\bar{4}$,

$$r_{ps} = r_{sp} = 0. \tag{9.67}$$

For $\bar{4}2m$ and $\bar{4}$

$$r_{ps} = -r_{sp} = \frac{2iF_1}{e(n_O + 1)^2}. \tag{9.68}$$

**Table 9.1** *The quantities $F_i$ in (9.13)–(9.17) for non-magnetic uniaxial crystals.*

| Class | $F_1$ | $F_2$ | $F_3$ | $F_4$ | $F_5$ |
|---|---|---|---|---|---|
| $4, 3, 6$ | $L_1$ | $L_2$ | $-L_2$ | $L_1$ | $L_3$ |
| $422, 32, 622$ | $L_1$ | $0$ | $0$ | $L_1$ | $L_3$ |
| $\bar{4}$ | $L_4$ | $-L_2$ | $-L_2$ | $-L_4$ | $0$ |
| $\bar{4}2m$ | $L_4$ | $0$ | $0$ | $-L_4$ | $0$ |
| $4mm, 3m, 6mm$ | $0$ | $L_2$ | $-L_2$ | $0$ | $0$ |
| $\bar{6}, \bar{6}m2$ | $0$ | $0$ | $0$ | $0$ | $0$ |

$$L_1 = \mu_0\omega\left(G'_{xx} + \frac{1}{2}\omega a_{xyz}\right), \qquad L_2 = -\mu_0\omega\left(G'_{xy} + \frac{1}{2}\omega a_{xxz} - \frac{1}{2}\omega a_{zxx}\right),$$

$$L_3 = (\varepsilon_0\omega n_E^2)^{-1}k_{1x}^2(G'_{zz} - \omega a_{xyz}), \qquad L_4 = \mu_0\omega\left(G'_{xx} - \frac{1}{2}\omega a_{xyz} + \frac{1}{2}\omega a_{zxy}\right)$$

## 9.4 Explicit results for non-magnetic cubic crystals

For cubic crystals $n_O = n_E$. The eigenpolarizations have been determined in Section 6.6, and based on these the reflection coefficients (9.38)–(9.41) can be evaluated for arbitrary angle of incidence on cubic crystals. The calculations are similar to the above for uniaxials, and we simply state the results. The symmetry classes possessing $G'_{ij}$ and $a_{ijk}$ are 23, 432, and $\bar{4}3m$. For these

$$r_{pp} = \frac{n_O \cos\theta - (1 - n_O^2 \sin^2\theta)^{1/2}}{n_O \cos\theta + (1 - n_O^2 \sin^2\theta)^{1/2}} \tag{9.69}$$

$$r_{ss} = \frac{\cos\theta - (n_O^2 - \sin^2\theta)^{1/2}}{\cos\theta + (n_O^2 - \sin^2\theta)^{1/2}} \tag{9.70}$$

$$r_{ps} = -r_{sp}$$
$$= \frac{-(iF/en_O^2)\sin^2\theta\cos\theta}{(n_O^2 - \sin^2\theta)^{1/2}[\cos^2\theta + (n_O^2 - \sin^2\theta)^{1/2}(1 + n_O^{-2})\cos\theta + n_O^{-2}(n_O^2 - \sin^2\theta)]}. \tag{9.71}$$

Here $F = 2\mu_0\omega G'_{xx}$ for the classes 23 and 432, and $F = 0$ for the class $\bar{4}3m$.

Again, simple results are obtained for normal incidence ($\theta = 0$). For the above classes

$$r_{pp} = -r_{ss} = \frac{n_O - 1}{n_O + 1} \tag{9.72}$$

$$r_{ps} = r_{sp} = 0. \tag{9.73}$$

## 9.5    Tests of translational and time-reversal invariance

We examine whether the reflection coefficients obtained in the previous three sections from transformed multipole theory for non-magnetic uniaxial and cubic crystals are consistent with the requirements of translational and time-reversal invariance.

1. *Translational invariance.* The calculation of reflection coefficients to electric quadrupole–magnetic dipole order is based on the translationally invariant multipole moment densities (8.69)–(8.74), see Section 9.2. The invariance of these densities carries over to the reflection coefficients. This follows from (3.66) and (3.69) which show that the five quantities $F_i$ in (9.13)–(9.17) are translationally invariant: therefore so are $\alpha_m$ and $\beta_m$ in (9.26) and (9.27), and also the $c_i$ and $d_i$ in (9.30)–(9.35). Thus *the Fresnel coefficients* (9.38)–(9.41) *obtained from transformed multipole theory are independent of the choice of origin of coordinates. This contrasts with the origin dependence of reflection coefficients (and hence of the reflected intensity) obtained from direct multipole theory* (Section 6.8). Translational invariance of the explicit formulas for Fresnel coefficients given in Sections 9.3 and 9.4 follows, by inspection, for the reason given above (translational invariance of the $F_i$, and of $F$ in (9.71)).

2. *Time-reversal invariance (reciprocity).* The polarizability tensors which enter the reflection coefficients (9.61)–(9.64) for uniaxials, and (9.69)–(9.71) for cubics, are all time even for non-magnetic crystals. Hence the time-reversed coefficients $r_{jk}(-t)$ (Section 6.2) are all equal to $r_{jk}(t)$. Thus the reciprocity relations (6.17) and (6.18), namely $r_{pp}(-t) = r_{pp}(t)$ and $r_{ss}(-t) = r_{ss}(t)$, are satisfied. Furthermore, we have already seen in (9.71) that for cubics $r_{sp}(t) = -r_{ps}(t)$; for each uniaxial class in Table 9.1 the same result can be deduced from (9.63) and (9.64). Thus, using also $r_{ps}(-t) = r_{ps}(t)$ (see above), we see that the reciprocity relation (6.19), namely $r_{ps}(-t) = -r_{sp}(t)$, is also satisfied.

## 9.6    Reflection from antiferromagnetic $Cr_2O_3$: first configuration

In this and the following section we consider a single example of reflection from a magnetic crystal, namely chromium oxide ($Cr_2O_3$) in its antiferromagnetic state. Because it exists at room temperature ($T_N = 307\,K$) and transmits light, this crystal has been much studied experimentally for its different properties. These experiments include reflection at normal incidence for two different configurations, the results of which allow comparison with the theory presented below. A theoretical advantage of $Cr_2O_3$ is its uniaxial point group $\bar{3}m$, for which, of its multipole polarizabilities up to electric quadrupole–magnetic dipole order, the following vanish relative to crystallographic axes [1]

$$\alpha'_{ij} = a_{ijk} = G'_{ij} = 0, \qquad (9.74)$$

while the non-vanishing tensors have only the components

$$\alpha_{xx} = \alpha_{yy}, \quad \alpha_{zz} \qquad (9.75)$$

$$G_{xx} = G_{yy}, \quad G_{zz} \qquad (9.76)$$

$$a'_{xxy} = a'_{xyx} = a'_{yxx} = -a'_{yyy}, \qquad a'_{xyz} = a'_{xzy} = -a'_{yzx} = -a'_{yxz}. \qquad (9.77)$$

Note that $G_{ij}$ and $a'_{ijk}$ are time-odd tensors (see Table 3.2).

In this and the next section we present the theory of reflection from $Cr_2O_3$ for two configurations which were studied experimentally at normal incidence by Krichevtsov *et al.* [2]. Their first experiment was on a crystal face perpendicular to the optic ($z$) axis. However, we initially consider incidence at an angle $\theta$ on this face, partly because this provides a more stringent test of the theory and also because an interesting feature emerges. The Fresnel coefficients for normal incidence are obtained from the limit $\theta = 0$, and it is these results which are compared with experiment (Section 9.8). We remark that $Cr_2O_3$ is coloured and therefore absorbs in the visible spectrum. In the following theory this absorption is taken into account by making the polarizability tensors complex, in the manner described in Section 2.8.

From (9.74) the transformed multipole moment densities in (8.69)–(8.74) are

$$P_i = \left(\alpha_{ij} + \frac{1}{2}S_{ijk}k_k\right)E_j + W_{ij}B_j \qquad (9.78)$$

$$Q_{ij} = iS_{ijk}E_k \qquad (9.79)$$

$$M_i = W_{ji}E_j, \qquad (9.80)$$

where from (8.56) and (8.60)

$$S_{ijk} = \frac{1}{3}(a'_{ijk} + a'_{jki} + a'_{kij}) \qquad (9.81)$$

$$W_{ij} = G_{ij} - \frac{1}{3}G_{ll}\delta_{ij} - \frac{1}{6}\omega\varepsilon_{jkl}a'_{kli}. \qquad (9.82)$$

The first configuration we treat is that described in Section 6.5, in which the crystallographic $xy$ and $xz$ planes of $Cr_2O_3$ are respectively the planes of reflection and incidence. For this configuration and from (6.43) and (9.75)–(9.82) the matching conditions in (6.69)–(6.74) for a harmonic plane wave have the following forms when laboratory axes coincide with crystallographic axes

$$E_{2x} = E_{1x} \tag{9.83}$$

$$E_{2y} = E_{1y} \tag{9.84}$$

$$E_{2z} = E_{1z} - \varepsilon_0^{-1}\alpha_{zz}E_{2z} + 2e^{-2}k_{2x}F_1E_{2y} \tag{9.85}$$

$$k_{2z}E_{2y} + F_1E_{2x} = k_{1z}E_{1y} \tag{9.86}$$

$$k_{2z}E_{2x} - F_1E_{2y} - k_{2x}E_{2z} = k_{1z}E_{1x} - k_{1x}E_{1z}. \tag{9.87}$$

Here

$$F_1 = \frac{1}{3}\mu_0\omega\left(G_{xx} - G_{zz} + \frac{1}{2}\omega a'_{xyz}\right) \tag{9.88}$$

is an origin-independent quantity, see (3.67) and (3.68). Using (9.85) to eliminate $E_{2z}$ from (9.87), one obtains

$$k_{2z}E_{2x} - (1 + 2e^{-2}n_E^{-2}k_{1x}^2)F_1E_{2y} = k_{1z}E_{1x} - k_{1x}(1 - n_E^{-2})E_{1z}. \tag{9.89}$$

The independent matching conditions for $Cr_2O_3$ in this configuration are then (9.83), (9.84), (9.86), and (9.89). These may be cast in the same form as (9.18)–(9.25), except that now

$$\alpha_m = k_{mz}E_{my} + F_1E_{mx} \tag{9.90}$$

$$\beta_m = k_{mz}E_{mx} - (1 + 2n_E^{-2}\sin^2\theta)F_1E_{my}. \tag{9.91}$$

Proceeding as in Section 9.2 we obtain the same expressions for the $c_i$ and $d_i$ in (9.30)–(9.35) and hence for the Fresnel reflection coefficients in (9.38)–(9.41). However, to determine these for $Cr_2O_3$ we require the eigenvalues and eigenvectors of the two propagating waves inside the crystal. These are found by solving the wave equation in (5.12). The elements of the secular determinant in (5.12) are obtained for $Cr_2O_3$ from (5.13), (5.5), and (9.74). To electric quadrupole–magnetic dipole order they are

$$a_{ij} = n^2\sigma_i\sigma_j - (n^2 - 1)\delta_{ij} + \varepsilon_0^{-1}\alpha_{ij} + \varepsilon_0^{-1}c^{-1}n\beta_{ij}^s, \tag{9.92}$$

where from (5.8)

$$\beta_{ij}^s = \sigma_k\left[-\varepsilon_{ikl}G_{jl} - \varepsilon_{jkl}G_{il} + \frac{1}{2}\omega(a'_{ijk} + a'_{jik})\right]. \tag{9.93}$$

As in (6.161) we take

$$\mathbf{j} = (c/\omega)\mathbf{k} = n\boldsymbol{\sigma}. \tag{9.94}$$

Then from (9.75)–(9.77), (9.88), and (9.92)–(9.94) the secular equation in (5.12) becomes

$$\begin{vmatrix} n_O^2 - j_z^2 & -e^{-1}j_x F_2 & j_x j_z \\ -e^{-1}j_x F_2 & n_O^2 - j_x^2 - j_z^2 & -3e^{-1}j_x F_1 \\ j_x j_z & -3e^{-1}j_x F_1 & n_E^2 - j_x^2 \end{vmatrix} = 0, \tag{9.95}$$

where from (3.67)

$$F_2 = \mu_0 \omega^2 a'_{yyy} \tag{9.96}$$

is an origin-independent quantity. To first order in $F_i$ the solutions of (9.95) are

$$j_{az}^2 = n_O^2 - j_x^2, \qquad \frac{E_{ax}}{E_{ay}} = \frac{e^{-1}[3j_x(n_O^2 - j_x^2)^{1/2}F_1 - (n_E^2 - j_x^2)F_2]}{j_x(n_O^2 - n_E^2)} \tag{9.97}$$

$$j_{bz}^2 = n_E^{-2}n_O^2(n_E^2 - j_x^2), \qquad \frac{E_{by}}{E_{bx}} = \frac{e^{-1}n_E[n_E(n_E^2 - j_x^2)^{1/2}F_2 - 3n_O j_x F_1]}{j_x(n_O^2 - n_E^2)n_E^2 - j_x^2)^{1/2}}. \tag{9.98}$$

The components $E_{az}$ and $E_{bz}$ are not required in the matching conditions.

From (9.97), (9.98), (9.90), and (9.91) the coefficients in (9.30)–(9.35) become

$$c_1 = -c_3$$
$$= [e\{n_E^{-1}n_O(n_E^2 - j_x^2)^{1/2} - (n_O^2 - j_x^2)^{1/2}\}E_{ay}E_{by} + F_1 E_{bx}E_{ay}]\cos\theta \sim O(F) \tag{9.99}$$

$$c_2 = -e[(n_O^2 - j_x^2)^{1/2} + \cos\theta]E_{bx}E_{ay} \sim O(F^0) \tag{9.100}$$

$$c_4 = -e[(n_O^2 - j_x^2)^{1/2} - \cos\theta]E_{bx}E_{ay} \sim O(F^0) \tag{9.101}$$

$$d_1 = e[n_E^{-1}n_O(n_E^2 - j_x^2)^{1/2}\cos\theta + n_E^{-2}(n_E^2 - j_x^2)]E_{bx}E_{ay} \sim O(F^0) \tag{9.102}$$

$$d_2 = d_4 = -[e\{n_E^{-1}n_O(n_E^2 - j_x^2)^{1/2} + (n_O^2 - j_x^2)^{1/2}\}E_{ax}E_{bx}$$
$$- n_E^2(n_E^2 + 2j_x^2)F_1 E_{bx}E_{ay}] \sim O(F) \tag{9.103}$$

$$d_3 = -e[n_E^{-1}n_O(n_E^2 - j_x^2)^{1/2}\cos\theta - n_E^{-2}(n_E^2 - j_x^2)]E_{bx}E_{ay} \sim O(F^0). \tag{9.104}$$

In each of (9.99)–(9.104) only the leading terms in $F_i$ are shown.

The Fresnel reflection coefficients for $Cr_2O_3$ in the configuration considered in this section follow from (9.55)–(9.58) and (9.99)–(9.104) and are

$$r_{pp} = \frac{n_O n_E \cos\theta - (n_E^2 - \sin^2\theta)^{1/2}}{n_O n_E \cos\theta + (n_E^2 - \sin^2\theta)^{1/2}} \tag{9.105}$$

$$r_{ss} = \frac{\cos\theta - (n_O^2 - \sin^2\theta)^{1/2}}{\cos\theta + (n_O^2 - \sin^2\theta)^{1/2}} \tag{9.106}$$

$$r_{ps} = 2g^{-1}[\{(2n_O^2 + n_E^2)(n_E^2 - \sin^2\theta)^{1/2} - 3n_O n_E(n_O^2 - \sin^2\theta)^{1/2}\}F_1$$
$$+ n_E(n_E^2 - \sin^2\theta)^{1/2}\{n_O(n_E^2 - \sin^2\theta)^{1/2} - n_E(n_O^2 - \sin^2\theta)^{1/2}\}F_2/\sin\theta]\cos\theta \tag{9.107}$$

$$r_{sp} = 2g^{-1}[\{(2n_O^2+n_E^2)(n_E^2-\sin^2\theta)^{1/2}-3n_On_E(n_O^2-\sin^2\theta)^{1/2}\}F_1$$
$$-n_E(n_E^2-\sin^2\theta)^{1/2}\{n_O(n_E^2-\sin^2\theta)^{1/2}-n_E(n_O^2-\sin^2\theta)^{1/2}\}F_2/\sin\theta]\cos\theta,$$
$$(9.108)$$

where

$$g = e(n_O^2 - n_E^2)\{(n_O^2 - \sin^2\theta)^{1/2} + \cos\theta\}\{(n_E^2 - \sin^2\theta)^{1/2} + n_On_E\cos\theta\}.$$
$$(9.109)$$

We remind the reader that because of absorption, the refractive indices $n_O$ and $n_E$, and the quantities $F_1$ and $F_2$ in the above formulas are all complex.

The translational invariance of the reflection coefficients (9.105)–(9.108) is obvious, bearing in mind the origin independence of $F_1$ and $F_2$. To show that the reciprocity relations (6.17)–(6.19) are satisfied by (9.105)–(9.108), we first consider the behaviour under time reversal of $\cos\theta$ and $\sin\theta$. Comparison of Fig. 6.3, which includes the time-reversed experiment, with Fig. 6.5, which displays the coordinate system of configuration 1, shows that $k_{iz}$ and $k_{rz}$ are time even and $k_{ix}$ and $k_{rx}$ are time odd. Thus from (6.77) $\cos\theta$ is time even, while from (6.78) $\sin\theta$ is time odd. It follows therefore that $r_{pp}$ and $r_{ss}$ satisfy reciprocity. In respect of $r_{ps}$ and $r_{sp}$, we note that $F_1$ and $F_2$ in (9.88) and (9.96) are time odd. However, although this property is sufficient to ensure that the term in $F_1$ satisfies reciprocity, this is not the case for the term in $F_2$. It is the combination of $F_2$ and $\sin\theta$ in this term that ensures reciprocity. This role of $\sin\theta$ is the interesting feature referred to earlier.

For normal incidence $(\theta = 0)$, (9.105)–(9.108) simplify to

$$r_{pp} = -r_{ss} = \frac{n_O - 1}{n_O + 1} \qquad (9.110)$$

$$r_{ps} = r_{sp} = -\frac{2F_1}{e(n_O + 1)^2}. \qquad (9.111)$$

These results for normal incidence have been obtained previously in a form appropriate to the Shelankov–Pikus reciprocity condition $r_{ij}(t) = r_{ji}(-t)$ for normal incidence, rather than ours in the **k**, **p**, **s** basis [3]

## 9.7    Reflection from antiferromagnetic Cr$_2$O$_3$: second configuration

In this section we present the theory for the second configuration that was studied in the reflection experiments on Cr$_2$O$_3$ by Krichevtsov et al. [2] and analyzed theoretically by these authors and by Graham and Raab [3]. Here the incident beam is parallel to one of the two-fold (C$_2$) rotation axes, which lie at right angles to the optic $(z)$ axis, while the reflection face is perpendicular to the incident beam. In the system of crystallographic axes adopted here (Section 3.6), on which Birss' tables are based [1], the chosen C$_2$ axis is $y$. (In Refs. 2 and 3 it is labelled $x$.)

The matching conditions in (6.69)–(6.74) apply when the outward normal to the reflection plane is along the laboratory $-z$-axis. To express these conditions for the second configuration, for which the outward normal is along the laboratory $-y$-axis, we make the cyclic replacement

$$\left.\begin{array}{c} x \\ y \\ z \end{array}\right\} \longrightarrow \left.\begin{array}{c} z \\ x \\ y \end{array}\right\} \tag{9.112}$$

and then arrange for corresponding laboratory and crystallographic axes to co-incide. Thus the required matching conditions, in the crystal's system and to order electric quadrupole–magnetic dipole, are now

$$E_{2z} = E_{1z} + \frac{1}{2}\varepsilon_0^{-1}\nabla_z Q_{yy} \tag{9.113}$$

$$E_{2x} = E_{1x} + \frac{1}{2}\varepsilon_0^{-1}\nabla_x Q_{yy} \tag{9.114}$$

$$E_{2y} = E_{1y} - \varepsilon_0^{-1}(P_y - \nabla_z Q_{zy} - \nabla_x Q_{xy}) \tag{9.115}$$

$$B_{2z} = B_{1z} - \mu_0 \left( \frac{1}{2}\dot{Q}_{xy} - M_z \right) \tag{9.116}$$

$$B_{2x} = B_{1x} + \mu_0 \left( \frac{1}{2}\dot{Q}_{xy} + M_x \right) \tag{9.117}$$

$$B_{2y} = B_{1y}. \tag{9.118}$$

The **E** and **B** components in these equations are field amplitudes.

By means of (6.43) and (9.75)–(9.82) the matching conditions reduce to

$$E_{2z} = E_{1z} \tag{9.119}$$

$$E_{2x} = E_{1x} \tag{9.120}$$

$$\left( k_{2y} + \frac{1}{2}F_2 \right) E_{2x} - 2F_1 E_{2z} = k_{1y} E_{1x} \tag{9.121}$$

$$k_{2y} E_{2z} - F_1 E_{2x} = k_{1y} E_{1z}, \tag{9.122}$$

where $F_1$ is given by (9.88) and $F_2$ by (9.96). As previously, we express the matching conditions in terms of the fields and wave vectors of the incident and reflected rays in the vacuum and of the two transmitted rays in the crystal. Thus

$$E_{az} + E_{bz} = E_{iz} + E_{rz} \tag{9.123}$$

$$E_{ax} + E_{bx} = E_{ix} + E_{rx} \tag{9.124}$$

$$\left( k_{ay} + \frac{1}{2}F_2 \right) E_{ax} + \left( k_{by} + \frac{1}{2}F_2 \right) E_{bx} - 2F_1(E_{az} + E_{bz}) = k_{iy} E_{ix} + k_{ry} E_{rx} \tag{9.125}$$

$$k_{ay} E_{az} + k_{by} E_{bz} - F_1(E_{ax} + E_{bx}) = k_{iy} E_{iz} + k_{ry} E_{rz}. \tag{9.126}$$

The components of $\mathbf{E}_i$ and $\mathbf{E}_r$ in (9.123)–(9.126) may be related to the corresponding fields $E_p$ and $E_s$ in Fig. 6.5 by using (9.112) in (6.114)–(6.116) and then setting $\theta = 0$ for normal incidence. Thus

$$E_{iz} = E_{ip}, \qquad E_{rz} = -E_{rp} \tag{9.127}$$
$$E_{ix} = E_{is}, \qquad E_{rx} = E_{rs} \tag{9.128}$$
$$E_{iy} = 0, \qquad E_{ry} = 0. \tag{9.129}$$

In a similar manner one obtains from (6.77)

$$k_{iy} = -k_{ry} = e, \tag{9.130}$$

while from Snell's law for normal incidence the only components of $\mathbf{k}_a$ and $\mathbf{k}_b$ in (6.75) are

$$k_{ay} = en_a, \qquad k_{by} = en_b. \tag{9.131}$$

From (9.127)–(9.131) the matching conditions in (9.123)–(9.126) can be expressed in the same form as (9.18)–(9.21) with

$$N_1 = E_{ip} - E_{rp} \tag{9.132}$$
$$N_2 = E_{is} + E_{rs} \tag{9.133}$$
$$N_3 = e(E_{is} - E_{rs}) \tag{9.134}$$
$$N_4 = e(E_{ip} + E_{rp}) \tag{9.135}$$
$$\alpha_m = \left(en_m + \frac{1}{2}F_2\right)E_{mx} - 2F_1E_{mz} \tag{9.136}$$
$$\beta_m = en_mE_{mz} - F_1E_{mx}, \tag{9.137}$$

where $m = a$ or $b$. Then from the equivalent of (9.18)–(9.21) one can obtain the $c_i$ and $d_i$ in (9.28) and (9.29) that apply to $Cr_2O_3$ in the second configuration. These are

$$c_1 = -c_3 = -(\alpha_aE_{bx} - \alpha_bE_{ax}) \tag{9.138}$$
$$c_2 = -\alpha_aE_{bz} + \alpha_bE_{az} + e(E_{az}E_{bx} - E_{ax}E_{bz}) \tag{9.139}$$
$$c_4 = -\alpha_aE_{bz} + \alpha_bE_{az} - e(E_{az}E_{bx} - E_{ax}E_{bz}) \tag{9.140}$$
$$d_1 = -(\beta_aE_{bx} - \beta_bE_{ax}) - e(E_{az}E_{bx} - E_{ax}E_{bz}) \tag{9.141}$$
$$d_2 = d_4 = -(\beta_aE_{bz} - \beta_bE_{az}) \tag{9.142}$$
$$d_3 = \beta_aE_{bx} - \beta_bE_{ax} - e(E_{az}E_{bx} - E_{ax}E_{bz}). \tag{9.143}$$

The reflection coefficients are expressed in terms of $c_i$ and $d_i$ by (9.38)–(9.41). To obtain explicit forms for them the secular equation in (5.12) must be solved for the field amplitudes and wave vectors of the two transmitted rays in the crystal. From (5.12), (5.13), (5.8), and (9.74)–(9.77)

$$\begin{vmatrix} n_O^2 - n^2 - e^{-1}nF_2 & 0 & 3e^{-1}nF_1 \\ 0 & n_O^2 + e^{-1}nF_2 & 0 \\ 3e^{-1}nF_1 & 0 & n_E^2 - n^2 \end{vmatrix} = 0, \tag{9.144}$$

the solutions of which to order $F_i$ are

$$n_a = n_O - \frac{1}{2}e^{-1}F_2, \qquad \frac{E_{az}}{E_{ax}} = \frac{3e^{-1}n_O F_1}{n_O^2 - n_E^2} \qquad (9.145)$$

$$n_b = n_E, \qquad \frac{E_{bx}}{E_{bz}} = -\frac{3e^{-1}n_E F_1}{n_O^2 - n_E^2}. \qquad (9.146)$$

From (9.145), (9.146), (9.136)–(9.143), and (9.38)–(9.41) the reflection coefficients to leading order in $F_i$ are

$$r_{pp} = \frac{n_E - 1}{n_E + 1} \qquad (9.147)$$

$$r_{ss} = -\frac{n_O - 1}{n_O + 1} \qquad (9.148)$$

$$r_{ps} = r_{sp} = \frac{2(2n_O - n_E)F_1}{e(n_O + n_E)(n_O + 1)(n_E + 1)}. \qquad (9.149)$$

The diagonal elements (9.147) and (9.148) agree with those published previously [2, 3] when account is taken of the different coordinate labelling of the $C_2$ axis as the propagation direction. The off-diagonal elements (9.149) agree in magnitude with those of Ref. 3.

## 9.8 Comparison with experiment for $Cr_2O_3$

In terms of the basis used in Fig. 6.1, the reflected field $\mathbf{E}_r$ of a harmonic plane wave is related to the incident field $\mathbf{E}_i$ by (6.1), namely

$$\begin{pmatrix} E_{rp} \\ E_{rs} \end{pmatrix} = \begin{pmatrix} r_{pp} & r_{ps} \\ r_{sp} & r_{ss} \end{pmatrix} \begin{pmatrix} E_{ip} \\ E_{is} \end{pmatrix}. \qquad (9.150)$$

When the incident wave is polarized parallel to $s$, for example, the components of $\mathbf{E}_r$ are $E_{rp} = r_{ps}E_{is}$ and $E_{rs} = r_{ss}E_{is}$. Thus a reflected wave is in general elliptically polarized with its major axis rotated by $\Delta\phi$ from the direction of the incident linear polarization. If the reflecting medium absorbs, then circular dichroism will also be manifest. Furthermore, in a magnetic crystal the signs of these two effects change for states of opposite spin.

In their reflection experiments on $Cr_2O_3$ at normal incidence on the two crystal orientations considered in Sections 9.6 and 9.7, Krichevtsov et al. [2] measured the rotation $\Delta\phi$ and the circular dichroism as a function of temperature. Below the Néel temperature of 307 K the time-reversed behaviour of these two effects is strikingly evident and allows a comparison to be made between their experimental values and the theoretical results of the previous two sections.

From the theory of Ref. 2 one can show that

$$\frac{\Delta\phi_z}{\Delta\phi_y} = \frac{(r_{ps})_z}{(r_{ps})_y}, \qquad (9.151)$$

where subscripts $z$ and $y$ denote the crystallographic axes used for light propagation. An estimate of this ratio made from the published graphs of Ref. 2 is

1.6 at a wavelength of 633 nm. From (9.111) and (9.149), and using $|n_O - n_E| = 5.8 \times 10^{-2}$ [2], we calculate a value of 2 for the ratio in (9.151). Considering the very small rotations measured ($< 10^{-4} \, rad$) and the difficulty of surface preparation, the agreement between experiment and theory is reasonable.

## 9.9 Uniqueness of fields

We have seen that through the matching conditions at an interface between two media, reflection imposes a requirement of uniqueness on the fields which does not arise in transmission theory. The latter is based in part on Maxwell's inhomogeneous equations which do not define the response fields $\mathbf{D}$ and $\mathbf{H}$ uniquely (Section 7.2). This stringency imposed in reflection may be illustrated by considering the fields at a boundary between a vacuum and a dielectric.

As mentioned earlier (Section 6.5), matching conditions at the interface involve field amplitudes. At a vacuum point adjacent to the boundary, the amplitudes of the fields $\mathbf{E}$, $\mathbf{B}$, $\mathbf{D} = \varepsilon_0 \mathbf{E}$, and $\mathbf{H} = \mu_0^{-1} \mathbf{B}$ are unique and origin independent. Hence the field amplitudes in the dielectric, and adjacent to the boundary, can be matched with the vacuum values only if the former are also origin independent. However, as shown in Chapter 4, in a dielectric the amplitudes of $\mathbf{D}$ and $\mathbf{H}$ obtained directly from multipole theory are origin dependent for multipole orders beyond electric dipole. It is therefore not possible to match them with the vacuum amplitudes. The unphysical consequences of attempting to enforce such a match have been demonstrated in Section 6.8.

This situation is avoided by use of the transformed multipole theory of Chapters 7 and 8. This theory yields unique, origin-independent forms for the fields $\mathbf{D}$ and $\mathbf{H}$ induced in a homogeneous dielectric by a harmonic plane wave. Use of the corresponding translationally invariant polarization densities in the reflection theory of Sections 9.2–9.7 ensures the origin independence of the multipole expression for an effect produced in reflection.

Further instances in which such uniqueness is required are the energy density $U = \frac{1}{2}(\mathbf{E} \cdot \mathbf{D} + \mathbf{B} \cdot \mathbf{H})$ and the Poynting vector $\mathbf{S} = \mathbf{E} \times \mathbf{H}$, if these are to be physically meaningful quantities. When a multipole description is used for the effect of a harmonic plane wave in a homogeneous dielectric, use of the transformed fields $\mathbf{D}$ and $\mathbf{H}$ of Chapter 8 will produce acceptable expressions for $U$ and $\mathbf{S}$.

## 9.10 Summary

This concludes our exposition of multipole theory in electromagnetism. Often the discussion has been rather intricate, and to assist the reader in forming a coherent picture of this subject we summarize the main features.

(i) Classical multipole theory for a charge distribution in vacuum is discussed in Sections 1.1–1.11, where multipole expansions are obtained for electrostatics, magnetostatics, and electrodynamics. Applications to topics such as energy, force, and torque are also presented.

(ii) Quantum multipole theory for a charge distribution in vacuum is considered in Sections 2.1–2.10 and used to obtain expressions for multipole moment operators and multipole polarizability-type (distortion) tensors. The latter describe the induction of multipole moments by the fields of a harmonic plane wave.

(iii) These classical and quantum aspects of multipole theory for a charge distribution are, in principle, straightforward and without any surprises.

(iv) Classical multipole theory is extended to macroscopic media in Sections 1.12 and 1.13, where the electromagnetic response fields $\mathbf{D}$ and $\mathbf{H}$ are introduced together with Maxwell's macroscopic equations. The formulation is in terms of primitive, rather than traceless, macroscopic multipole moment densities (Section 1.15). In Section 2.11 it is described how macroscopic multipole moments and polarizabilities are obtained from the quantum theory of a charge distribution.

(v) The translational properties (behaviour under translation of the coordinate system) of multipole operators and polarizability tensors are derived in Section 3.7. These important results enable us to determine the translational behaviour of physical properties calculated from multipole theory.

(vi) In Chapter 4 we deduce linear constitutive relations for the response fields $\mathbf{D}$ and $\mathbf{H}$ from the multipole theory of Chapters 1 and 2 (the standard formulation of multipole theory). Three conditions are identified which should be satisfied by the calculated constitutive tensor, namely, origin independence (translational invariance), symmetries for non-dissipative media, and the Post constraint. We then consider the extent to which results obtained from the standard formulation of multipole theory satisfy these conditions (Section 4.5). Somewhat surprisingly it turns out that, for contributions beyond electric dipole order, the constitutive tensor does not satisfy translational invariance, and sometimes violates the Post constraint. Thus, for example, multipole expressions for the magnetoelectric tensors are unphysical in that they are origin dependent, and this occurs even in the electric dipole–magnetic dipole approximation (Section 4.6). Violation of the Post constraint depends on the multipole order and whether the medium is non-magnetic or magnetic.

(vii) Multipole theory is applied to a variety of transmission and scattering phenomena in Chapter 5. It is emphasized that constitutive relations for $\mathbf{D}$ and $\mathbf{H}$ with origin-dependent material constants (Chapter 4) lead to a wave equation which is, nevertheless, origin independent. Consequently, in its application to transmission effects, multipole theory yields physically acceptable results.

(viii) Multipole theory is applied to reflection phenomena in Chapter 6, where it is shown that for reflection of electromagnetic radiation from a crystal surface this theory produces unphysical results, notably reflected intensities which depend on the choice of coordinate origin.

(ix) The results of our analysis of the standard formulation of multipole theory for macroscopic media are depicted schematically in Fig. 9.1 Here symbols placed outside the indicated domain represent phenomena or properties for which the results of multipole theory are unphysical; those placed inside are physically acceptable. It is clear that although the standard formulation of multipole theory for macroscopic media is successful in certain instances, it is essentially an unphysical theory.

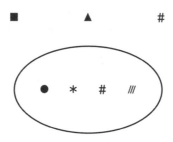

■ Constitutive relations

● Transmission and scattering effects

▲ Reflection effects

\# Post constraint

\* Reciprocity (time-reversal invariance)

/// Symmetries of material constants (non-dissipative media)

FIG. 9.1. Schematic illustration of the physical acceptability of results obtained from the standard formulation of multipole theory for macroscopic media.

(x) In Chapters 7 and 8 we consider how multipole theory should be modified in order to remove its unphysical features. This modification involves a transformation theory based on the inherent non-uniqueness of the response fields **D** and **H** in Maxwell's electrodynamics. The transformation theory is devised to make the multipole constitutive tensor translationally invariant and to satisfy the symmetries for a non-dissipative medium.

(xi) The most general electromagnetic fields to which our analyses of the standard and transformed multipole theories apply are fields represented by complex harmonic plane waves. We emphasize that this restriction allows consideration also of dissipative media: for these media the wave vector **k** is complex, and the plane waves are homogeneous if the real and imaginary parts of **k** are parallel, and inhomogeneous otherwise.

(xii) Consideration of a non-dissipative limit, and consequent symmetries of the constitutive tensor, play an important role in the transformation theory (Chapter 8).

(xiii) The Post constraint is automatically satisfied in the transformed theory (Chapter 8).

(xiv) Application of the transformed multipole theory to transmission and reflection phenomena is presented in Chapter 9. The wave equation for propagation of electromagnetic radiation is invariant under the transformations considered, and consequently for transmission phenomena the transformed theory yields the same results as the standard theory. For reflection phenomena the results are different and those of the transformed theory satisfy requirements of both translational and time-reversal invariance. Thus the theory of reflection from a crystal surface provides a sensitive test of multipole theory in macroscopic electromagnetism.

(xv) We conclude that the theory of Chapters 7 and 8 yields a unique, transformed multipole theory for which all symbols in Fig. 9.1 fall within the physical domain.

Part of the presentation in this book relies heavily on the use of transformations, notably space–time transformations of physical quantities (Chapter 3) and more abstract transformations of the response fields and their constitutive relations (Chapters 7 and 8). The reader will have noticed that the role played by simple translation of the coordinate system is more significant than one may have anticipated. We feel, therefore, that it is appropriate to conclude our work with some brief comments on the historical development of this topic.

The fact that multipole moments, beyond the leading non-vanishing order, are origin dependent (Sections 1.2 and 1.7) has been appreciated for a long time [4]. The values of such origin-dependent observables have meaning only with respect to a specified origin of coordinates. How this may be done is a nontrivial question that was first answered by Buckingham and Longuet-Higgins in connection with measurements of the electric quadrupole moment of dipolar molecules. Their theory shows how to extract values of an origin-dependent observable (quadrupole moment) from measurements of an origin-independent quantity (birefringence), and was later used in numerical calculations of these moments by Rizzo *et al.* [5] and Coriani *et al.* [6] (Section 5.11).

With regard to origin-independent observables, multipole theory yields expressions for these as combinations of polarizabilty tensors which are generally origin dependent (Chapters 3–6). It is therefore an essential task to demonstrate the origin independence of such expressions. To our knowledge, this was first done by Van Vleck for the *dc* magnetizability (Section 2.4), and later by Buckingham and others in studies of certain effects in transmission (Chapter 5) and light scattering [7], and of transition intensities [8].

These developments gave strong support for the applicability of multipole theory, and it therefore came as a surprise that the theory yields unphysical results for response fields and certain reflection effects (Chapters 4 and 6). The resolution of this situation provided part of the motivation for the present work.

# References

[1] Birss, R. R. (1966). *Symmetry and magnetism*. North-Holland, Amsterdam.

[2] Krichevtsov, B. B., Pavlov, V. V., Pisarev, R. V., and Gridnev, V. N. (1993). Spontaneous non-reciprocal reflection of light from antiferromagnetic $Cr_2O_3$. *Journal of Physics: Condensed Matter*, **5**, 8233–8244.

[3] Graham, E. B. and Raab, R. E. (1997). Macroscopic theory of reflection from antiferromagnetic $Cr_2O_3$. *Journal of Physics: Condensed Matter*, **9**, 1863–1869.

[4] Scaife, B. K. P. (1989). *Principles of dielectrics*, Ch. 10 Clarendon, Oxford.

[5] Rizzo, A., Coriani, S., Halkier, A., and Hättig, C. (2000). *Ab initio* study of the electric-field-gradient-induced birefringence of a polar molecule: CO. *Journal of Chemical Physics*, **113**, 3077–3087.

[6] Coriani, S., Halkier, A., Jonsson, D., Gauss, J., Rizzo, A., and Christiansen, O. (2003). On the electric field gradient induced birefringence and electric quadrupole moment of CO, $N_2O$, and OCS. *Journal of Chemical Physics*, **118**, 7329–7339.

[7] Barron, L. D. and Buckingham, A. D. (1971). Rayleigh and Raman scattering from optically active molecules. *Molecular Physics*, **20**, 1111–1119.

[8] Stephens, P. J. (1970). Theory of magnetic circular dichroism. *Journal of Chemical Physics*, **52**, 3489–3516.

# APPENDIX A

## TRANSFORMATIONS INVOLVING **J**

In this appendix we present results that are useful in transforming integrals involving the current density **J**, which occur in both the magnetostatic and the electrodynamic expansions of the vector potential **A** (see Chapter 1).

From Gauss' theorem

$$\int_V \nabla_i (J_i T_{jk\ldots})\, dv = \oint_S J_i n_i T_{jk\ldots}\, ds = 0. \tag{A.1}$$

Here $T_{jk\ldots}$ is some function of **r** and possibly of $t$, $ds$ is an element of area on the surface $S$ enclosing the volume $V$, of which $dv$ is an element, and **n** is the outward normal to $ds$. In (A.1) we have used the boundary condition $\mathbf{J}\cdot\mathbf{n} = 0$ on $S$.

In addition, we use the equation of continuity

$$\nabla\cdot\mathbf{J}(\mathbf{r},t) = -\dot{\rho}(\mathbf{r},t) \tag{A.2}$$

and the following identities

$$J_j r_i - J_i r_j = \varepsilon_{ijk}(\mathbf{r}\times\mathbf{J})_k \tag{A.3}$$

$$\nabla_j(r_i J_j) = J_i - r_i \dot{\rho} \tag{A.4}$$

$$\nabla_k(r_i r_j J_k) = 2J_i r_j + \varepsilon_{ijk}(\mathbf{r}\times\mathbf{J})_k - r_i r_j \dot{\rho} \tag{A.5}$$

$$\nabla_l(r_i r_j r_k J_l) = 3J_i r_j r_k + \varepsilon_{ijl}(\mathbf{r}\times\mathbf{J})_l r_k + \varepsilon_{ikl}(\mathbf{r}\times\mathbf{J})_l r_j - r_i r_j r_k \dot{\rho}. \tag{A.6}$$

Equation (A.3) follows from $(\mathbf{r}\times\mathbf{J})_k = \varepsilon_{klm} r_l J_m$ and the relation

$$\varepsilon_{ijk}\varepsilon_{klm} = \delta_{il}\delta_{jm} - \delta_{im}\delta_{jl}. \tag{A.7}$$

Integrating (A.4)–(A.6) over $V$ and using (A.1) yields

$$\int_V J_i\, dv = \int_V r_i \dot{\rho}\, dv \tag{A.8}$$

$$\int_V J_i r_j\, dv = -\frac{1}{2}\varepsilon_{ijk}\int_V (\mathbf{r}\times\mathbf{J})_k\, dv + \frac{1}{2}\int_V r_i r_j \dot{\rho}\, dv \tag{A.9}$$

$$\int_V J_i r_j r_k\, dv = -\frac{1}{3}\varepsilon_{ijl}\int_V (\mathbf{r}\times\mathbf{J})_l r_k\, dv$$
$$-\frac{1}{3}\varepsilon_{ikl}\int_V (\mathbf{r}\times\mathbf{J})_l r_j\, dv + \frac{1}{3}\int_V r_i r_j r_k \dot{\rho}\, dv. \tag{A.10}$$

Equations (A.8)–(A.10) can be expressed in terms of the multipole moments
(1.5)–(1.7), (1.44), and (1.46) (with $\rho(\mathbf{r})$ and $\mathbf{J}(\mathbf{r})$ replaced by $\rho(\mathbf{r}, t)$ and $\mathbf{J}(\mathbf{r}, t)$
in the electrodynamic case) as

$$\int_V J_i(\mathbf{r}, t)\, dv = \dot{p}_i(t) \tag{A.11}$$

$$\int_V J_i(\mathbf{r}, t) r_j\, dv = -\varepsilon_{ijk} m_k(t) + \frac{1}{2}\dot{q}_{ij}(t) \tag{A.12}$$

$$\int_V J_i(\mathbf{r}, t) r_j r_k\, dv = -\frac{1}{2}\varepsilon_{ijl} m_{lk}(t) - \frac{1}{2}\varepsilon_{ikl} m_{lj}(t) + \frac{1}{3}\dot{q}_{ijk}(t). \tag{A.13}$$

The magnetostatic forms of (A.11)–(A.13) are obtained from these equations
by suppressing the time-derivative terms

$$\int_V J_i(\mathbf{r})\, dv = 0 \tag{A.14}$$

$$\int_V J_i(\mathbf{r}) r_j\, dv = -\varepsilon_{ijk} m_k \tag{A.15}$$

$$\int_V J_i(\mathbf{r}) r_j r_k\, dv = -\frac{1}{2}\varepsilon_{ijl} m_{lk} - \frac{1}{2}\varepsilon_{ikl} m_{lj}. \tag{A.16}$$

Equations (A.15) and (A.16) yield (1.43) and (1.45), respectively. Thus equations
(A.14)–(A.16) enable us to express the expansion (1.40) in the convenient form
of (1.47). Equations (A.11)–(A.13) enable us to express (1.77) in the form (1.78).

# APPENDIX B

## MAGNETOSTATIC FIELD

We prove that the curl of the vector potential (1.47) yields the magnetostatic field (1.48). Using (1.47) in

$$B_i = (\mathbf{\nabla} \times \mathbf{A})_i = \varepsilon_{ijk}\nabla_j A_k, \tag{B.1}$$

we have

$$B_i = \frac{\mu_0}{4\pi}\,\varepsilon_{ijk}\varepsilon_{kln}\nabla_j\left[\frac{m_l R_n}{R^3} + \frac{3R_s R_n - R^2\delta_{sn}}{2R^5}\,m_{ls} + \cdots\right]. \tag{B.2}$$

Now

$$\nabla_j\frac{R_n}{R^3} = -\frac{3R_n R_j - R^2\delta_{jn}}{R^5} \tag{B.3}$$

$$\nabla_j\frac{3R_s R_n - R^2\delta_{sn}}{R^5} = \frac{3}{R^7}\left[-5R_j R_n R_s + R^2(R_j\delta_{ns} + R_n\delta_{sj} + R_s\delta_{jn})\right]. \tag{B.4}$$

Equations (B.2)–(B.4) and (A.7) yield (1.48).

# APPENDIX C

## MAGNETOSTATIC FORCE

We express the total force on a steady current distribution in a slowly varying external magnetic field (equation (1.58)) in terms of the magnetic multipole moments (equation (1.59)). The leading term in the expansion of the force (1.58) is from (A.15)

$$\varepsilon_{ijk}(\nabla_l B_k) \int_V J_j \, r_l \, dv = -\varepsilon_{ijk}\varepsilon_{jlm}(\nabla_l B_k)\, m_m$$
$$= (\nabla_i B_j)\, m_j. \qquad (C.1)$$

In the last step leading to (C.1) we have used (A.7) and $\nabla \cdot \mathbf{B} = 0$.

The next term in (1.58) becomes from (A.16)

$$\frac{1}{2}\varepsilon_{ijk}(\nabla_m \nabla_l B_k) \int_V J_j \, r_l r_m \, dv = -\frac{1}{4}\varepsilon_{ijk}(\nabla_m \nabla_l B_k)(\varepsilon_{jln} m_{nm} + \varepsilon_{jmn} m_{nl})$$
$$= -\frac{1}{2}\varepsilon_{ijk}\varepsilon_{jln}(\nabla_m \nabla_l B_k)\, m_{nm}$$
$$= \frac{1}{2}(\nabla_j \nabla_i B_k)\, m_{kj}. \qquad (C.2)$$

Because the sources of the external field $\mathbf{B}$ are outside the current distribution and $\mathbf{B}$ is magnetostatic, $\nabla \times \mathbf{B} = 0$. Thus in (C.2)

$$\nabla_i B_j = \nabla_j B_i. \qquad (C.3)$$

From (C.1)–(C.3) the force in (1.58) can be written

$$F_i = m_j \nabla_j B_i + \frac{1}{2} m_{jk} \nabla_k \nabla_j B_i + \cdots, \qquad (C.4)$$

which is (1.59).

# APPENDIX D

## MAGNETOSTATIC TORQUE

We express the total torque on a steady current distribution in a slowly varying magnetic field (equation (1.63)) in terms of the magnetic multipole moments (equation (1.64)). The leading term in the expansion of the torque in (1.63) is from (A.15)

$$\varepsilon_{ijk}\varepsilon_{klm}B_m \int_V J_l\, r_j\, dv = -\varepsilon_{ijk}\varepsilon_{klm}\varepsilon_{ljn}B_m m_n$$

$$= \varepsilon_{ijk}m_j B_k, \tag{D.1}$$

where we have used (A.7).

The next term in (1.63) is from (A.16)

$$\varepsilon_{ijk}\varepsilon_{klm}(\nabla_n B_m) \int_V J_l\, r_j r_n\, dv = -\frac{1}{2}\varepsilon_{ijk}\varepsilon_{klm}(\nabla_n B_m)(\varepsilon_{ljp}m_{pn} + \varepsilon_{lnp}m_{pj})$$

$$= \frac{1}{2}\varepsilon_{ijk}(m_{jl} + m_{lj})\nabla_l B_k, \tag{D.2}$$

where we have used (A.7), (C.3), and $\nabla \cdot \mathbf{B} = 0$. From (D.1) and (D.2) the torque in (1.63) can be written

$$N_i = \varepsilon_{ijk}\left[m_j B_k + \frac{1}{2}(m_{jl} + m_{lj})\nabla_l B_k + \cdots\right], \tag{D.3}$$

which is (1.64).

# APPENDIX E

## INTEGRAL TRANSFORMATIONS

We outline the calculations by which the scalar and vector potentials (1.95) and (1.96) for a macroscopic medium are transformed into (1.99) and (1.100). The calculations are based on the following five identities:

$$\nabla_i \left( \frac{P_i}{R} \right) = \frac{1}{R} \nabla_i P_i + P_i \nabla_i \frac{1}{R} \tag{E.1}$$

$$\nabla_i \left\{ \left( \nabla_j \frac{1}{R} \right) Q_{ij} \right\} - \nabla_i \left( \frac{1}{R} \nabla_j Q_{ij} \right) = Q_{ij} \nabla_j \nabla_i \frac{1}{R} - \frac{1}{R} \nabla_j \nabla_i Q_{ij}, \quad \text{if } Q_{ji} = Q_{ij} \tag{E.2}$$

$$\nabla_i \left\{ \left( \nabla_j \nabla_k \frac{1}{R} \right) Q_{ijk} \right\} - \nabla_i \left\{ \left( \nabla_k \frac{1}{R} \right) \nabla_j Q_{ijk} \right\} + \nabla_i \left( \frac{1}{R} \nabla_k \nabla_j Q_{ijk} \right)$$

$$= Q_{ijk} \nabla_k \nabla_j \nabla_i \frac{1}{R} + \frac{1}{R} \nabla_k \nabla_j \nabla_i Q_{ijk}, \quad \text{if } Q_{jik} = Q_{ijk} \text{ and } Q_{kji} = Q_{ijk} \tag{E.3}$$

$$\nabla_j \left( \frac{1}{R} T_{ij} \right) = \frac{1}{R} \nabla_j T_{ij} + T_{ij} \nabla_j \frac{1}{R} \tag{E.4}$$

$$\nabla_j \left\{ \left( \nabla_k \frac{1}{R} \right) T_{ijk} \right\} - \nabla_k \left\{ \frac{1}{R} \nabla_j T_{ijk} \right\} = T_{ijk} \nabla_k \nabla_j \frac{1}{R} - \frac{1}{R} \nabla_k \nabla_j T_{ijk}. \tag{E.5}$$

Note that the electric multipole moment densities $Q_{ij}$ and $Q_{ijk}$ possess full permutation symmetry (Section 1.1) and therefore the symmetry conditions for (E.2) and (E.3) are satisfied.

We integrate equations (E.1)–(E.5) over the volume $V$ of the medium and use Gauss' theorem. This gives

$$\int_V P_i \nabla_i \frac{1}{R} \, dv = - \int_V \frac{1}{R} \nabla_i P_i \, dv + \oint_S \frac{1}{R} P_i n_i \, ds \tag{E.6}$$

$$\int_V Q_{ij} \nabla_j \nabla_i \frac{1}{R} \, dv = \int_V \frac{1}{R} \nabla_j \nabla_i Q_{ij} \, dv - \oint_S \frac{1}{R} \nabla_j Q_{ij} n_i \, ds - \oint_S \frac{R_j}{R^3} Q_{ij} n_i \, ds \tag{E.7}$$

(In the last term of (E.7) we have used (1.92).)

$$\int_V Q_{ijk} \nabla_k \nabla_j \nabla_i \frac{1}{R} \, dv = - \int_V \frac{1}{R} \nabla_k \nabla_j \nabla_i Q_{ijk} \, dv + \oint_S \frac{1}{R} \nabla_k \nabla_j Q_{ijk} n_i \, ds$$

$$+ \oint_S \frac{R_k}{R^3} \nabla_j Q_{ijk} n_i \, ds + \oint_S \frac{3R_j R_k - R^2 \delta_{jk}}{R^5} Q_{ijk} n_i \, ds$$

$$(\text{E.8})$$

(In the last two terms of (E.8) we have used (1.92) and (1.93).)

$$\int_V T_{ij} \nabla_j \frac{1}{R} \, dv = - \int_V \frac{1}{R} \nabla_j T_{ij} \, dv + \oint_S \frac{1}{R} T_{ij} n_j \, ds \qquad (\text{E.9})$$

$$\int_V T_{ijk} \nabla_k \nabla_j \frac{1}{R} \, dv = \int_V \frac{1}{R} \nabla_k \nabla_j T_{ijk} \, dv - \oint_S \frac{1}{R} \nabla_j T_{ijk} n_k \, ds - \oint_S \frac{R_k}{R^3} T_{ijk} n_j \, ds.$$

$$(\text{E.10})$$

(In the last term of (E.10) we have used (1.92).) Here $ds$ is an element of area on the surface $S$ enclosing $V$, and $\mathbf{n}$ is the outward normal to $ds$.

Using (E.6)–(E.8) in (1.95) we obtain (1.99). Using (E.9) and (E.10) in (1.96) we obtain (1.100).

# APPENDIX F

## ORIGIN DEPENDENCE OF A POLARIZABILITY TENSOR

We derive here the origin dependence of the polarizability $L_{ijk}$ as an example of how to obtain those in (3.64)–(3.79). Using the notation in (3.55) for a quantity relative to origin $\bar{O}$, and from the expression for $L_{ijk}$ in (2.127), we have

$$
\begin{aligned}
\Delta L_{ijk} &= \bar{L}_{ijk} - L_{ijk} \\
&= 2\hbar^{-1} \sum_s \omega_{sn} Z_{sn} \mathcal{R}e\left\{\langle \bar{q}_{ij}\rangle_{ns}\langle \bar{m}_k\rangle_{sn} - \langle q_{ij}\rangle_{ns}\langle m_k\rangle_{sn}\right\} \\
&= 2\hbar^{-1} \sum_s \omega_{sn} Z_{sn} \mathcal{R}e\left\{\langle q_{ij}+\Delta q_{ij}\rangle_{ns}\langle m_k+\Delta m_k\rangle_{sn} - \langle q_{ij}\rangle_{ns}\langle m_k\rangle_{sn}\right\} \\
&= 2\hbar^{-1} \sum_s \omega_{sn} Z_{sn} \mathcal{R}e\left\{\langle q_{ij}\rangle_{ns}\langle \Delta m_k\rangle_{sn}+\langle \Delta q_{ij}\rangle_{ns}\langle m_k\rangle_{sn}+\langle \Delta q_{ij}\rangle_{ns}\langle \Delta m_k\rangle_{sn}\right\} \\
&= 2\hbar^{-1} \sum_s \omega_{sn} Z_{sn} \mathcal{R}e\left\{\langle q_{ij}\rangle_{ns}\left[-\frac{i}{2}\varepsilon_{klm}d_l\omega_{sn}\langle p_k\rangle_{sn}\right]\right. \\
&\qquad \left. +[-d_i\langle p_j\rangle_{ns}-d_j\langle p_i\rangle_{ns}]\langle m_k\rangle_{sn}+[-d_i\langle p_j\rangle_{ns}-d_j\langle p_i\rangle_{ns}]\times\left[-\frac{i}{2}\varepsilon_{klm}d_l\omega_{sn}\langle p_m\rangle_{sn}\right]\right\}
\end{aligned}
\tag{F.1}
$$

in which we used (3.57), (3.59), the canonical commutation relation, and $\langle q^{(\alpha)}\rangle_{ns} = q^{(\alpha)}\langle n|s\rangle = 0$ for orthogonal kets. Then from (2.75) and (2.131), (F.1) becomes

$$
\begin{aligned}
\Delta L_{ijk} &= \frac{1}{\hbar}\varepsilon_{klm}d_l \sum_s (1+\omega^2 Z_{sn})\mathcal{I}m\left\{\langle q_{ij}\rangle\langle p_k\rangle_{sn}\right\} \\
&\quad -\frac{2}{\hbar} \sum_s \omega_{sn} Z_{sn} \mathcal{R}e\left\{d_i\langle p_j\rangle_{ns}\langle m_k\rangle_{sn}+d_j\langle p_i\rangle_{ns}\langle m_k\rangle_{sn}\right\} \\
&\quad -\frac{1}{\hbar}\varepsilon_{klm}d_l \sum_s (1+\omega^2 Z_{sn})\mathcal{I}m\left\{[d_i\langle p_j\rangle_{ns}+d_j\langle p_i\rangle_{ns}]\langle p_m\rangle_{sn}\right\}.
\end{aligned}
\tag{F.2}
$$

From the closure relation

$$
\sum_s \langle q_{ij}\rangle_{ns}\langle p_k\rangle_{sn} = \langle q_{ij}p_k\rangle_{nn}
\tag{F.3}
$$

$$
\sum_s \langle p_i\rangle_{ns}\langle p_j\rangle_{sn} = \langle p_i p_j\rangle_{nn}.
\tag{F.4}
$$

The product operators in the expectation values in (F.3) and (F.4) are Hermitian because their factors are commuting Hermitian operators. Thus (F.3) and (F.4) are real. Using this, and (2.116), (2.118), and (2.119), we can express (F.2) as

$$\Delta L_{ijk} = \frac{1}{2}\omega\varepsilon_{klm}d_l a'_{kij} - d_i G_{jk} - d_j G_{ik} + \frac{1}{2}\omega\varepsilon_{klm}d_l(d_i\alpha'_{jm} + d_j\alpha'_{im})$$

$$= -d_i G_{jk} - d_j G_{ik} + \frac{1}{2}\omega\varepsilon_{klm}d_l(a'_{kij} + d_i\alpha'_{jm} + d_j\alpha'_{im}). \tag{F.5}$$

This is (3.76).

# APPENDIX G

## INVARIANCE OF TRANSFORMED TENSORS

We prove the origin independence of the tensors $V_{ij}$ (8.32), $W_{ij}$ (8.56), and $S_{ijk}$ (8.60) which appear in the transformed material constants (8.65)–(8.68):

(i) From (8.32), (3.66), and (3.69) the change in $V_{ij}$ for a shift of origin $\mathbf{d}$ is

$$
\begin{aligned}
\Delta V_{ij} &= \frac{1}{2}\omega\varepsilon_{jkl}(d_k\alpha_{il} + d_i\alpha_{kl} + d_l\alpha_{ki}) \\
&= \frac{1}{2}\omega\varepsilon_{jkl}(d_k\alpha_{il} + d_l\alpha_{ki}) \\
&= \frac{1}{2}\omega\varepsilon_{jkl}d_k\alpha_{il} + \frac{1}{2}\omega\varepsilon_{jlk}d_k\alpha_{li} \\
&= 0,
\end{aligned} \tag{G.1}
$$

where we have used the symmetry (2.115) of $\alpha_{ij}$.

(ii) In (8.56) we use the identity

$$
G_{ll}\delta_{ij} = G_{ij} + \varepsilon_{ikl}\varepsilon_{jkm}G_{ml}
$$

and (3.67) and (3.68). Then

$$
\begin{aligned}
\Delta W_{ij} &= \Delta G_{ij} - \frac{1}{3}(\Delta G_{ij} + \varepsilon_{ikl}\varepsilon_{jkm}\Delta G_{ml}) - \frac{1}{6}\omega\varepsilon_{jkl}\Delta a'_{kli} \\
&= \frac{1}{6}\omega\varepsilon_{jkl}(d_i\alpha'_{kl} + d_l\alpha'_{ki} - 2d_k\alpha'_{il}) + \frac{1}{6}\omega\varepsilon_{ikl}\varepsilon_{jkm}\varepsilon_{lpq}d_p\alpha'_{mq} \\
&= \frac{1}{6}\omega\varepsilon_{jkl}[d_i(\alpha'_{kl} + \alpha'_{lk}) + d_l\alpha'_{ki} - d_k(2\alpha'_{il} + \alpha'_{li})] \\
&= \frac{1}{6}\omega\varepsilon_{jkl}d_k(-2\alpha'_{il} - \alpha'_{li} - \alpha'_{li}) \\
&= 0,
\end{aligned} \tag{G.2}
$$

where we have used the antisymmetry (2.116) of $\alpha'_{ij}$.

(iii) From (8.60) and (3.67)

$$
\Delta S_{ijk} = -\frac{1}{3}[d_k(\alpha'_{ij} + \alpha'_{ji}) + d_j(\alpha'_{ik} + \alpha'_{ki}) + d_i(\alpha'_{jk} + \alpha'_{kj})] = 0 \tag{G.3}
$$

because of the antisymmetry of $\alpha'_{ij}$.

# GLOSSARY OF SYMBOLS

1.*Roman letters*

| | |
|---|---|
| $\mathbf{a}$ | Acceleration, 67 |
| $a_i$ | Components of $\mathbf{a}$, 69 |
| $\mathbf{A}$ | Vector potential, 10 |
| $A_i$ | Components of $\mathbf{A}$, 11 |
| $A^\mu$ | 4-vector potential, 87 |
| $A_{ij}$ | Permittivity tensor, 85. Transformed permittivity tensor, 175 |
| $A_{ij}^M$ | Direct multipole form of $A_{ij}$, 92 |
| $a_{ij}$ | Coefficients in secular equation, 102 |
| $a_{ijk}$ | Polarizability tensor, 35. Polarizability density, 53, 85 |
| $a'_{ijk}$ | Polarizability tensor, 44. Polarizability density, 53, 85 |
| $\tilde{a}_{ijk}$ | Invariant polarizability density, 187 |
| | Non-invariant polarizability density, 188 |
| $A_{ijk}$ | 163 |

| | |
|---|---|
| $\mathbf{b}$ | Magnetic field, 23 |
| $\mathbf{B}$ | Magnetic field, 11 |
| $B_i$ | Components of $\mathbf{B}$, 11 |
| $\mathbf{B}_a,\ \mathbf{B}_b$ | Magnetic fields of two refracted waves, 158 |
| $b_{ij}$ | 103 |
| $b_{ijkl}$ | Polarizability tensor, 7. Polarizability density, 53, 85 |
| $b'_{ijkl}$ | Polarizability tensor, 48. Polarizability density, 53, 85 |

| | |
|---|---|
| $c$ | Speed of light in vacuum, 15 |
| $c_1,\dots,c_4$ | 161, 194 |
| $c_s$ | Coefficient, 34 |
| $c_s(t)$ | Mixing coefficient, 41 |

| | |
|---|---|
| $d_1,\dots,d_4$ | 161, 194 |
| $\mathbf{d}$ | Displacement of an origin, 5 |
| $\mathbf{D}$ | Response field, 24 |
| $D_i$ | Components of $\mathbf{D}$, 24 |
| $\mathbf{D}_0$ | Amplitude of $\mathbf{D}$, 96 |
| $D_i^F$ | Effect of a Faraday transformation on $\mathbf{D}_i$, 174 |
| $\mathbf{D}(\mathbf{k},\omega)$ | Fourier transform of $\mathbf{D}$, 173 |

221

| | |
|---|---|
| $dq$ | Element of charge, 11 |
| $ds$ | Surface element, 10 |
| $dS$ | Surface element, 21 |
| $dv$ | Volume element, 1 |
| $dV$ | Volume element, 19 |
| $d_{ijkl}$ | Polarizability tensor, 7. Polarizability density, 54, 85 |
| $d'_{ijkl}$ | Polarizability tensor, 48. Polarizability density, 54, 85 |
| | |
| $e$ | 157 |
| $\mathbf{e}$ | Electric field, 23 |
| $\mathbf{E}$ | Electric field, 4 |
| $E_i$ | Components of $\mathbf{E}$, 4 |
| $\mathbf{E}_a, \mathbf{E}_b$ | Electric fields of two refracted waves, 158 |
| $\mathbf{E}_0$ | Electric field amplitude, 48 |
| $E_{ij}(t), E_{ijk}(t)$ | Electric field gradients, 46 |
| $E_{ip}, E_{is}$ | Components of $\mathbf{E}$ for an incident wave, 146 |
| $E_{rp}, E_{rs}$ | Components of $\mathbf{E}$ for a reflected wave, 146 |
| | |
| $f$ | Dispersion line-shape function, 50 |
| $F$ | Faraday transformation, 174 |
| $F_1, \ldots, F_8$ | 160 |
| $F_1, \ldots, F_5$ | 193 |
| $\mathbf{F}$ | Force, 8. Vector representing $\mathbf{E}$ or $\mathbf{B}$, 85 |
| $F_i$ | Components of $\mathbf{F}$, 7 |
| $\mathbf{F}^{(\alpha)}$ | Force on particle $\alpha$, 8 |
| $F^{\mu\nu}$ | Electromagnetic field tensor for $\mathbf{E}$ and $\mathbf{B}$, 88 |
| | |
| $g$ | Absorption line-shape function, 50 |
| $G$ | Gauge transformation, 173 |
| $g^{(\alpha)}$ | $g$-factor of particle $\alpha$, 39 |
| $g_{\mu\nu}$ | Metric tensor, 87 |
| $G_{ij}, G'_{ij}$ | Polarizability tensors, 44. Polarizability densities, 53, 85 |
| $\tilde{G}_{ij}$ | Invariant polarizability density, 187 |
| | Non-invariant polarizability density, 188 |
| $G_{\mu\nu}$ | Field tensor for $\mathbf{D}$ and $\mathbf{H}$, 88 |
| | |
| $\hbar$ | Planck's constant, 33 |
| $H$ | Hamiltonian operator, 33 |
| $\mathbf{H}$ | Response field, 24 |
| $H_i$ | Components of $\mathbf{H}$, 24 |
| $H_i^F$ | Effect of a Faraday transformation on $H_i$, 174 |
| $H_i^G$ | Effect of a gauge transformation on $H_i$, 173 |

| | |
|---|---|
| $H^{(0)}$, $\langle H^{(0)}\rangle_{sn}$ | Zeroth-order perturbation Hamiltonian and its matrix elements, 33, 34 |
| $H^{(1)}$, $\langle H^{(1)}\rangle_{sn}$ | First-order perturbation Hamiltonian and its matrix elements, 33, 34 |
| $H^{(2)}$, $\langle H^{(2)}\rangle_{sn}$ | Second-order perturbation Hamiltonian and its matrix elements, 33, 34 |
| $H_{ijk}$, $H'_{ijk}$ | Polarizability tensors, 48. Polarizability densities, 53, 85 |
| $\mathbf{H}(\mathbf{k},\omega)$ | Fourier transform of $\mathbf{H}$, 173 |
| | |
| $I_1,\ldots,I_{14}$ | 166 |
| $I_i$, $I_r$ | Incident and reflected intensities, 148 |
| $I'_i$, $I'_r$ | Incident and reflected intensities in a time-reversed experiment, 148 |
| | |
| $\mathbf{j}$ | Current density, 23 |
| $j$ | 163 |
| $\mathbf{J}$ | Current density, 10 |
| $J_i$ | Components of $\mathbf{J}$, 10 |
| $\mathbf{J}_b$ | Bound current density, 22 |
| $J_{bi}$ | Components of $\mathbf{J}_b$, 22 |
| $\mathbf{J}_f$ | Free current density, 22 |
| $J_{fi}$ | Components of $\mathbf{J}_f$, 23 |
| $\mathbf{J}_f(\mathbf{k},\omega)$ | Fourier transform of $\mathbf{J}_f$, 173 |
| | |
| $\mathbf{k}$ | Wave vector, 41 |
| $k_i$ | Components of $\mathbf{k}$, 85 |
| $\mathbf{k}_a$, $\mathbf{k}_b$ | Wave vectors of two refracted waves, 157 |
| $\mathbf{k}_i$, $\mathbf{k}_r$ | Incident and reflected wave vectors, 146 |
| $K$ | Complex conjugation, 66 |
| $\mathbf{K}$ | Bound surface current density, 151 |
| $K_i$ | Components of $\mathbf{K}$, 152 |
| $K_{bi}$ | Components of bound surface current density, 22 |
| | |
| $L_1,\ldots,L_5$ | 164 |
| $\mathbf{L}$ | Angular momentum, 60 |
| $\mathbf{l}^{(\alpha)}$ | Angular momentum of particle $\alpha$, 12 |
| | Angular momentum operator, 39 |
| $l_i^{(\alpha)}$ | Components of $\mathbf{l}^{(\alpha)}$, 12 |
| $l_{ij}$ | Direction cosines, 60 |
| $L_{ij}$ | Coefficients, 147 |
| $L_{bij}$ | Bound surface magnetic dipole moment current density, 23 |
| $L_{ijk}$, $L'_{ijk}$ | Polarizability tensors, 48. Polarizability densities, 54, 85 |

| | |
|---|---|
| $\mathbf{m}$ | Magnetic dipole moment, 11 |
| $m_i$ | Components of $\mathbf{m}$, 11. Magnetic dipole moment operator, 38 |
| $m^{(\alpha)}$ | Mass of particle $\alpha$, 12 |
| $\mathbf{m'}$ | Perturbed magnetic dipole moment operator, 40 |
| $m'_i$ | Components of $\mathbf{m'}$, 40 |
| $m_i^{(0)}$ | Permanent magnetic dipole moment, 39 |
| $M_i$ | Macroscopic magnetic dipole moment density, 19 |
| | Transformed magnetic dipole moment density, 187 |
| $m_{ij}$ | Magnetic quadrupole moment, 11 |
| | Magnetic quadrupole moment operator, 46 |
| $\bar{m}_{ij}$ | Magnetic quadrupole moment relative to a displaced origin, 12 |
| $M_{ij}$ | Macroscopic magnetic quadrupole moment density, 19 |
| $m'_{ij}$ | Perturbed magnetic quadrupole moment operator, 47 |
| $m_{ijk}$ | Magnetic octopole moment operator, 46 |
| $m'_{ijk}$ | Perturbed magnetic octopole moment operator, 47 |
| $m_{ijk...z}$ | Magnetic $2^n$-pole operator, 47 |
| | |
| $\mathbf{n}$ | Unit vector along outward normal, 10 |
| $n_j$ | Components of $\mathbf{n}$, 10 |
| $n$ | Refractive index, 101 |
| $n_1, n_2$ | Refractive indices, 104 |
| $n_L, n_R$ | Refractive indices, 104 |
| $n_O, n_E$ | Refractive indices, 108 |
| $n_+, n_-$ | Refractive indices, 110 |
| $n_x, n_y$ | Refractive indices, 114 |
| $N$ | Number of particles, 3 |
| $N_1,...,N_4$ | 160, 193 |
| $\mathbf{N}$ | Torque, 13 |
| $N_i$ | Components of $\mathbf{N}$, 8 |
| $|n\rangle$ | Perturbed eigenvector, 34 |
| $|n^{(0)}\rangle$ | Unperturbed eigenvector, 34 |
| $|n(t)\rangle$ | Time-dependent eigenvector, 41 |
| | |
| $p_i$ | Electric dipole moment, 3 |
| $\mathbf{P}$ | Macroscopic electric dipole moment density, 121 |
| $P_i$ | Components of $\mathbf{P}$, 19 |
| | Transformed electric dipole moment density, 187 |
| $p_i^{(0)}$ | Permanent electric dipole moment, 6 |
| $\hat{\mathbf{p}}$ | Unit vector, 146, 148 |
| | |
| $q$ | Electric charge, 2 |
| $q^{(\alpha)}$ | Electric charge of particle $\alpha$, 3 |
| $q_{ij}$ | Electric quadrupole moment, 3 |

| | |
|---|---|
| $\bar{q}_{ij}$ | Electric quadrupole moment relative to a displaced origin, 6 |
| $Q_{ij}$ | Macroscopic electric quadrupole moment density, 18 |
| | Transformed electric quadrupole moment density, 187 |
| $q_{ij}^{(0)}$ | Permanent electric quadrupole moment, 6 |
| $q_{ijk}$ | Electric octopole moment, 3 |
| $q_{ij}$ | Electric quadrupole moment, 3 |
| $q_{ijk}^{(0)}$ | Permanent electric octopole moment, 6 |
| $Q_{ijk}$ | Macroscopic electric octopole moment density, 19 |
| $\mathbf{r}$ | Position vector, 1 |
| $r_i$ | Components of $\mathbf{r}$, 2 |
| $\mathbf{r}^{(\alpha)}$ | Position vector of particle $\alpha$, 3. Position operator, 33 |
| $r_i^{(\alpha)}$ | Components of $\mathbf{r}^{(\alpha)}$, 8 |
| $R$ | Reflection matrix, 146 |
| $R'$ | Time-reversed reflection matrix, 148 |
| $\mathbf{R}$ | Position vector, 1 |
| $R_i$ | Components of $\mathbf{R}$, 2 |
| $r_{pp}, r_{ps}, r_{sp}, r_{ss}$ | Fresnel coefficients of reflection, 146 |
| $r'_{pp}, r'_{ps}, r'_{sp}, r'_{ss}$ | Time-reversed Fresnel coefficients, 148 |
| $r_{jk}, r'_{jk}$ | Fresnel coefficients, 150 |
| | |
| $S$ | Closed surface, 10. Scalar, 63 |
| $S'$ | Transformed scalar, 63 |
| $S_a, S_b$ | Scale factors, 160 |
| $\hat{\mathbf{s}}$ | Unit vector, 146, 148 |
| $|s^{(0)}\rangle$ | Unperturbed eigenvector, 34 |
| $S_{ijk}$ | Translationally invariant tensor, 185 |
| | |
| $t$ | Time, 15 |
| $t'$ | Retarded time, 15 |
| $T$ | Time reversal operator, 66 |
| $\tilde{t}_{ij}, t_{ij}^a, t_{ij}^s$ | Coefficients, 101 |
| $T_{ij}$ | 20 |
| $T_{ij}$ | Second-rank tensor, 62. Magnetoelectric tensor, 85 |
| | Transformed magnetoelectric tensor, 175 |
| $T_{ij}^M$ | Direct multipole form of magnetoelectric tensor $T_{ij}$, 92 |
| $T'_{ij}$ | Transformed tensor, 62 |
| $T_{ijk}$ | 20 |
| $T_{ijk}$ | Third-rank tensor, 63 |
| $T'_{ijk}$ | Transformed tensor, 63 |
| | |
| $u$ | Beam profile function, 128 |
| $u(z)$ | Unit step function, 150 |

| | |
|---|---|
| $\mathbf{u}$ | Velocity, 11 |
| $U$ | Unitary operator, 66 |
| $\lvert u\rangle,\ \lvert u'\rangle$ | Ket and time-reversed ket, 66 |
| $U_{ij}$ | Magnetoelectric tensor, 85 |
| | Transformed magnetoelectric tensor, 175 |
| $U_{ij}^{M}$ | Direct multipole form of $U_{ij}$, 92 |
| $U_{ij}^{G}$ | 175 |

| | |
|---|---|
| $v$ | 119 |
| $V$ | Volume, 2. Potential energy operator, 33 |
| $\mathbf{V}$ | Vector, 60 |
| $V_{i}$ | Components of $\mathbf{V}$, 60 |
| $V_{i}'$ | Components of transformed vector, 60 |
| $V_{\alpha\beta}$ | Two-body potential, 62 |
| $\lvert v\rangle,\ \lvert v'\rangle$ | Ket and time-reversed ket, 66 |
| $V_{ij}$ | Translationally invariant tensor, 182 |

| | |
|---|---|
| $W$ | Electrostatic potential energy, 8. Energy eigenvalue, 37 |
| | Magnetostatic potential energy, 13. Energy of a molecule, 121 |
| $W_{n}$ | Energy eigenvalue, 34 |
| $W_{n}^{(0)}$ | Unperturbed energy eigenvalue, 34 |
| $W_{n}^{(1)},\ W_{n}^{(2)}$ | First- and second-order perturbation contributions to $W_{n}$, 34 |
| $\lvert w\rangle,\ \lvert w'\rangle$ | Ket and time-reversed ket, 66 |
| $W_{ij}$ | Translationally invariant tensor, 185 |

| | |
|---|---|
| $x_{i},\ x_{i}'$ | Cartesian axes, 59 |
| $x^{\mu}$ | Position 4-vector, 87 |
| $X_{ij}$ | Inverse permeability tensor, 85 |
| | Transformed inverse permeability tensor, 175 |
| $X_{ij}^{M}$ | Direct multipole form of $X_{ij}$, 92 |
| $X_{ij}^{G}$ | 175 |

| | |
|---|---|
| $Y_{ij}^{F}$ | 174 |

| | |
|---|---|
| $Z_{sn}$ | 49 |
| $Z_{ij}^{F}$ | 174 |

## 2. Greek letters

| | |
|---|---|
| $\alpha_{1},\dots,\alpha_{4}$ | Phases, 149 |
| $\alpha_{a},\ \alpha_{b}$ | 160 |
| $\alpha_{ij}$ | Polarizability tensor, 7. Polarizability density, 53, 85 |
| $\alpha_{ij}'$ | Polarizability tensor, 42. Polarizability density, 53, 85 |
| $\tilde{\alpha}_{ij}$ | Coefficients, 101. Polarizability density, 187 |
| $\alpha_{ij}^{a},\ \alpha_{ij}^{s}$ | Coefficients, 101 |

| | |
|---|---|
| $\beta_a$, $\beta_b$ | 160 |
| $\beta_i$ | Coefficients, 179 |
| $\beta_1$, $\beta_2$, $\beta$ | 113 |
| $\beta_{ij}^a$, $\beta_{ij}^s$ | Coefficients, 101 |
| $\beta_{ijk}$ | Hyperpolarizability tensor, 7 |
| $\gamma_i$ | Coefficients, 179 |
| $\gamma_1, \ldots, \gamma_6$ | 117 |
| $\gamma'$, $\gamma''$ | 119 |
| $\gamma_7$, $\gamma_8$ | 119 |
| $\tilde{\gamma}_{ij}$, $\gamma_{ij}^a$, $\gamma_{ij}^s$ | Coefficients, 101 |
| $\gamma_{ijkl}$ | Hyperpolarizability tensor, 7 |
| $\Gamma_{sn}$ | Width of a Lorentzian curve, 51 |
| $\delta(\mathbf{r})$ | Dirac delta function, 3 |
| $\Delta$ | 161 |
| $\delta_{ij}$ | Kronecker delta function, 2 |
| $\Delta V$ | Macroscopic volume element, 54 |
| $\varepsilon_0$ | Permittivity of free space, 1 |
| $\tilde{\varepsilon}_{ij}$ | Complex dynamic permittivity, 26 |
| $\varepsilon_{ijk}$ | Levi–Civita tensor, 8 |
| $\eta$ | Charge density, 23 |
| $\theta$ | Angle of incidence, 146 |
| $\theta_{ij}$ | 60 |
| $\Theta_{ij}$ | Traceless electric quadrupole moment, 4 |
| $\lambda$ | Wavelength, 17 |
| $\mu_0$ | Permeability of free space, 10 |
| $\xi_{ijk}$ | Polarizability tensor, 52 |
| $\mathbf{\Pi}^{(\alpha)}$ | Generalized momentum operator, 33 |
| $\rho$ | Charge density, 1 |
| $\rho_b$ | Bound charge density, 21 |
| $\rho_f$ | Macroscopic ("free") charge density, 19 |
| $\sigma$ | Surface charge density, 151 |
| $\boldsymbol{\sigma}_a$, $\boldsymbol{\sigma}_b$ | Unit-wave-normals of two refracted waves, 157 |
| $\sigma_b$ | Bound surface charge density, 21 |
| $\sigma_i$ | Components of $\boldsymbol{\sigma}$, 101 |
| $\boldsymbol{\sigma}$ | Unit-wave-normal, 101 |
| $\tau$ | Orientational variable, 121 |
| $\phi$ | Angle of rotation, 69 |
| $\Phi$ | Electrostatic potential, 1. Scalar potential, 15 |
| $\chi$ | Gauge function, 172 |
| $\chi_{ij}$ | $DC$ magnetic susceptibility, 39 |
| | Polarizability tensor (magnetic susceptibility), 48 |
| | Magnetic susceptibility operator, 46 |
| $\chi'_{ij}$ | Polarizability tensor, 48 |

| | |
|---|---|
| $\chi_{ijk}, \chi_{ijkl}$ | Magnetic susceptibility operators, 46 |
| $\chi_{\nu\mu\lambda\rho}$ | Covariant constitutive tensor, 90 |
| $\lvert\Psi\rangle$ | Eigenket, 37 |
| $\omega$ | Angular frequency, 26 |
| $\omega_{sn}$ | Angular frequency of a transition, 34 |
| $\Omega$ | Operator, 34 |
| $\Omega^{(0)}$ | Unperturbed operator, 47 |
| $\Omega'$ | Perturbed operator, 47. Time-reversed operator, 66 |
| $\langle\Omega\rangle_{sn}$ | Matrix elements of $\Omega$, 34 |

3. *Other letters*

| | |
|---|---|
| $\mathfrak{a}_{ijk}$ | Polarizability tensor, 6. Polarizability density, 54 |
| $\mathfrak{b}_{ijkl}$ | Polarizability tensor, 6 |
| $\mathcal{B}$ | Magnetic field of a light wave, 121 |
| | Surface contribution to $\mathbf{B}$, 153 |
| $\mathcal{E}$ | Electric field of a light wave, 121 |
| | Surface contribution to $\mathbf{E}$, 153 |
| $\mathcal{H}_{ijk}, \mathcal{H}'_{ijk}$ | Polarizability tensors, 48 |
| $\mathcal{L}$ | Lagrangian density, 87 |
| $\mathcal{L}'$ | Legendre transformation of $\mathcal{L}$, 88 |
| $\mathcal{L}_{ijk}, \mathcal{L}'_{ijk}$ | Polarizability tensors, 48 |
| $\mathcal{P}, \mathcal{P}_i$ | Bound surface electric dipole moment density, 151, 152 |
| $\mathcal{P}_{bi}$ | Bound surface electric dipole moment density, 21 |
| $\mathcal{Q}_{bij}$ | Bound surface electric quadrupole moment density, 22 |
| $\mathcal{V}$ | Volume, 20 |

# INDEX

absorption 50-51, 105
  coefficient 111
action principle 87
additional wave 116, 119
  absence of 138
  neglect of 156
angular momentum 12, 38, 39, 46, 60, 62
anti-unitary operator 66
average
  Boltzmann 121
  isotropic 109, 122, 123
  macroscopic 19, 23-25
  orientational 108, 121-123
axes
  Cartesian 59
  fast and slow 111, 113
axial tensor, *see* tensor
axial vector, *see* vector

Barron-Gray gauge, *see* gauge
beam profile function 128, 133, 134
birefringence
  circular 104, 106, 120, 140
  electric-field-gradient-induced 127-140
  gyrotropic 55, 112-115
  in $Cr_2O_3$ 115
  induced 120, 127
  Jones 55, 111, 112, 114, 115
  linear 55, 106, 111, 112, 124
  Lorentz 55, 115
  multipole orders of 118
  of quantum vacuum 112
bound
  surface charge density of multipole
        moments 21, 151-153
  surface current density of multipole
        moments 22, 151-153
  volume charge density 21
  volume current density 22-23
Buckingham effect 135
Buckingham's
  derivation 37
  quadrupole experiment 8, 131

Cartesian
  axes 59
  basis 5
  tensor, *see* tensor
charge density 1, 15, 17, 19, 21-24

bound 21-23, 151-153
free 19, 23, 28
macroscopic 19
surface contributions 151-153
chiral effect 81
chiral medium 109
circular birefringence, *see* birefingence
circular dichroism, *see* dichroism
closure relation 43, 218
commutator 33, 41, 43, 218
computer calculation
  of electric quadrupole moment 135-136
constitutive relations for response fields
  and wave equation 100-101, 191-192
  covariant form 90
  from multipole theory 84-85, 207
  transformed 174-176, 178-189
constitutive tensor 90-92
  transformed 175
continuity equations 10, 150-153, 211
convergence in macroscopic theory 24, 98
Coulomb gauge, *see* gauge
crystal classes 71, 105-107, 115-119,
        163-170, 197-199
crystal systems 71
cubic crystals 71, 72, 106, 115-120, 156,
        158, 163-166, 193, 194, 197-198
current density 10, 15
  bound 22-23, 25
  free 22-24, 28
  surface contributions 151-153

depolarization 55
dichroism
  circular 105, 110, 111, 140, 205
differential scattering 55, 81
dipole moment, *see* multipole moments
Dirac $\delta$ function 3
  derivative of 150
direction cosines 60
dissipation
  effect on material constants 95-96, 189
  effect on polarizability tensors 50-51, 79
  effect on wave vector 95, 208
  in optically active medium 110

effective quadrupole center 135, 136
eigenpolarizations 103, 104, 110, 113, 114,
        116, 117, 120, 163-165

eigenvalue 37
eigenvector 34, 41
eigenvector theory 103
electric dipole
  approximation 42
  effects 55, 103, 118, 121
  moment, *see* multipole moments
  moment density 19, 53
  order 25, 50, 69, 92, 97, 100, 102, 106,
      162, 178-180
electric dipole-magnetic dipole
      approximation 25-26, 97
electric dipole moment operator 34, 35,
      41, 46
  expectation value of 35, 42, 48
electric displacement, *see* response fields
electric field
  defining relation 68
  gradients 7, 35-37, 43-47, 54-56,
      129-131, 136-139
  in Buckingham's experiment 131, 134,
      138-140
  in reflection, *see* reflection
  in transmission 100, 136-140; *see* also
      transmission effects
  macroscopic 23
  matching conditions for, *see* matching
      conditions
  microscopic 23
  radiated by a lamina 126-127, 130-131
  radiated by a molecule 124-125, 128-129
  space-time nature of 68-69, 175, 176
  wave equation for 101, 137
electric-field-gradient-induced
      birefringence 127-140
electric octopole-magnetic quadrupole
  effects 55, 78, 115-120
  order 1, 17, 24, 25, 27-28, 48, 50, 69,
      78, 84, 85, 94-95, 97, 100-102,
      116, 118, 153, 155, 189, 191
electric octopole moment, *see* multipole
      moments
electric octopole moment operator 34, 35,
      46
  expectation value of 35, 48
electric quadrupole contribution
  to circular birefringence 109
  to $\mathbf{D}$ 24-26, 111, 136
  to gyrotropic birefringence 112
electric quadrupole-magnetic dipole
  effects 25, 55, 106, 112, 118, 128-139
  order 24, 25, 69, 93, 97, 100, 102,106,
      118, 136, 137, 151-156, 157,
      162, 163, 167, 180-189, 192,
      199, 200

electric quadrupole moment, *see*
      multipole moments
  calculated values 135-136
  measurements of 8, 128, 135-136
  necessity for inclusion 25-26, 43, 109,
      111, 187
electric quadrupole moment operator 34,
      35, 46
  expectation value of 35, 48
electromagnetic field tensor 88
electrostatic energy 9, 36-37
electrostatic field 4
  in Buckingham's experiment 131
electrostatic perturbation 33-37
electrostatic potential 1
energy density 206
Euler–Lagrange equations 87
expectation value 34-35, 40, 42, 47, 48

Faraday effect
  and symmetry 79-81
  induced 79-81
  intrinsic 55, 103-105, 118-120
Faraday transformations 174-176, 179,
      182
  effect on constitutive tensor 175-176
far-zone 17-18, 27, 135
force
  on a charge distribution 7-8
  on a current distribution 13
forward scattering
  by a lamina 125-126, 130-131
four-current density 87
four-vector potential 87
  and gauge invariance 172-173
Fresnel
  coefficients of reflection 146, 148-150,
      161-162, 165-169, 194-198,
      201-202, 205-206
  integrals 126, 133
functional field 91

gauge
  Barron–Gray 45-46, 51
  Coulomb 38, 45
  Lorenz 15, 17, 45
  standard 40
gauge transformations
  effect on constitutive tensor 175-176
  for 4-vector potential 172-173
  for response fields 173
gyrotropic birefringence, *see* birefringence

Hamiltonian
  perturbed 33, 38, 41, 45-46

semi-classical 32-33, 47
harmonic plane wave 41, 47, 85, 101, 139,
        157, 173
    distorted 140
    homogeneous 95, 208
    inhomogeneous 95, 208
Hellmann–Feynman theorem 38
Helmholtz's theorem 173
Hermitian multipole moment operators
        35, 39, 44, 46, 51, 52, 219
hierarchy of multipole contributions 25,
        187
hyperpolarizability 7, 52, 53, 122

induced multipole moments 6-7, 122-123,
        129-130
integral relation for response fields 86
intensity 147-149, 168-170, 198
International convention 71, 73, 75
intrinsic symmetry, see symmetry
invariance
    of equations of physics 61-62
    translational, see translational
        invariance
inverse permeability tensor 86
    transformed 186
inversion
    of coordinate axes 59-61, 67, 175
isotropic average 109, 122, 123
isotropic tensor, see tensor

Jones
    birefringence, see birefringence
    calculus 111, 140
    matrix 111
    platelets 111, 113-114

Kerr effect 7, 120
    forward scattering theory of 124-127
    wave theory of 120-124
Kleinman's conjecture 53
Kramers–Kronig relations 51
Kronecker $\delta$ function 2, 64, 68-69, 72

Lagrangian density 87-91, 95
Legendre transformation 88
Levi–Civita tensor 8, 64, 68-69, 72
linear birefringence, see birefringence
Lorentz birefringence, see birefringence
Lorentzian 51
Lorenz gauge, see gauge
Lorenz–Lorentz equation 127

macroscopic media 18-29, 53-54

macroscopic multipole moment densities
        19-26, 28-29, 53-54, 84-85,
        157-158, 170
    transformed 187, 192, 199
magnetic crystals
    reflection from 170, 199-206
magnetic dipole moment, see multipole
        moments
magnetic field
    defining relation 68
    gradient 45-48, 51-52, 54-55
    macroscopic 23
    matching conditions for, see matching
        conditions
    microscopic 23
    space-time nature of 68-69, 175, 176
magnetic media 71, 92-95, 97, 103, 110,
        112, 118, 170, 180, 183-186,
        199-206
magnetic object 71, 73-75
magnetic quadrupole moment, see
        multipole moments
magnetic susceptibility operators 46
magnetizability 38, 39
magnetoelectric effect 25, 55, 187
magnetoelectric tensors 86
    transformed 186
magnetostatic energy 14, 39
magnetostatic field 11
magnetostatic perturbation 38-40
matching conditions 153-156, 158-160,
        192-194, 200, 203
material constants
    from multipole theory 85
    properties of 86-97
    transformed 175, 179-186
matrix element 34, 41
    behaviour under time reversal 67, 69
Maxwell's equations
    microscopic 23
    macroscopic 23-26, 100-101, 154-155
Maxwell-type relations 89
measurements of electric quadrupole
        moment 8, 128, 135-136
multipole expansion
    hierarchy for 25
    of $\mathbf{D}$ and $\mathbf{H}$ 24
    of dynamic scalar and vector potentials
        15-21
    of electrostatic field 4
    of electrostatic force 8
    of electrostatic potential 2-5
    of electrostatic potential energy 9
    of electrostatic torque 8
    of magnetostatic field 11

of magnetostatic force 13
of magnetostatic potential energy 14
of magnetostatic torque 13-14
of magnetostatic vector potential 10-11
multipole moment densities, *see*
      macroscopic multipole moment
      densities
multipole moment operators
  classical limit of 46
  electric dipole 34, 35, 41, 46
  electric octopole 34, 35, 46
  electric quadrupole 34, 35, 46
  Hermitian property of 35, 39, 44, 46,
    51, 52, 219
  magnetic dipole 38, 39, 46-47
  magnetic octopole 46-47
  magnetic quadrupole 46-47
  magnetic $2^n$-pole 47
  origin dependence of 75-78
  perturbed magnetic 40, 47
multipole moments
  electric dipole 3, 6, 7, 9, 122
  electric monopole 2
  electric octopole 3, 6, 7, 9, 27, 28
  electric quadrupole 3-9, 27, 28, 135-136
  induced 6-7, 122-123, 129-130
  macroscopic 19-20, 28-29, 53-54, 84-85,
    158, 170, 187, 192, 199
  magnetic dipole 11-12
  magnetic quadrupole 11-12
  of order $2^l$ 5, 12
  origin dependence of, *see* origin
    dependence
  permanent 6, 35
  primitive 4, 5, 12, 20, 24, 26-29, 96, 135
  space-time properties of 68-69
  symmetric 5, 12, 27
  total 6, 7, 37, 122
  traceless, *see* traceless multipole
    moments

near-zone 17-20, 27
Neumann's principle 72, 74
non-dissipative media
  material constants for 92-95
  symmetries for 90, 92-96, 175, 179, 182,
    185-189
non-magnetic crystals 71
  reflection from 156-169, 192-199
non-magnetic media 72, 92-94, 97, 106,
    107, 115-117, 156, 158-169
non-magnetic object 73-74
  vanishing of polarizability tensors for
    72, 92
non-polarizable distribution 7, 9, 14

normal incidence, *see* reflection

oblique incidence, *see* multipole moments
Onsager reciprocal relations 147, 150
operator, *see* multipole moment operators
  generalized momentum 33
  time-even 66
  time-odd 66
  time-reversal 66
optical activity
  and symmetry 69-70, 188
  condition for 107
  in a fluid of aligned molecules 109, 188
  in anisotropic media 55
  in fluids and cubic crystals 55
  microwave measurements of 109
  natural 106
optical rotary dispersion 110
orientational average 108, 121-123
origin dependence
  of amplitudes of response fields 96-97,
    206
  of electric multipole moments 5, 98,
    128, 135-136
  of Fresnel coefficients 168-170
  of magnetic multipole moments 12
  of multipole expansions 98
  of multipole material constants 93-97
  of multipole moment operators 75-78
  of polarizability tensors 75-79, 102
  of reflected intensity in multipole
    theory 168-170, 207
origin-dependent
  tensor 65
  vector 61, 62
origin independence
  imposition of in transformation theory
    175, 181-182, 184-187
  of circular birefringence 105
  of field amplitudes 206
  of Fresnel coefficients 198, 202
  of linear birefringence 111, 117
  of Lorentz birefringence 117
  of magnetizability 39, 209
  of material constants 86, 97, 208
  of Maxwell's equations 26, 29
  of natural optical activity 106, 109
  of polarizability densities 187, 192
  of reflection coefficients and intensities
    in transformed multipole
    theory 198, 202, 205, 209
  of transformed material constants 175,
    179-187
  of transformed tensors 220
  of wave equation 102, 191-192

origin-independent
  observable 78, 86
  tensor 65
  vector 61-62
orthogonality relations 63
orthonormality 63

parity 67, 70, 79-81
perfect differentials 89, 96
permittivity tensor 86
  complex dynamic 26
  transformed 186
perturbation theory
  time-dependent 41
  time-independent 34
phenomenology
  of wave-matter interaction 54-56
pictorial approach to symmetry
      conditions 79-82
point groups
  magnetic 71-75
  non-magnetic 71-72
polar tensor, see tensor
polar vector, see vector
polarizability densities 53-54, 85, 174-176
  invariant 188, 192
  non-invariant 188
polarizability tensors
  dipole 6, 101, 103, 106, 107, 108, 110,
      112, 113, 115
  for absorptive medium 50-51
  for harmonic plane waves 47-50
  for non-magnetic objects 68
  in transformation theory 178, 180, 183,
      186-189
  macroscopic 53-54
  magnetic 39, 51-52
  multipole order of 50, 69
  origin dependence of 75-79
  octopole 6, 116-118
  quadrupole 6, 107-114
  quantum-mechanical expressions for
      42-44, 49-50
  space-time properties of 68-69
  static 35-36, 39, 51-52
polarization 104, 120, 132, 133, 134, 139,
      140, 147-149, 205; see also
      eigenpolarizations
Post constraint 90-92
  in multipole theory 93-97, 207-209
  in transformed multipole theory 175,
      183, 186, 187
potential energy
  of a charge distribution 8-9
  of a current distribution 14

of a molecule 121-122
Poynting vector 206
primitive multipole moments 4, 5, 12, 20,
      24, 26-29, 96, 135
  number of independent components 5
principle of reciprocity 147-150; see also
      time-reversal invariance
profile function 128, 133, 134
property tensor 70, 72-75

quadrupole moment, see multipole
      moments
quantum-mechanical polarizability
      tensors 35-36, 39, 42-44, 49-50
quantum theory
  of multipole moments and
      polarizabilities 32-56

reciprocity principle, see principle of
      reciprocity
reflection
  from magnetic crystals 170, 199-206
  from non-magnetic crystals 156-169,
      192-198
  matrix 146, 148, 150
  normal incidence 162, 165, 167-168,
      197, 199, 202, 204-205
  oblique incidence 157, 162-168, 192-198,
      202
  of coordinate system 59-60
  of harmonic plane wave 145-146
reflection coefficients, see Fresnel
      coefficients of reflection
refracted rays 156-157
refractive index 101, 104-114, 116-120
  effect of an electric field gradient 134,
      139
  in Kerr effect 124, 127
response fields $\mathbf{D}$ and $\mathbf{H}$
  and the wave equation 101, 123, 136
  constitutive relations for, see
      constitutive relations
  for Kerr effect 121, 123
  for the quadrupole experiment 136
  material constants for, see material
      constants
  multipole expansions of 24-26
  non-uniqueness of 26, 173
  origin dependence of amplitudes 96-97,
      206
  space-time properties of 68-69, 175, 176
  transformations of 173-176, 186
retarded time 15, 16, 20, 124, 128
rotation
  by optically active medium 69

intrinsic in antiferromagnetics 55
of coordinate system 59-62

scalar
    axial 63, 70
    polar 63
    time-even 68-70
scalar potential
    in Barron–Gray gauge 45
    in electrodynamics 15-21, 27, 33
    in electrostatics 2-5, 33
scaling factors 160, 194
scattering theory
    of electric-field-gradient-induced
            birefringence 127-135
    of Kerr effect 124-127
Schönflies convention 71, 73
secular equation 102, 104, 106, 108, 110,
            112, 113, 114, 116, 119, 123
senkrecht 146
Shelankov–Pikus reciprocity condition 202
small parameter in macroscopic theory 24
Snell's law 157
space-time properties
    of electromagnetic quantities 68-69, 176
    of mechanical quantities 67-69
    of multipole quantities 68-69, 179, 180,
            183
spatial dispersion 54
spin
    inclusion of 38-39, 78-79
structural field 91
summation convention 3, 87
surface density, see bound
symmetry
    and Faraday effect 79-81
    and Neumann's Principle 72, 74
    and property tensors 70-75
    classes 71-75, 80, 105, 107-110, 112-119,
            163-169, 197-198
    intrinsic 7, 39, 44, 46, 52-53, 74, 94, 96,
            102, 109, 122, 138, 179, 180,
            183, 184
    in wave equation 53, 101-102
    of an object 70-75
    of material constants 53, 86-97, 175,
            179, 182, 185-189
    of multipole moments 5, 12
    permutation 50-53, 123
    pictorial approach to 79-82
    point-group 53, 70-75, 82
    space-time 53, 70-75

Tellegen media 92
tensor 3, 62-65
    axial 62, 67-70, 75, 107, 118, 176, 179,
            180, 183
    constitutive 90-92
    equation 64
    for wave equation 101-102
    inverse permeability 86, 186
    isotropic 64-65, 68, 115
    Kronecker $\delta$ 2, 64, 68-69, 72
    Levi–Civita 8, 64, 68-69, 72
    magnetoelectric 86, 186
    metric 87
    of electromagnetic field 88
    origin dependent 65
    origin independent 65
    permittivity 86, 186
    polar 62, 67-69, 75, 107, 118, 176, 179,
            180, 188
    polarizability, see polarizability tensor
    property tensor 70, 72-75
    space and time nature of 67-70
    symmetric 188
    time-even 65, 67-69, 102, 103, 107, 110,
            118, 150, 176, 179, 180
    time-odd 65, 67-69, 102, 103, 105, 110,
            118, 150, 176, 180, 183, 188
    traceless 4, 5, 12, 135, 186
time-antisymmetric objects 71
time reversal 65-70, 72, 79-81, 102,
            147-150, 175, 183
time-reversal invariance 147, 168-169,
            198, 202, 209
time-even and time-odd tensors, see
            tensor
time-even operator 66
time-odd operator 66
time-reversal operator 66
time-symmetric objects 71
torque
    on a charge distribution 8
    on a current distribution 13-14
traceless multipole moments
    and material constants 96
    electric octopole 5, 27-28
    electric quadrupole 4, 5, 27-28, 135
    magnetic quadrupole 12, 27
    number of independent components 5,
            12
transformation
    improper 59-63
    matrix 60
    proper 59-63
transformation theory of constitutive
            relations 174-176
    to electric dipole order 178-180

to electric octopole-magnetic
        quadrupole order 189
to electric quadrupole-magnetic dipole
        order 180-189
transformed
    material constants 175, 179-186
    multipole moment densities 187, 192,
        199
    polarizability densities 187
translation
    of coordinate system 61-62, 65; *see also*
        origin independence, origin
        dependence
translational invariance 61-62, 168-169,
        198, 202, 209; *see also* origin
        independence
transmission effects
    natural 103-120
    induced 120-140

uniaxial crystal 71, 103-110, 112, 156,
        158, 163-169, 193, 194-198
uniqueness of fields 206
unit step function 150

Van Vleck
    and origin independence 39, 209

vector 60
    axial 61, 67-69
    origin dependent 61-62
    origin independent 61-62
    polar 61, 67-69
    pseudo 61
vector potential
    four-vector 87, 172
    in Barron–Gray gauge 45, 51
    in electrodynamics 15-21, 27, 41
    in magnetostatics 10-11, 38

wave equation
    and transformation theory 191-192
    for a homogeneous medium 100-103
    for an inhomogeneous medium 137
    origin independence of 102, 191-192
    solutions of 138-140, 162-165; *see also*
        secular equation
wave theory
    of electric-field-gradient-induced
        birefringence 136-140
    of Kerr effect 120-124
wave vector 85, 95, 208